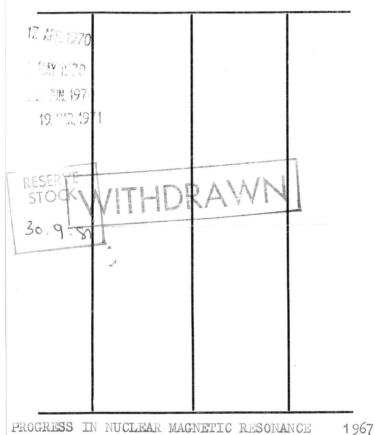
PROGRESS IN NUCLEAR MAGNETIC RESONANCE   1967
SPECTROSCOPY.VOL.3.
ed.by J.W.Emsley,J.Feeney&L.H.Sutcliffe.

Progress in

NUCLEAR MAGNETIC RESONANCE SPECTROSCOPY

Volume 3

Progress in

NUCLEAR MAGNETIC RESONANCE SPECTROSCOPY

Volume 3

Edited by

J. W. EMSLEY
The University, Durham City

J. FEENEY
Varian Associates Ltd., Walton-on-Thames

L. H. SUTCLIFFE
The University, Liverpool

THE QUEEN'S AWARD
TO INDUSTRY 1966

PERGAMON PRESS

OXFORD · LONDON · EDINBURGH · NEW YORK
TORONTO · SYDNEY · PARIS · BRAUNSCHWEIG

Pergamon Press Ltd., Headington Hill Hall, Oxford
4 & 5 Fitzroy Square, London W.1
Pergamon Press (Scotland) Ltd., 2 & 3 Teviot Place, Edinburgh 1
Pergamon Press Inc., 44–01 21st Street, Long Island City, New York 11101
Pergamon of Canada, Ltd., 6 Adelaide Street East, Toronto, Ontario
Pergamon Press (Aust.) Pty. Ltd., Rushcutters Bay,, Sydney, New South Wales
Pergamon Press S.A.R.L., 24 rue des Écoles, Paris 5ᵉ
Vieweg & Sohn GmbH, Burgplatz 1, Braunschweig

First edition 1967

Library of Congress Catalog Card No. 66–17931.

PRINTED IN GREAT BRITAIN BY ADLARD AND SON, LTD., DORKING, SURREY.
08 003322 9

CONTENTS

ERRATUM

Line 3 on page 38 should read:

$$\mid J_{BX} - [J_{AB}^2 + (\delta_{AB} + \tfrac{1}{2}J_{BX})^2]^{1/2} + [J_{AB}^2 + (\delta_{AB} - \tfrac{1}{2}J_{BX})^2]^{1/2} \mid\; < \Delta_{1/2}$$

EMSLEY, FEENEY and SUTCLIFFE:
Progress in N.M.R. Spectroscopy, Volume 3.

PREFACE

HIGH resolution NMR is one of the most rapidly expanding branches of chemistry. The expansion is taking place over a wide front, embracing not only the fundamentals of the subject but also instrumentation and extended applications of the technique to structural and analytical problems. It is not surprising, therefore, that both the newcomer to NMR and the established worker are finding it increasingly difficult to keep abreast of new developments, and although useful attempts have been made to distribute information quickly among the many active groups, we believe that the situation is improved by making available a continuous supply of up-to-date authoritative reviews dealing with topics of current interest. We aim to satisfy the demand for reviews of this type with this Progress series.

We are indebted to the contributors for their co-operation and sacrifice of precious time.

J. W. EMSLEY
J. FEENEY
L. H. SUTCLIFFE

vi

CHAPTER 1

SUB-SPECTRAL ANALYSIS

P. Diehl,[†] R. K. Harris[‡] and R. G. Jones[**]

CONTENTS

† Physikalisches Institut, Universität Basel, Basel, Switzerland.
‡ School of Chemical Sciences, University of East Anglia, Norwich, England.
** Department of Chemistry, School of Physical Sciences, University of Essex, Colchester, England.

1

GLOSSARY OF SYMBOLS

AB, AA'BB', AA'X_nX_n' etc.	NMR spin system notation. Primes denote magnetic non-equivalence
ab, aa'bb', a_n etc.	NMR sub-spectral notation (primes as above)
J_{AB}, J_{ab}	Coupling constant and effective coupling constant
ν_A, ν_a	Larmor frequency and effective Larmor frequency
δ_{AB}, δ_{ab}	Chemical shift and effective chemical shift
N, K, L, M	Parameters for spin systems of the types AA'BB', AA'XX' etc.
n, k, l, m	Parameters for aa'bb' sub-spectra
$R(= J/\delta)$	Factor determining A_aB_b spectra
$\mathscr{H}$	Nuclear spin Hamiltonian

A'_{mn}, B'_{mn}	General elements of the Hamiltonian
A_{mn}, a_{mn}	Hamiltonian elements ($A'_{mn} = A_{mn} + a_{mn}$)
F^2, I^2, F_x, F_z	Spin operators
F, I	Quantum numbers associated with F^2 and I^2
S, D, T etc	Composite particle states
m	Eigenvalue of F_z
m_T	Sum of m taken over a complete spin system
m_x	Sum of m taken over a group of magnetically equivalent nuclei
$m(XX')$	Sum of m taken over a group of chemically equivalent but magnetically non-equivalent nuclei
ψ	Spin basis function
ϕ_p, ϕ_q	Spin eigenfunctions
Z_{pq}	Relative intensity of transitions between ϕ_p and ϕ_q
g	Statistical weight of spin state
χ, q	Parameters in $AA'X_nX'_n$ spin systems
ϵ	Off-set factor in determining sub-spectral transformations
$\Delta_{1/2}$	Full band width at half peak height
i	Minimum observable intensity
C_{2v}, D_{2h}, P_n	Point groups
E, σ, i, C_n, P	Symmetry elements
A_1, B_2, E	Symmetry species (irreducible representations).

1. INTRODUCTION

1.1. *General*

The techniques devised to analyze high resolution NMR spectra in principle simplify the procedure. These techniques include the use of computers and the manipulation of theoretical aids, or both in conjunction with one another.

The computer may be programmed to diagonalize the secular determinant and to match the experimental data by an iterative procedure. This tends to obliterate the logical sequence of steps inherent in the process. It is the purpose of this article to describe, with examples, the alternative methods of analysis, applicable in many cases, which initially do not use computers. The ultimate object of all these methods is to derive the maximum information in the most convenient way.

The approach to be considered can be broadly classified under the heading "Sub-spectral analysis". This implies that the systems studied consist of a number of simpler systems which contribute independently to the overall spectrum. It will become obvious later that many but not all of the simpler systems can be identified with known independent systems.

1.2. *Development*

The occurrence of repeated spacings in NMR spectra have formed the basis for analysis in the past.[1] Group theory has been used to simplify the analysis of NMR spectra of systems possessing symmetry.[2, 3] The secular determinant factorizes into a number of lower order determinants if the matrix elements of the Hamiltonian are constructed from functions which form the basis for an irreducible representation of the symmetry group of the spin system. The literature contains many examples of the principles[4] and the application[5] of this very useful tool. More recent developments have extended the application of group theory to non-rigid molecules[6] and in particular to the analysis of the NMR spectra of non-rigid systems.[6, 7].

The Hamiltonian can be factorized using the total spin quantum number of individual groups in those cases where the system is composed of groups of magnetically equivalent nuclei[4, 8–10] (this is frequently called the composite particle method).

The "X approximation" [3] and other algebraic simplifications[11, 12] have been used extensively to facilitate maximum factorization of the secular determinant. A combination of the X approximation and magnetic equivalence has led to the development of the effective Larmor frequency method for cases involving magnetic equivalence[13, 14]. This method has more recently been extended to systems containing groups of nuclei which are chemically equivalent but magnetically non-equivalent.[15, 16] Further, sub-spectral transformations and effective coupling constants have been introduced to describe and characterize sub-spectra.[17]

The effective Larmor frequency breakdown and the composite particle method are in fact special cases of sub-spectral analysis but it has been shown that the generalization of the method has limitations of applicability.[17, 18, 22]

1.3. *Tools of Sub-spectral Analysis*

1.3.1. *Symmetry*

Rigid systems may possess planes of symmetry (σ), centres of inversion (i), rotation axes (C) and rotation–reflection axes (S). The effects of these elements of symmetry on the molecular configuration are called covering operations and the appropriate symmetry classification for the NMR problem includes all covering operations which do not alter the NMR parameters but merely interchange identical parameters. It is usually possible to assign the correct symmetry classification quite quickly by enumerating the possible covering operations and checking the list against the character tables which can be found in the many books on group theory.[19]

The symmetry classification of non-rigid molecules is not as straightforward. In general the appropriate symmetry classification includes all those permutations of nuclei which do not alter the NMR parameters.[6, 7] However, in the case of restricted rotation in molecules there may be some doubt about the time-averaged values of some coupling constants. This problem can usually be resolved by considering the complexity of the experimental spectrum, if available.

The steps in constructing basic symmetry wave functions in the general case have been outlined.[4, 15] They lead to the initial factorization of the secular determinant into sub-determinants characterized by wave functions which transform according to the respective symmetry species. Examples follow in later sections.

1.3.2. Total spin quantum number

The composite particle approach depends upon two features common to any group of magnetically equivalent nuclei:[3, 4, 45]

(1) Coupling between spins within the group does not contribute to the spectrum, even when there is more than one unique coupling (as in the case of the four magnetically equivalent equatorial fluorine nuclei of SF_5 compounds).

(2) There is no mixing between wave functions having a different eigenvalue of F^2 (the square of the total spin angular momentum of the group). The eigenvalues permitted are of the form $\hbar^2 F(F + 1)$, where F is the total spin quantum number associated with F^2. Moreover, no transitions between the eigenstates with different F are allowed ($\Delta F = 0$).

The first feature means that all terms involving coupling between magnetically equivalent spins may be omitted from the Hamiltonian. Attention has been drawn[20] to the distinction in theory between a magnetically equivalent group of nuclei which has equal coupling between all nuclei of the group (referred to as equally coupled magnetic equivalence) and a magnetically equivalent group where this does not hold. Such a distinction is not important if only transition energies and intensities are required. It is necessary, however, to make the distinction if true eigenfunctions and energies are required (for discussion of relaxation processes for example). This amounts to using the correct equation (1) for unequally coupled magnetically equivalent groups rather than the related equation (2):

$$\mathscr{H} = \sum_i \left\{ \nu_i F_z(i) + \tfrac{1}{2} \sum_{j \neq k} J_{ij_ik} I_j \cdot I_k + \tfrac{1}{2} \sum_{l \neq i} J_{il} F_i \cdot F_l \right\} \tag{1}$$

$$\mathscr{H} = \sum_i \left\{ \nu_i F_z(i) + \tfrac{1}{2} \sum_{l \neq i} J_{il} F_i \cdot F_l \right\} \tag{2}$$

The summations labelled i and l are over the magnetically equivalent *groups* of nuclei.

The justification for the usual composite particle procedure using equation (2) in all cases is that the second term of equation (1) commutes with $\mathcal{H}$ and with $\sum_i \mathbf{F}_x(i)$.

The second feature implies that each group of magnetically equivalent nuclei may be considered as a single composite spin particle of fixed maximum total spin F_{max}, which may exist in different "spin states" with differing eigenvalues of $\mathbf{F}^2$. The only difference between such a group of magnetically equivalent spins and a single spin particle of the same total spin $I = F_{max}$ is that the single particle has only one possible eigenvalue of $\mathbf{I}^2$ and therefore exists in only a single spin state. Thus the same type of calculation may be used to evaluate the transition energies and frequencies of spin systems containing magnetically equivalent nuclei and those containing single spins of $I > \frac{1}{2}$. For a group of n equivalent spin $\frac{1}{2}$ nuclei, the maximum value of F is $n/2$ and permitted values of F are $n/2, n/2 - 1, \ldots, \frac{1}{2}$ for n odd, and $n/2$, $n/2 - 1, \ldots, 0$ for n even. Consequently, there are $(n + 1)/2$ spin states for n odd and $(n/2) + 1$ spin states for n even. There are $(2F + 1)$ possible eigenvalues m of the z component of the spin angular momentum $\mathbf{F}_z$, of the group, for a spin state of total spin quantum number F. These eigenvalues m may be F, F $- 1$, F $- 2, \ldots, -$F. By analogy with standard practice for electronic spectroscopy, $(2F + 1)$ is known as the multiplicity of the spin state. The notations introduced to describe the possible spin states are S, D, T, Q, Qt, Sx, Sp, $\ldots$ for states with eigenvalues $\frac{1}{2}, 1, \frac{3}{2}, \ldots$ of the total spin.[10] Each spin state has an associated statistical weight, g, which is the degeneracy of the representation in the appropriate symmetry group.† Equation (3) applies for a state with quantum number F of a group of n nuclei.

$$g = \frac{(2F + 1)n!}{(n/2 - F)!\,(n/2 + F + 1)!} \tag{3}$$

This may readily be seen by drawing up a table of the total number of spin wave functions possible for each value of m[4]. Basis spin wave functions are normally chosen to be eigenfunctions of both $\mathbf{F}^2$ and $\mathbf{F}_z$, although this is not the only possible choice (or necessarily the one that yields the optimum initial diagonalization[9]) when there is more than one group of nuclei. Two notations have been introduced for such spin functions. The more straightforward simply lists values of F and m as $|F, m\rangle$. The notation F^m using S, D, T, etc., rather than numbers for F offers certain advantages in clarity and will therefore be used here. The two basis functions of a doublet state (F $= \frac{1}{2}$)

---

† It has been pointed out that this factor is not, strictly speaking, to be regarded as a degeneracy of the total wave function when the space part is included[21].

would be written $D_G^{1/2}$ and $D_G^{-1/2}$, for example, the subscript designating the type of nucleus referred to.

1.3.3. X *approximations and algebraic simplifications*

The secular matrices of systems which contain weakly coupled groups ($J \ll \delta$) of nuclei can be treated using second-order perturbation theory. In those cases where the second-order perturbation contributions of the type (J^2/δ) are negligible the term X approximation is used. The consequence of this approach is that the magnetic quantum numbers m of the individual weakly coupled groups are good quantum numbers. For example, transitions within the AB part of an ABX system are defined by the selection rules $\Delta m(AB) = +1$, $\Delta m(X) = 0$.

Special values of, or relationships between, molecular parameters lead to additional simplifications (for example, when some coupling constants are zero). The example of the $AA'X_3X_3'$ system, including the special case $J_{XX'} = 0$ where simpler sub-systems are apparent, will be dealt with later.

1.3.4. *Effective Larmor frequency*

A combination of the X approximation and magnetic equivalence gives rise to a general formulation of the analysis of strongly coupled parts for cases where there exist strongly coupled groups of chemically equivalent nuclei weakly coupled to other groups of magnetically equivalent nuclei which have to be weakly coupled.[13, 14] The magnetic quantum numbers, m, of the groups of magnetically equivalent nuclei are then good quantum numbers and the spectrum of the strongly coupled part breaks down completely into sub-spectra. These sub-spectra are characterized by the coupling constants of the strongly coupled part but the Larmor frequencies of the strongly coupled nuclei are replaced by effective Larmor frequencies ν_i. Thus for an AB_nX_q system the AB_n part will be a superposition of separate ab_n sub-spectra† with equal coupling constants but different effective Larmor frequencies given in general by,

$$\nu_a(X) = \nu_A + m_x J_{AX}$$
$$\nu_b(X) = \nu_B + m_x J_{BX}$$

where m_x is used to denote the sum of eigenvalues of F_z over the group of q magnetically equivalent X nuclei.

It is possible to obtain a partial analysis based on effective Larmor frequencies limited to the extreme values of $\sum m$, (the sum of eigenvalues of F_z over any strongly coupled group of nuclei) for systems consisting of

---

† Small letters are used to denote nuclei in sub-spectra.

groups of magnetically non-equivalent nuclei. A system such as $AA'A''A'''XX'$ will contain two X sub-spectra of type x_2 with effective Larmor frequencies $\nu_X + \frac{1}{2}\sum_A J_{AX}$ and $\nu_X - \frac{1}{2}\sum_A J_{AX}$, where the summations are over all A type nuclei.

1.3.5. *Sub-spectral transformations*

The method of sub-spectral transformations[17] makes use of all the tools mentioned above in order to obtain maximum factorization of the Hamiltonian. The sub-Hamiltonians which are obtained are identified, where possible, with Hamiltonians of simple NMR systems. The necessary transformations which relate the parameters of the simpler sub-system (effective Larmor frequencies and effective coupling constants) to the parameters of the complex system are then derived in those cases where the identity of Hamiltonians is valid.

The applications of these tools are demonstrated in the following sections.

2. SUB-SPECTRAL TRANSFORMATIONS

2.1. *Factorization of the Hamiltonian*

The principles of sub-spectral analysis are explained below in a discussion of the relatively simple case of an $AA'XX'$ spin system. The arguments are based on a rigid system of nuclei. An example of a non-rigid system will be given later.

The appropriate symmetry elements which can be used to characterize the symmetry classification of the rigid $AA'XX'$ spin system are E, the identity operation, and either C_2, a two-fold axis of symmetry or σ, a plane of symmetry.[2] The basic symmetry wave functions of the individual pairs of chemically equivalent nuclei AA' and XX' are analogous and well known.[2] They are the symmetrical functions $\alpha\alpha$, $(\alpha\beta + \beta\alpha)2^{-1/2}$, $\beta\beta$ (A species of the C_2 point group) and the antisymmetric function $(\alpha\beta - \beta\alpha)2^{-1/2}$ (B species). These wave functions are eigenfunctions of the appropriate Hamiltonian giving eigenvalues directly. This is summarized in an abbreviated notation below:

Basic symmetry wave function	$m(AA'$ or $XX')$	Notation	Symmetry
$\alpha\alpha$	1	1	*A*
$(\alpha\beta + \beta\alpha)2^{-1/2}$	0	1	*A*
$\beta\beta$	−1	1	*A*
$(\alpha\beta - \beta\alpha)^{-1/2}$	0	1	*B*

or briefly

m	A species	B species
1	1	
0	1	1
−1	1	

The notation of bold numbers denotes the order of the sub-determinant (or sub-matrix).

The 2^n molecular basic symmetry product wave functions can be constructed by forming all possible products of the basic group symmetry wave functions and the symmetry species of the products are derived as shown in Table 1. (An equivalent result can be obtained by constructing basic molecular symmetry wave functions directly.[5])

TABLE 1. THE BASIC MOLECULAR SYMMETRY PRODUCT WAVE FUNCTIONS FOR THE AA′XX′ SYSTEM CLASSIFIED UNDER THE C_2 POINT GROUP

m	XX′ part			m	AA′ part		
	Species				Species		
	A	B			A	B	
1	1		$\times$	1	1		$=$
0	1	1		0	1	1	
−1	1			−1	1		

m_T	Species products							
	$A \times A = A$			$B \times B = A$		$A \times B = B$		$B \times A = B$
2	1							
1	1	1				1		1
0	1	1	1	1			1	1
−1		1	1				1	1
−2			1					
$m(XX′)$	1	0	−1	0		1	0	−1	0

The molecular product wave functions which result from the operations illustrated in Table 1 are regrouped into the appropriate symmetry species, A or B. The result is summarized in Table 2 where it is obvious that not all of the molecular product wave functions are stationary state wave functions. Off-diagonal matrix elements exist between pairs of basic symmetry wave functions where the wave functions are characterized by the same value of $m(AA′XX′)$ (written m_T) and belong to the same symmetry species

TABLE 2. FACTORIZATION OF THE SECULAR MATRIX OF THE AA'XX'
SYSTEM USING SYMMETRY AND GOOD QUANTUM NUMBERS
(the X approximation)

m_T	A species			B species		
2	1					
1	1	1		1	1	
0	1	2	1		2	
−1		1	1		1	1
−2			1			
$m(XX')$	1	0	−1	1	0	−1

($m(XX') = 0$, A and B species). However, the four-spin Hamiltonian which could have a maximum order of sixteen has been factorized into twelve sub-matrices of order 1 and two sub-matrices of order 2 using symmetry and the good quantum number $m(XX')$ (the quantum number $m(AA')$ is also a good quantum number and can be used to isolate the transitions within the XX' part).

The methods used to achieve the maximum factorization of the secular matrix of the AA'XX', illustrated in Table 2, are well known and form a firm foundation for the method of sub-spectral analysis, where they give the most convenient factorization. The next step which involves the evaluation of the matrix elements in terms of the molecular parameters ν_A, ν_X, $J_{AA'}$, $J_{XX'}$, J_{AX} and $J_{AX'}$ is also well known[3] and the result is the similarly factorized secular determinant.[2] The sub-spectral analysis approach now goes one step further, introducing the new concept of sub-spectral transformations. This new concept involves the *attempt* to discuss the factorized Hamiltonian as a linear superposition of sub-Hamiltonians which can be attributed to well known and simple sub-systems.[17] Table 2 can be written as a sum of four independent sub-systems of wave functions (diagonalization of the secular determinant gives four independent sub-systems of energy levels) which belong to one of the following types:

$$\mathbf{1 : 1 : 1} \ (A \text{ species}) + \mathbf{1} \ (B \text{ species}),$$

or

$$\mathbf{1 : 2 : 1} \ (A \text{ and } B \text{ species}).$$

These sub-systems can be superficially identified with a_2 and ab sub-systems and therefore the AA' part of an AA'XX' spectrum, i.e. AA'(AA'XX'), can be considered to be a superposition of four sub-spectra,

$$AA'(AA'XX')_{m(XX')} = (a_2)_{+1} + (a_2)_{-1}$$
$$+ (ab)_0 \ A \text{ species}$$
$$+ (ab)_0 \ B \text{ species}.$$

However, it has to be demonstrated that the sub-systems given above $(1:1:1+1$ and $1:2:1)$ can be formally identified as a_2 and ab sub-systems, respectively, by deriving the relations between sub-spectral parameters ν_a, ν_b, J_{ab} and the parameters of the AA'XX' system, ν_A, ν_X, $J_{AA'}$, $J_{XX'}$, J_{AX} and $J_{AX'}$ in a unique way.

Such relations can be found by direct comparison of the two Hamiltonians element by element in the case of effective Larmor frequency spectra (see later) but this is not possible in the ab sub-systems arising within the AA'XX' system. Consequently the method of sub-spectral transformations was developed.

2.2. The Derivation of Sub-spectral Transformations

Sub-spectral transformations in general transform parts of Hamiltonians which are sub-Hamiltonians of complex problems into Hamiltonians of simpler sub-systems. An example is provided by comparison of respective matrices of the $1:2:1$ (A species) sub-system (of the AA'XX' system) and the ab system as in Table 3.

TABLE 3. COMPARISON OF MATRICES IN SYSTEMS AA'XX' AND ab

$$\text{AA'XX'}(m(\text{XX'}) = 0, \quad A \text{ species})$$

$$\begin{bmatrix} \nu_A + \frac{1}{4}(J_{AA'} + J_{XX'}) & \\ -\frac{3}{4}(J_{AA'} + J_{XX'}), & -\frac{1}{2}(J_{AX} - J_{AX'}) \\ -\frac{1}{2}(J_{AX} - J_{AX'}), & \frac{1}{4}(J_{AA'} + J_{XX'}) \\ & -\nu_A + \frac{1}{4}(J_{AA'} + J_{XX'}) \end{bmatrix}$$

$$\text{ab}$$

$$\begin{bmatrix} \frac{1}{2}\nu_a + \frac{1}{2}\nu_b + \frac{1}{4}J_{ab} & \\ \frac{1}{2}\nu_a - \frac{1}{2}\nu_b - \frac{1}{4}J_{ab}, & \frac{1}{2}J_{ab} \\ \frac{1}{2}J_{ab}, -\frac{1}{2}\nu_a + \frac{1}{2}\nu_b - \frac{1}{4}J_{ab} \\ & -\frac{1}{2}\nu_a - \frac{1}{2}\nu_b + \frac{1}{4}J_{ab} \end{bmatrix}$$

The two systems must have the same transition energies, when diagonalized, if indeed the sub-Hamiltonian of the AA'XX' system gives rise to an ab type spectrum. However, the respective eigenvalues may be shifted relative to each other by a constant amount ϵ. The two matrices must have the same eigenvalues after compensation of the possible off-set (ϵ may be zero) by addition of ϵ to all the diagonal elements of the AA'XX' matrix so that the coefficients of their secular polynomials must be equal. This argument provides four relations for the four unknowns, ν_a, ν_b, J_{ab} and ϵ. General elements A'_{mn} are used to formulate the approach rather than the specific elements of the AA'XX' Hamiltonian given in Table 3. The conditions set out above then become

$$A'_{11} + \epsilon = \tfrac{1}{2}\nu_a + \tfrac{1}{2}\nu_b + \tfrac{1}{4}J_{ab}$$

$$C'_{11} + \epsilon = -\tfrac{1}{2}\nu_a - \tfrac{1}{2}\nu_b + \tfrac{1}{4}J_{ab} \qquad (4)$$

$$B'_{11} + B'_{22} + 2\epsilon = -\tfrac{1}{2}J_{ab}$$

$$\epsilon^2 + \epsilon(B'_{11} + B'_{22}) + B'_{11}B'_{22} - (B'_{12})^2 = -3/16J^2_{ab} - \tfrac{1}{4}(\nu_a - \nu_b)^2$$

The solutions are:

$$
\left.\begin{aligned}
\epsilon &= -\tfrac{1}{4}(A'_{11} + B'_{11} + B'_{22} + C'_{11}) \\
\nu_a &= \tfrac{1}{2}(A'_{11} - C'_{11}) \\
&\pm \sqrt{[\tfrac{1}{4}(A'_{11} + C'_{11})(2(B'_{11} + B'_{22}) - (A'_{11} + C'_{11})) + (B'_{12})^2 - B'_{11}B'_{22}]} \\
\nu_b &= \tfrac{1}{2}(A'_{11} - C'_{11}) \\
&\mp \sqrt{[\tfrac{1}{4}(A'_{11} + C'_{11})(2(B'_{11} + B'_{22}) - (A'_{11} + C'_{11})) + (B'_{12})^2 - B'_{11}B'_{22}]} \\
J_{ab} &= (A'_{11} + C'_{11}) - (B'_{11} + B'_{22})
\end{aligned}\right\} \quad (5)
$$

These general relations prove that every sub-Hamiltonian consisting of an energy level diagram with grouping $(1 : 2 : 1)$ can be transformed into an ab sub-spectrum. The argument is also valid for the intensities of transitions because there is a unique relation between shift and intensity for this type of NMR spectrum.

The following sub-spectral transformations can be derived by applying the relations (5) to the AA′XX′ sub-Hamiltonian of Table 3.

A species

$$
\left.\begin{aligned}
\nu_a &= \nu_A + \tfrac{1}{2}(J_{AX} - J_{AX'}) \\
\nu_b &= \nu_A - \tfrac{1}{2}(J_{AX} - J_{AX'}) \\
J_{ab} &= J_{AA'} + J_{XX'}
\end{aligned}\right\} \quad (6)
$$

B species

$$
\left.\begin{aligned}
\nu_a &= \nu_A + \tfrac{1}{2}(J_{AX} - J_{AX'}) \\
\nu_b &= \nu_A - \tfrac{1}{2}(J_{AX} - J_{AX'}) \\
J_{ab} &= J_{AA'} - J_{XX'}
\end{aligned}\right\} \quad (7)
$$

Similarly the a_2 type sub-spectral transformations can be obtained by inspection of the corresponding matrix elements. They are as follows:

$$
\nu_a = \nu_A \pm \tfrac{1}{2}(J_{AX} + J_{AX'}) \quad (8)
$$

A theoretical illustration of the contributing sub-spectra in an AA′XX′ system is given in Fig. 1.

Whereas the a_2 relations are of the effective Larmor frequency type the general sub-spectral transformations (as, for example, ab) lead to effective Larmor frequency and effective coupling constants. The extreme values of m of the weakly coupled groups always characterize effective Larmor frequency sub-spectra (for example, $m(XX') = \pm 1$ in AA′XX′ systems). The procedure for obtaining sub-spectral transformations can be further simplified by invoking the condition that the sub-spectral transformations degenerate properly into effective Larmor frequency transformations in the case of magnetic equivalence. These are always linear combinations of Larmor frequencies and weak coupling constants. Therefore effective Larmor fre-

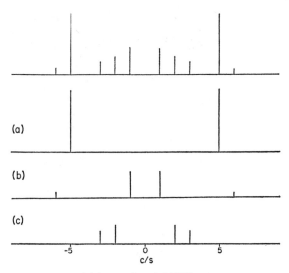

FIG. 1. AA' part of an AA'XX' spectrum.

$J_{AA'} = 3 \cdot 0 \, \text{c/s}, \quad J_{XX'} = 2 \cdot 0 \, \text{c/s}, \quad J_{AX} = 2 \cdot 5 \, \text{c/s}, \quad J_{AX'} = 7 \cdot 5 \, \text{c/s}.$

    (a) Both a_2 sub-spectra (transformation (8)).
    (b) Species A ab sub-spectrum (transformation (6)).
    (c) Species B ab sub-spectrum (transformation (7)).

quencies should never contain a contribution involving strong coupling constants. Similarly, the effective coupling constants in the case of magnetic equivalence are strong coupling constants so that it is highly likely, but it has not yet been proved rigorously, that the effective coupling constants should in general contain exclusively strong coupling contributions (relations (6) and (7)).

The relations (4) can be reformulated using capital letters for weak coupling and shift contributions and small letters for strong coupling contributions $(A'_{11} = A_{11} + a_{11})$. Furthermore, it has been shown that the constant offset ϵ can only contain strong coupling contributions.[22] These considerations lead to the relations (9) below,

$$\left.\begin{array}{ll} A_{11} = \frac{1}{2}\nu_a + \frac{1}{2}\nu_b, & a_{11} + \epsilon = \frac{1}{4}J_{ab}, \\ C_{11} = -\frac{1}{2}\nu_a - \frac{1}{2}\nu_b, & C_{11} + \epsilon = \frac{1}{4}J_{ab}, \\ B_{11} + B_{22} = 0, & b_{11} + b_{22} + 2\epsilon = -\frac{1}{2}J_{ab}, \\ B_{11}B_{22} - B_{12}^2 = -\frac{1}{4}(\nu_a - \nu_b)^2 \\ \epsilon^2 + \epsilon(b_{11} + b_{22}) + b_{11}b_{22} - b_{12}^2 = -3J_{ab}^2/16 \\ \epsilon(B_{11} + B_{22}) + B_{11}b_{22} + b_{11}B_{22} - 2B_{12}b_{12} = 0. \end{array}\right\} \quad (9)$$

There are nine equations for the four unknowns so that five equations must contain special conditions which a system must fulfil in order to allow the use of the working hypothesis. These special conditions are:

$$\left.\begin{array}{c} A_{11} = -C_{11}, \quad a_{11} = c_{11} \\ B_{11} = -B_{22}, \\ B_{11}(b_{11} - b_{22}) = -2B_{12}b_{12} \\ (b_{11} - a_{11})(b_{22} - a_{11}) - b_{12}^2 = 0 \end{array}\right\} \quad (10)$$

It can be seen that spin $\frac{1}{2}$ systems generally will have the properties required by the special conditions (10). The first three conditions result from the symmetry of the Hamiltonian with respect to an inversion of spins. The fourth condition states that the off-diagonal elements should not contain "weak" and "strong" contributions at the same time. The last condition fixes the magnitude of the off-diagonal elements.

The simplified sub-spectral transformations are:

$$\left.\begin{array}{c} \nu_a = A_{11} \pm \sqrt{(B_{11}^2 + B_{12}^2)} \\ \nu_b = A_{11} \mp \sqrt{(B_{11}^2 + B_{12}^2)} \\ J_{ab} = 2a_{11} - (b_{11} + b_{22}) \\ \epsilon = -\frac{1}{4}(2a_{11} + b_{11} + b_{22}) \end{array}\right\} \quad (11)$$

2.3. *Applications of* ab *Sub-spectral Transformations*

The general sub-spectral transformation procedure contains the effective Larmor frequency method as well as the composite particle approach as special cases. This will now be demonstrated in the ab sub-spectra of ABX and AB$_3$ systems.

2.3.1. *The* ABX *system*

It has been shown[13, 14] that the AB part of the ABX spectrum can be broken down into two ab sub-spectra characterized by $m_X = \pm\frac{1}{2}$:

$$AB(ABX)_{m_X} = (ab)_{+1/2} + (ab)_{-1/2}$$

The following contributions to the ABX sub-Hamiltonian are obtained using the symbols of relations (9):

$$A_{11} = \frac{1}{2}[\nu_A + \nu_B + m_X(J_{AX} + J_{BX})]; \qquad a_{11} = \frac{1}{4}J_{AB}$$
$$B_{11} = \frac{1}{2}[\nu_A - \nu_B + m_X(J_{AX} - J_{BX})]; \qquad b_{11} = -\frac{1}{4}J_{AB}$$
$$B_{22} = -\frac{1}{2}[\nu_A - \nu_B + m_X(J_{AX} - J_{BX})]; \qquad b_{22} = -\frac{1}{4}J_{AB}$$
$$B_{12} = 0; \qquad b_{12} = \frac{1}{2}J_{AB}$$
$$C_{11} = -\frac{1}{2}[\nu_A + \nu_B + m_X(J_{AX} + J_{BX})]; \qquad c_{11} = -\frac{1}{4}J_{AB}$$

The following sub-spectral transformations can be derived directly using the relations (11) because the conditions (10) are fulfilled.

$$\left.\begin{array}{l} \nu_a = \nu_A + m_X J_{AX} \\ \nu_b = \nu_B + m_X J_{BX} \end{array}\right\} \text{ or } \left\{\begin{array}{l} \nu_B + m_X J_{BX} \\ \nu_A + m_X J_{AX} \end{array}\right.$$
$$J_{ab} = J_{AB}$$

These relations can be recognized to be of the effective frequency type (see later illustrations).

2.3.2. The AB_3 system

The relevant parts of the Hamiltonian for $ab(AB_3)$ derived in terms of the symbols of the relations (9) are

$$\begin{array}{ll} A_{11} = \tfrac{1}{2}(\nu_A + \nu_B), & a_{11} = \tfrac{1}{4}J_{AB} \\ B_{11} = \tfrac{1}{2}(\nu_A - \nu_B), & b_{11} = -\tfrac{1}{4}J_{AB} \\ B_{22} = -\tfrac{1}{2}(\nu_A - \nu_B), & b_{22} = -\tfrac{1}{4}J_{AB} \\ B_{12} = 0, & b_{12} = \tfrac{1}{2}J_{AB} \\ C_{11} = -\tfrac{1}{2}(\nu_A + \nu_B), & c_{11} = \tfrac{1}{4}J_{AB} \end{array}$$

The conditions (10) are fulfilled so that the relations (11) give the trivial transformations

$$\begin{array}{l} \nu_a = \nu_A \\ \nu_b = \nu_B \end{array} \quad J_{ab} = J_{AB}$$

It is not necessary to pursue this approach for such systems composed of groups of magnetically equivalent nuclei since composite particle methods imply identity transformations (see Section 4).

2.4. ab_2 Sub-spectra

The ab_2 sub-spectral transformations corresponding to the ab transformations can be derived from the symmetrical part of the AB_2 Hamiltonian. There are six equations for four unknowns so that two conditions must be obeyed for the transformations to be valid. These conditions will not be listed; they can be shown to be general properties of spin $\tfrac{1}{2}$ systems.[22] The six equations can be divided into fourteen independent equations if the separation of the parts of the Hamiltonian as explained above is invoked. There are then ten conditions which must be satisfied in order to use the transformation formula and these again are general properties of spin $\tfrac{1}{2}$ systems. The ab_2 sub-spectral transformations are:

$$\epsilon = -\tfrac{1}{3}(a_{11} + b_{11} + b_{22})$$
$$\nu_a = 2A_{11} - 2(B_{11} + B_{22})$$
$$\nu_b = B_{11} + B_{22}$$
$$J_{ab} = \tfrac{1}{3}[4a_{11} - 2(b_{11} + b_{22})]$$

The transformations for an ab_2 sub-spectrum in the AA'A"XX'X" system[16, 17] are given as an example. The breakdown for the AA'A"XX'X" system (D_3 effective symmetry) is as follows:

$$AA'A''(AA'A''XX'X'') = (a_3)_{+3/2} + (a_3)_{-3/2} \qquad A + E \text{ species}$$
$$+ (ab_2)_{+1/2} + (ab_2)_{-1/2} \quad A_1 + A_2 \text{ species}$$
$$+ (abc)_{+1/2} + (abc)_{-1/2} \quad E \text{ species}$$

The Hamiltonian contributions for the ab_2 system have the following values:[16]

$$A_{11} = \tfrac{3}{2}\nu_A + \tfrac{1}{4}N, \qquad\qquad a_{11} = \tfrac{3}{4}K$$
$$B_{11} = \tfrac{1}{2}\nu_A + \tfrac{1}{12}N, \qquad\qquad b_{11} = \tfrac{3}{4}K$$
$$B_{22} = \tfrac{1}{2}\nu_A + \tfrac{1}{12}(2N - 3L), \quad b_{22} = -\tfrac{3}{4}K$$

where $K = J_{AA'} + J_{XX'}$, $N = 2J_{AX} + J_{AX'}$ and $L = 2J_{AX} - J_{AX'}$.

Therefore,

$$\epsilon = -\tfrac{1}{4}K = -\tfrac{1}{4}(J_{AA'} + J_{XX'})$$
$$\nu_a = \nu_A + \tfrac{1}{2}L = \nu_A + \tfrac{1}{2}(2J_{AX} - J_{AX'})$$
$$\nu_b = \nu_A + \tfrac{1}{12}(2N - 3L) = \nu_A + \tfrac{1}{2}J_{AX'}$$
$$J_{ab} = K = J_{AA'} + J_{XX'}$$

The sub-spectra of an AA'A"XX'X" system are shown in Fig. 2.

2.5. abc *Sub-spectra*

The sub-spectral transformations of abc systems are rather complex because they involve the manipulation of third-order matrices. There are eight equations for the seven unknown parameters ν_a, ν_b, ν_c, J_{ab}, J_{ac}, J_{bc} and ϵ. Therefore the Hamiltonian elements of the complex sub-spectrum must meet the requirements imposed by one condition which is

$$A'_{11} - (B'_{11} + B'_{22} + B'_{33}) = D'_{11} - (C'_{11} + C'_{22} + C'_{33})$$

This trace relation will be fulfilled by all spin $\tfrac{1}{2}$ systems containing the sub-spectral pattern (**1, 3, 3, 1**) because the elements A'_{11} and B'_{mn} can be derived from the elements D'_{11} and C'_{mn}, respectively, by spin inversion. Although this proves the existence of sub-spectral relations for the abc-type sub-spectra, these relations are usually non-analytical in terms of the sub-spectral parameters.[22] It is therefore necessary to use the separation of the Hamiltonian into strongly coupled, weakly coupled and mixed contributions in order to derive the analytical transformations. The following transformations

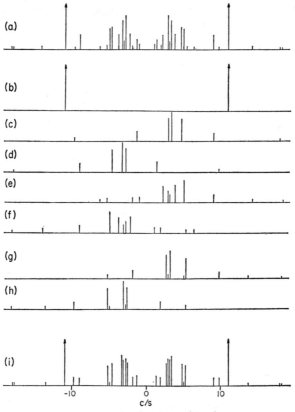

Fig. 2. An AA′A″XX′X″ spectrum and its composite sub-spectra.

$$J_{AA'} = 2 \cdot 0 \text{ c/s}, \quad J_{XX'} = 8 \cdot 0 \text{ c/s}, \quad J_{AX} = 10 \cdot 0 \text{ c/s}, \quad J_{AX'} = 2 \cdot 0 \text{ c/s}.$$

(a) The XX′X″ part (superposition of (b), (c), (d), (e) and (f)).
(b) Both x_3 ($\equiv a_3$) sub-spectra.
(c), (d) xy_2 ($\equiv ab_2$) sub-spectra.
(e), (f) xyz sub-spectra.
(g), (h) abc sub-spectra.
(i) The AA′A″ part (superposition of (b), (c), (d), (g) and (h)).
The arrows indicate that the lines have been reduced to one-half the actual intensity.

have been obtained[17] for abc sub-spectra of the AA′A″XX′X″ system as an example:

$$\nu_a = \nu_A \pm \tfrac{1}{2}(2J_{AX} - J_{AX'})$$
$$\nu_b = \nu_A \pm \tfrac{1}{2}J_{AX'}$$
$$\nu_c = \nu_A \pm \tfrac{1}{2}J_{AX'}$$
$$J_{ab} = J_{AA'} + \tfrac{1}{2}(\sqrt{3} - 1)J_{XX'}$$
$$J_{ac} = J_{AA'} - \tfrac{1}{2}(\sqrt{3} + 1)J_{XX'}$$
$$J_{bc} = J_{AA'} - \tfrac{1}{2}J_{XX'}$$

The abc sub-spectra are illustrated for a theoretical AA′A″XX′X″ system in Fig. 2.

The sub-spectral transformations for the xyz sub-spectra can be derived by interchange of the A, A′, A″ and X, X′, X″ nuclei. The coupling constants obviously change their values as a consequence of this operation, when $J_{AA'} \neq J_{XX'}$. Thus the abc and xyz sub-spectra can be quite different, unlike the ab_2 and xy_2 sub-spectra which are always identical. The sub-spectral transformation method alone has made it possible to see the precise analytical reason for the differences observed in the AA′A″ and XX′X″ parts of the AA′A″XX′X″ spectrum. The identical appearance of the AA′ and XX′ parts of an AA′XX′ spectrum can be explained in a similar way. The signs of J_{ab} in the respective anti-symmetrical ab sub-spectra are opposed but this sign reversal cannot be observed (see relations (7)).

2.6. aa′bb′ Sub-spectra and the Limitations of Sub-spectral Transformations

The sub-spectral transformations for an aa′bb′ sub-spectrum[17, 18] can be derived from the symmetrical (1, 2, 4, 2, 1) part of the sub-Hamiltonian. This results in ten equations for six unknowns: ν_a, ν_b, $k = J_{aa'} + J_{bb'}$, $n = J_{ab} + J_{ab'}$, $l = J_{ab} - J_{ab'}$ and ϵ. The parameter $m = J_{aa'} - J_{bb'}$ only affects the antisymmetrical (2, 2, 2) part. Two equations not needed to calculate the transformations lead to general conditions which are fulfilled in every spin $\frac{1}{2}$ system. The other two give special conditions which have been shown to be violated in the (1, 2, 4, 2, 1) sub-pattern of the general AA′A″A‴XX′ system, for example. Therefore, this sub-pattern cannot be identified with a real aa′bb′ sub-spectrum. In particular however, the requirements of the special conditions *are* satisfied when $J_{XX'} = 0$ in the AA′A″A‴XX′ system.

The number of transformation conditions rapidly increases for even more complex spectra and the general feasibility of sub-spectral transformations involving known simple spin systems cannot be guaranteed for sub-spectral patterns other than (1, 2, 1 ≡ ab), (1, 2, 2, 1 + 1, 1 ≡ ab_2) and (1, 3, 3, 1 ≡ abc). Consequently, the individual conditions which may be complex algebraic expressions have to be tested in every case. In such cases the practicability of sub-spectral analysis methods must be in doubt.

2.7. Approximate Validity of Sub-spectral Transformations

The (1, 2, 4, 2, 1) sub-Hamiltonian of the AA′A″A‴XX′ system considered in Section 2.6 can still be approximately treated as a part of an aa′bb′ spectrum. This is due to the fact that the conditions which are not satisfied stem from the two highest order secular polynomial coefficients of the fourth-order matrix. A spin system can be constructed which has eigenvalues in the first- and second-order matrices identical with those of the real spectrum using the sub-spectral transformations. The eigenvalues of the fourth-order matrix only fulfil the condition that their sum and their sum of products taken two at a time is identical with the real spectrum, whereas the sum of products taken three and four at a time respectively are different. The precision of such an approximation depends upon the magnitude of $J_{XX'}$.

3. EFFECTIVE LARMOR FREQUENCY SPECTRA

The principles of this method are straightforward[13, 14] and have been described in Section 1.3.4. Examples provide the best illustrations of the method.

3.1. *The* AB_2X *System*

The strongly coupled part of the AB_2X system can be written in the notation of sub-spectral breakdown as,

$$AB_2(AB_2X) = (ab_2)_{+1/2} + (ab_2)_{-1/2}$$

The effective Larmor frequencies are

$$\nu_a = \nu_A \pm \tfrac{1}{2}J_{AX}$$
$$\nu_b = \nu_B \pm \tfrac{1}{2}J_{BX}, \quad \text{respectively.}$$

The information immediately available in the proton spectrum of 3,5-dichlorofluorobenzene is illustrated in Fig. 3 where the sub-spectra are shown independently.

3.2. *The* AB_2X_2 *System*

The strongly coupled part of the AB_2X_2 system can be summarized in sub-spectral notation as

$$AB_2(AB_2X_2) = (ab_2)_{+1} + 2(ab_2)_0 + (ab_2)_{-1}$$

The spectrum consists of three ab_2 sub-spectra one of which has twice the intensity of each of the other two and has effective Larmor frequencies identical to the principal Larmor frequencies because of the non-magnetic state of the X nuclei, $m_X = 0$ (see Section 4.3.2).

An illustration of an AB_2X_2 system and its composite sub-spectra is given in Fig. 4.

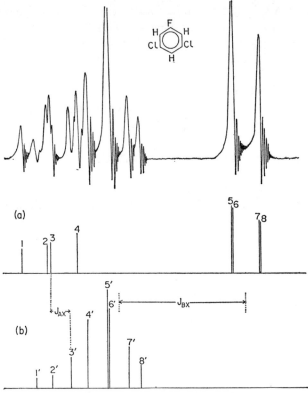

F_{IG}. 3. Proton resonance spectrum at 60 Mc/s of 3,5-dichlorofluorobenzene, an
AB$_2$X system. (a) and (b) ab$_2$ sub-spectra. The relevant data are indicated.

$$J_{ab} = J_{AB} = 1/3[(1 - 4) + (6 - 8)] = 1/3[(1' - 4') + (6' - 8')].$$
$$\delta_{AB} = 1/4(5 + 7 + 5' + 7') - 1/2(3 + 3')$$

From the fact that $1/2(5 + 7) > 1/2(5' + 7')$ but $3 < 3'$ it may be derived that J_{AX}
and J_{BX} have opposite signs.

4. COMPOSITE PARTICLE APPROACH

4.1. *Calculation Procedure*

When the composite particle approach is used alone, the sub-spectra are
simply due to different spin states of the molecule, and sub-spectral analysis
consists of recognizing patterns attributable to individual spin states (see
Section 1.3.2). A variety of spin systems may give rise to a sub-spectrum of a
common type as in the more general cases of sub-spectral analysis discussed
above. Essentially, however, the sub-spectra in composite particle theory are
equivalent to complete spectra for spin systems containing values $I > \frac{1}{2}$ as
well as $I = \frac{1}{2}$. The process of matrix transformation is no longer particularly

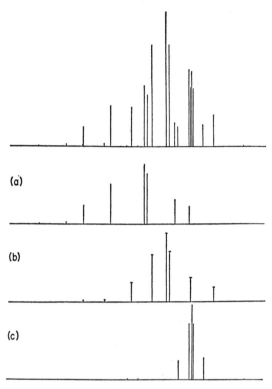

FIG. 4. An AB_2X_2 spectrum and its composite sub-spectra.

$$J/\delta = 0\cdot 8, \quad (J_{AX} - J_{BX})/\delta = 0\cdot 6.$$

(a) $ab_2(AB_2X_2)$, $J/\delta = 0\cdot 5$, $m_X = +1$.
(b) $ab_2(AB_2X_2)$, $J/\delta = 0\cdot 8$, $m_X = 0$ (singlet state) (one-half the intensity is shown).
(c) $ab_2(AB_2X_2)$, $J/\delta = 2\cdot 0$, $m_X = -1$.

useful since identity transformations are the general rule, but rather it is necessary to know the simple spectral patterns associated with individual spin states, and to know how to calculate a complete spectrum using composite particle ideas. The general calculation procedure may be divided into four parts, as follows:

1. A table of the possible overall spin states of the molecule, together with the appropriate statistical weights, is drawn up. The possible spin states are simply the products of the individual group spin states and the statistical weights are also simple products. Thus for the A_2B_6 spin system $F_B = $ Sp, Qt, T and S with $g_B = 1, 5, 9$ and 5 respectively (from equation (1)), and $F_A = $ T and S with $g_A = 1$ in each case. The possible overall spin states are given below:

Spin state	$S_A S_B$	$S_A T_B$	$S_A Q_{tB}$	$S_A S_{pB}$	$T_A S_B$	$T_A T_B$	$T_A Q_{tB}$	$T_A S_{pB}$
Statistical weight	5	9	5	1	5	9	5	1
Number of basis functions	1	3	5	7	3	9	15	21

The third column gives the number of basis functions $\prod F^m$ in each secular sub-determinant (neglecting the effect of g). This number is again obtained as a simple product of the multiplicities.

2. The matrix representation of the Hamiltonian for each overall spin state is then written down using the usual expression (2) where the summations are over all groups of nuclei, and ν_i is the Larmor frequency of group i given by $\nu_i = \gamma_i H_0(1 - \sigma_i)/2\pi$. It is necessary, of course, to use the Pauli spin matrices[50] appropriate to the values of F. The diagonal elements of $\mathscr{H}$ are then given quite simply by (12a).

$$\sum_i \{\nu_i m_i + \tfrac{1}{2} \sum_{l \neq i} J_{il} m_i m_l\} \tag{12a}$$

the quantum numbers m_i and m_l being the group values. The off-diagonal elements of $\mathscr{H}$ are non-zero only if the wave functions involved differ in quantum numbers m of two composite particles i and l only, and then are given by equation (12b):

$$\langle F_i^{m_i \pm 1} F_l^{m_l \mp 1} | \mathscr{H} | F_i^{m_i} F_l^{m_l} \rangle$$
$$= \tfrac{1}{2} J_{il} \{ [F_i(F_i + 1) - m_i(m_i \pm 1)][F_l(F_l + 1) - m_l(m_l \mp 1)] \}^{1/2} \tag{12b}$$

It is important to realize that at no stage is it necessary to define the spin wave functions $F_i^{m_i}$ in terms of basic spin $\tfrac{1}{2}$ product functions of the type $\alpha\beta\alpha\beta \ldots$ (for example).

3. Each matrix is factorized according to the total z component of the spin angular momentum, that is m_T. It is diagonalized to obtain eigenvalues and eigenfunctions in the usual way. The transition energies are obtained by subtraction of the eigenvalues using the selection rule $\Delta m_T = +1$.

4. The relative transition intensities may be obtained from the expression (13A):

$$Z_{pq} = 4g \langle \phi_p | \sum_i F_x(i) | \phi_q \rangle^2 \tag{13a}$$

where Z_{pq} is the relative intensity of the transition between eigenfunctions ϕ_p and ϕ_q which are expressed in terms of the basis functions. The appropriate Pauli matrices must again be used for $\sum_i F_x(i)$, but it is clear that this operator has a non-zero matrix element only between two basis functions which differ by ± 1 in the value of m for one of the composite particles alone. In such cases, the matrix element is given by equation (13b):

$$\langle F_i^{m_i \pm 1} | F_x(i) | F_i^{m_i} \rangle = \tfrac{1}{2} \{ F_i(F_i + 1) - m_i(m_i \pm 1) \}^{1/2} \tag{13b}$$

Transitions between the different overall spin states determined in step 1 are completely forbidden and the total eigenspectrum problem may indeed be treated as the superposition of several separate eigenspectrum problems. The composite particles therefore provide another example of sub-spectral analysis. The expression (13a) is normalized to give a total spectral intensity of $n2^{n-1}$ for a spin system containing n spin $\frac{1}{2}$ nuclei. The intensity due to each composite particle for a given spin state is readily calculated in the first-order case and is given by (14a). Thus in general the total intensity attributable to each spin state is given by (14b):

$$\tfrac{2}{3}g F_i \, (F_i + 1) \prod_j \, (2F_j + 1) \qquad (14a)$$

$$\tfrac{2}{3}g \sum_i \, [F_i(F_i + 1)] \prod_j \, (2F_j + 1) \qquad (14b)$$

Each spin state yields the number of distinct first-order lines given by (15), and extra combination lines are normally of negligible intensity in systems of only two composite particles (but see Section 4.3.1).

$$\sum_i \, [2F_i \prod_{j \neq i} \, (2F_j + 1)] \qquad (15)$$

The appearance of g in equation (13a) is the only effect of the number of nuclei in the various composite particles on the spectrum of a given spin state. Therefore, except for g, the equations derived for the transition frequencies and intensities of, say, the DT spin state of an AB_4 system are general for the DT spin state of any spin system (for example, for an AB_6 or A_3B_4 case). They are equally valid for corresponding states of spin systems involving nuclei with spin greater than $\frac{1}{2}$, for example the DT spin state of an AB_3 spin system with $I_B = 1$.

4.2. *Systems of Two Composite Particles*

The general case of spin systems of the type A_aB_b has been discussed[4, 9, 46] but the illustrations were limited to a single example of the AB_b system.[9] The specific cases A_2B_6[10] and AB_4[23] have been described with examples. The AB_b case is of particular interest because only second-order sub-matrices need to be diagonalized and consequently expressions for line positions and intensities can be obtained in closed form. The general appearance of the spectrum is easy to recognize and the coupling constant J_{AB} and the chemical shift δ_{AB} may be obtained using simple equations and line separations measured from the spectrum. It is not necessary to turn to a computer to completely analyze the spectrum, although a least-squares fitting of the data will be much easier if a computer is used. Consequently it is worth while to consider specific examples and to make some general comments on the spectra.

The AB_2 case contains all the general features of the AB_{2r} spin system (there are slight differences for AB_{2r+1} systems), with the minimum number of nuclei. The features of this spin system have been described several times[3] from a group theoretical approach (as in Section 3) which correlates readily with the composite particle approach used here (see Section 5). The lone A nucleus can only exist with $F = \frac{1}{2}$, that is in a doublet D state. The composite particle of two B nuclei may have $F = 1$ or 0 corresponding to T or S states respectively. In this case all statistical weights are unity. There are therefore two separate overall spin states, DS and DT, i.e.

$$AB_2 = DS + DT$$

It should be emphasized that the composite particle technique features a sub-spectral breakdown of the whole spectrum and not just the part assigned to strongly coupled nuclei (as is the case for the Effective Larmor Frequency approach considered in Section 3). The matrix for the DS state is simply as given in Table 4. It is clear from Table 4 that there is a single transition at ν_A.

TABLE 4. MATRIX FOR THE DS STATE OF THE AB_2 SYSTEM

m_T	Spin function	Diagonal energy terms
$\frac{1}{2}$	$D^{1/2}S^0$	$\frac{1}{2}\nu_A$
$-\frac{1}{2}$	$D^{-1/2}S^0$	$-\frac{1}{2}\nu_A$

It has unit intensity. This provides an example of the fact that a group of nuclei in a singlet spin state does not contribute to the energies of the transitions (or the intensities when the factor g is omitted). Therefore particles in singlet spin states can be ignored (see Section 4.3.2), i.e. in this instance the DS state reduces effectively to a D state. The matrix for the DT state is given in Table 5. Eight transitions are possible between eigenstates of the DT

TABLE 5. MATRIX FOR THE DT STATE OF THE AB_2 SYSTEM

m_T	Spin function	Energy terms	
		Diagonal	Off-diagonal
$\frac{3}{2}$	$D^{1/2}T^1$	$\frac{1}{2}\nu_A + \nu_B + \frac{1}{2}J_{AB}$	
$\frac{1}{2}$	$D^{-1/2}T^1$ $D^{1/2}T^0$	$-\frac{1}{2}\nu_A + \nu_B - \frac{1}{2}J_{AB}$ $\frac{1}{2}\nu_A$	$2^{-1/2}J_{AB}$
$-\frac{1}{2}$	$D^{-1/2}T^0$ $D^{1/2}T^{-1}$	$-\frac{1}{2}\nu_A$ $\frac{1}{2}\nu_A - \nu_B - \frac{1}{2}J_{AB}$	$2^{-1/2}J_{AB}$
$-\frac{3}{2}$	$D^{-1/2}T^{-1}$	$-\frac{1}{2}\nu_A - \nu_B + \frac{1}{2}J_{AB}$	

spin state. In the AX_2 first-order limit three of the lines are in the vicinity of ν_A and may be denoted as A lines, all three having unit intensity. Four other lines may be assigned to the B nuclei and each has relative intensity two. The final line occurs at $\nu_A + 2(\nu_A - \nu_B)$ and is forbidden in the first-order limit; it may be designated a combination line.

Several features of the AB_2 spectrum, together with their generalizations, may be noted at this point.

1. The spectrum, apart from a scaling factor depends only on one variable, namely the ratio $R = J_{AB}/\delta_{AB}$ but the resolvability of the lines depends upon the magnitudes of J_{AB} and δ_{AB}. This feature is common to all A_aB_b spin systems.

2. One pair of B lines are symmetrical in frequency (though not in intensity) about ν_B. In general for a D_AF_B state, there will be $(2F_B - 1)$ such pairs of lines.

3. The DS line remains constant in position, at ν_A, and in intensity regardless of the value of R. Such a line is always present for the A resonance of any A_aB_b spin system if b is even. It has fractional intensity $g_{DS}/n2^{n-1}$, where $n = a + b$.

4. As R increases, all A lines on the side of ν_A remote from ν_B decrease in intensity and migrate away from the remaining lines. Similar behaviour is shown by certain B lines which are most remote from ν_A (in the AB_2 case there is only one line of this type).

5. These features make assignment of the spectrum relatively straightforward, and closed-form expressions may be obtained relating J_{AB} and δ_{AB} to measured line separations.† Clearly it is best in the general case to use the simplest sub-spectra (i.e. DD or DT) for this purpose.

The application of composite particle theory to the AB_4 case has been described in detail.[23] It is particularly useful for pentafluorosulphur[48, 49] compounds SF_5X, where X is an effectively non-magnetic atom or group, since in these compounds there is a unique axial fluorine nucleus and four magnetically equivalent equatorial fluorine nuclei. There are three overall spin states D_AQt_B, D_AT_B and D_AS_B with statistical weights one, three and two respectively, which can be expressed alternatively by,

$$AB_4 = DQt + 3DT + 2DS$$

This means that there are the same spin states as for the AB_2 case (giving the same expressions for energies and intensities of transitions, but with different statistical weights) plus the D_AQt_B state. The D_AQt_B state yields thirteen first-order lines, giving a total of twenty-one lines (excluding combination lines) for the complete AB_4 spectrum. This number is to be compared with the

† The expression for δ_{AB} given for the AB_4 spin system in reference (23) contains an error. It should read
$$\delta_{AB} = (E_7 - E_1) + \tfrac{1}{2}(E_4 - E_7)$$

eighty first-order lines of the general five spin $\frac{1}{2}$ system. The simplifications due to symmetry can be seen to be very appreciable. The schematic energy level diagram for the AB_4 spin system is shown in Table 6. It should be

TABLE 6. SCHEMATIC ENERGY LEVEL DIAGRAM
FOR THE AB_4 SYSTEM

m_T	Spin state		
	DQt	DT($g = 3$)	DS($g = 2$)
$\frac{5}{2}$	1		
$\frac{3}{2}$	2	1	
$\frac{1}{2}$	2	2	1
$-\frac{1}{2}$	2	2	1
$-\frac{3}{2}$	2	1	
$-\frac{5}{2}$	1		

noted at this stage that the three sets of energy levels for the DT state are not necessarily degenerate (they are only so if there is a single magnitude for coupling within the magnetically equivalent group, a condition not satisfied by SF_5X compounds). However, the transition energies *are* degenerate and it is therefore legitimate to use a higher degree of symmetry than that displayed by the molecule itself (this is implicit in the composite particle approach; the point will be discussed in Section 5.1). Figure 5 shows the composite sub-spectra of AB_4 spin systems for various values of R. The effects mentioned in the discussion of AB_2 spectra can be clearly seen. Once the lines are assigned to sub-spectra, values of J_{AB} and δ_{AB} can be obtained from the AB_2 closed-form expressions.

The A_3B_2 case is one of importance because of the common occurrence of ethyl groups.[45] There are four overall spin states, namely QT, DT, QS and DS with statistical weights 1, 2, 1 and 2, respectively,
i.e.

$$A_3B_2 = QT + 2DT + QS + 2DS$$

The transition energies and intensities (except for g) of the DT and DS states are identical with those already discussed for the AB_2 spin system. The QS state gives a single line of relative intensity ten. Thus the only state for which further calculations need to be performed is the QT state, which in the first order limit gives rise to nine A lines (total relative intensity, thirty) and eight B lines (total relative intensity, sixteen). Unfortunately this state supplies one-half of the B intensity and five-eights of the A intensity. Further, closed-form expressions can only be given for four of the lines. Calculation of transition energies for the remaining thirteen lines requires the diagonalization of third-order sub-matrices.

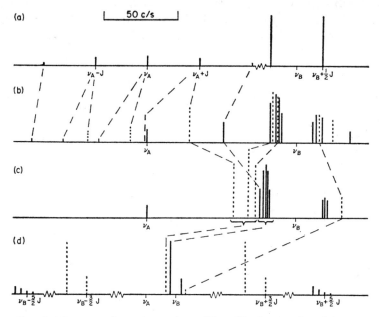

FIG. 5. AB$_4$ spectra for various values of R $= J/\delta$, showing the sub-spectra.

(a) $\delta \to \infty$, $J = 35$ c/s, R $\to 0$ (AX$_4$ case).
(b) $\delta = 100$ c/s, $J = 35$ c/s, R $= 0.35$.
(c) $\delta = 100$ c/s, $J = 200$ c/s, R $= 2.00$.
(d) $\delta = 20$ c/s, $J \to \infty$, R $\to \infty$.

The sub-spectra are not differentiated in (a), as lines coincide. In (b), (c) and (d) the line at ν_A is the DS sub-spectrum; the other full lines belong to the DQt sub-spectrum, while the DT lines are dotted. The intensities of (a) and the central lines of (d) have been reduced by a factor of 4, while the "satellite" lines near $\nu_B \pm 3J/2$ and $\nu_B \pm 5J/2$ of (d) have been increased by a factor of $64/625R^2$. The lines off-scale in (c) are those corresponding to the "satellite" lines of (d), and are very weak.

4.3. *Special Cases*

4.3.1. *Limiting Cases of* A$_a$B$_b$ *Spectra*

The appearance of spectra of the A$_a$B$_b$ type depends only on the ratio R $= J_{AB}/\delta_{AB}$ and limiting cases are obtained as R approximates either to zero or infinity. The former case gives, ultimately, first-order spectra ($\delta \gg J$); the latter, ultimately, a single line if $\delta = 0$ (but see Section 7.1). However, cases where $J \gg \delta$ but $\delta \neq 0$ are of some interest. Methyl acetylene has been treated[24] as an AB$_3$ system under such conditions using a perturbation approach based on the quantum numbers F_A, F_B, F (associated with $\mathbf{F} = \mathbf{F_A} + \mathbf{F_B}$) and m_T. The results are expressed in terms of R. Table 7 gives the

B

values for the AB_4 case, transposed into the notation used here, and Fig. 5(d) shows a calculated spectrum. The values in Table 7 may also be obtained from the general expressions for AB_4 transitions[23] by making approximations suitable for $R \to 0$.

TABLE 7. CALCULATED AB_4 LINE POSITIONS AND RELATIVE INTENSITIES FOR THE LIMITING CASE $R(J/\delta) \to 0$

	State	Line position[a]	Relative intensity	
Principal lines	DS	δ	2	
	DT	$-\delta/3$	3	
		$\delta/3$	30[b]	
	DQt	$-\delta/5$	10[b]	
		$\delta/5$	35[b]	
Satellite lines[c]	DT	$3J/2$	1[d]	
	DT	$3J/2 + 2\delta/3$	3	$\times\ 8/27R^2$
	DQt	$5J/2 - 2\delta/5$	1[d]	
		$5J/2$	3[d]	
		$5J/2 + 2\delta/5$	6[d]	$\times\ 8/625R^2$
		$5J/2 + 4\delta/5$	10	

  [a] Relative to ν_B, with $\delta = \nu_A - \nu_B$.
  [b] These intensities arise from superposition of 3, 3 and 5 lines respectively of the general AB_4 case.
  [c] Corresponding lines occur at positions obtained by reversing the sign of the terms in J_{AB}.
  [d] The lines corresponding to these are actually combination lines in the general AB_4 case.

4.3.2. Singlet states

It has been pointed out in Section 4.2 that the DS state of an AB_2 spin system gives rise to a single line at ν_A, as would be the case in the absence of the B nuclei. Thus the singlet state has no effect on the spectrum. This illustrates the general principle that a group of magnetically equivalent nuclei in an S state is effectively non-magnetic. This means that sub-spectra for molecular spin states containing an S group are not dependent on the chemical shift of that group or any coupling constants to it. Such S states only occur for a magnetically equivalent group containing an *even* number of spin $\frac{1}{2}$ nuclei. If such a group is present, the spin system *minus* that group will always be a sub-system and will give particularly simple sub-spectra. Thus an A_3B_2C system, as in 1,1-dichloropropane, contains an a_3c sub-system since the methylene protons may be in a singlet spin state. However, the degeneracy factor, g, to be used for each spin state, remains that appropriate to the complete spin system, as always. If a spin system contains more than one magnetically equivalent group with an even number of nuclei, there will, of course, be further simplifications. Thus the $A_3B_2C_2D$ system contains an a_3d sub-system. If there is only one magnetically equivalent group with an *odd* number of nuclei, there will always be a line at the Larmor frequency of those nuclei. This has been pointed out in Section 4.2 for the AB_{2r} spin system. Similarly the $A_3B_2C_2$ system contains an a_3 sub-system because of the possibility of singlet spin states for B_2 and C_2; hence there is a line at ν_A. These principles are particularly useful for the AB_2X_2 spin system discussed in Section 3 since:

$$AB_2X_2 = DTT + DTS + DST + DSS$$

Only the DTT state involves more than two effectively magnetic groups of nuclei.

4.3.3. *Simultaneous existence of magnetic and chemical equivalence*

If additional symmetry is present, over and above that included in the composite particle notation, it must be dealt with explicitly. However it is not necessary to consider such additional symmetry *between different* composite particle spin states.[25] For instance, the $AA'B_2B_2'$ system contains states as follows:

$$AA'B_2B_2' = DDSS + DDST + DDTS + DDTT$$

The DDTT and DDSS states may be further factorized in accordance with the symmetry. The DDTS and DDST states could in principle be mixed to form symmetrical and anti-symmetrical combinations. However, there is no point in doing this since composite particle principles indicate that the DDTS and DDST states are degenerate and do not mix. Thus the transitions for these states are best derived using *un*-symmetrical wave functions, and the total spectrum may be summarized as:

$$AA'B_2B_2' = DDSS(s) + DDSS(a) + 2DDTS + DDTT(s) + DDTT(a),$$

where (*s*) and (*a*) indicate symmetrical and antisymmetrical wave functions, respectively.

Since this type of spin system, that is $AA'B_nB_n'$ or $AA'X_nX_n'$, is met in later sections, it is appropriate at this point to comment on the parameters involved. There are four coupling constants that affect the spectrum (as is well known from the AA'BB' case [3]). These are indicated in the diagram:

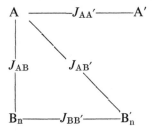

The coupling between B nuclei *within* the B_n and B_n' groups does not, of course, lead to any spectral changes. The (A, B) coupling constants are often more conveniently described by the linear combinations $N = J_{AB} + J_{AB'}$ and $L = J_{AB} - J_{AB'}$. Substitution of X for B gives the notation for the first-order case.

5. THE RELATION BETWEEN GROUP THEORY AND COMPOSITE PARTICLE APPROACHES

5.1. *General*

The symmetry group appropriate to the NMR problem comprises all permutations of the nuclei which do not alter the NMR parameters (coupling constants and chemical shifts).[7] Such a definition assumes that averaging is carried out over all chemical and conformational changes that occur rapidly with respect to the NMR time scale at the temperature considered. The appropriate symmetry for a group of n magnetically equivalent nuclei is that of a permutation group P_n because coupling within such a group of nuclei does not effect line positions or intensities.† Such a classification will frequently result in the use of higher effective symmetry than that determined by a rigid molecular geometry. Thus four magnetically equivalent nuclei may be classified according to P_4, irrespective of whether the molecule has D_{2h} symmetry (as for ethylene), D_{2d} (allene), C_{4v} (SF_5Cl and, due to exchange, PF_4Cl) or T_d (methane) symmetry. Similarly, the fluorine nuclei in PF_5 may be classified according to P_5 due to intramolecular exchange, although there is no simple geometrical counterpart. It has been shown that in these circumstances, for spin $\frac{1}{2}$ nuclei, there is a one-to-one correlation between the total spin F of the group and the irreducible representation of the permutation group.[21] This is the basis of the composite particle approach, and the result means it is not necessary to consider explicitly the symmetry of magnetically equivalent groups of nuclei. Additional symmetry must be considered explicitly as stated in Section 4.3.3.

It is perhaps instructive at this stage to mention that for several groups of magnetically equivalent nuclei, with no additional symmetry, the appropriate NMR group is $P_{n_1} \times P_{n_2} \times \ldots$, where the direct product is over all groups i of n_i nuclei.

The simplification that may be achieved by using the composite particle approach (that is, permutation groups) is well illustrated by the case of four

---

† Usually no account is taken of this simplying condition when only the classical symmetry groups are considered. Hence the matrix transformations must be carried out before it becomes apparent that certain sets of transitions are degenerate. This illustrates the power of the composite particle approach.

magnetically equivalent nuclei of local symmetry D_4 or D_{4h}, mentioned above. The use of D_4 symmetry alone produces the energy level scheme of Table 8(a), while Table 8(b) is the result of the composite particle approach (an equivalent breakdown can be achieved using the T_d rigid point group). Table 8(a)

TABLE 8. FACTORIZATION OF THE SECULAR MATRIX FOR FOUR NUCLEI USING (a) D_4 SYMMETRY AND (b) COMPOSITE PARTICLE APPROACH (P_4 SYMMETRY)

m_T	(a) D_4 symmetry				(b) Composite particle		
	A_1	B_1	B_2	E†	S†	T‡	Q†
2	1						1
1	1		1	1		1	1
0	2	1	1	1	1	1	1
−1	1		1	1		1	1
−2	1						1

† Doubly degenerate.   ‡ Triply degenerate.

does not make it clear that the B_2 functions give transition energies degenerate with those of the E species (these two correlate with the T composite particle state) until the matrix transformations are carried out. Moreover, it is not immediately obvious from Table 8(a) that one of the A_1 functions with $m_T = 0$ does not mix with the other; nor that no transitions are allowed which involve this level, nor that for the purposes of calculating transition energies in the full molecular spin system it is degenerate with the B_1 level. The further information that all transition energies are degenerate for any isolated group of equivalent nuclei is needed to make these points clear.

In principle, of course, it should be noted that use of the permutation symmetry groups as defined above[7] leads to identical factorization of the Hamiltonian matrix as does the composite particle method, for spin $\frac{1}{2}$ nuclei. The calculations are, however, much simpler using the latter.

5.2. The Non-rigid AB₂X₂ System

The non-rigid AB_2X_2 system has been mentioned[15] as a further system in which the special conditions of effective frequencies apply because of the magnetic equivalence of the two B nuclei and the two X nuclei as a result of intramolecular torsional oscillations. However, the basic symmetry wave function diagram was not derived because the appropriate point group was not determined.

It is the intention here to show the way in which it is possible to elucidate the symmetry point group for the non-rigid AB_2X_2 system (from first principles) with particular regard to the specific requirements of the NMR problem involved.

The basic ideas and definitions necessary for the extrapolation of our ideas about the symmetry point groups of rigid molecules to those of non-rigid molecules have been set out.[6] The clearest and most useful definition of the molecular symmetry point group is that it represents all feasible permutations of the positions and spins of identical nuclei or any product of such permutations, including E, the identity, and all feasible P^* ($PE^* = E^*P$) not necessarily including E^*; where E^* is defined as the inversion of all particle positions through the centre of mass.

The simple product spin functions used as a representation of an NMR problem are invariant under E^*.[6] They are, therefore, transformed by P^* in the same way as by P. Hence it is only necessary to use the P elements of the group to characterize the symmetry point group for the NMR problem. This point group is usually a sub-group of the molecular symmetry point group.

The feasible P for the non-rigid AB_2X_2 system can be defined using the numbering notation of Fig. 6.

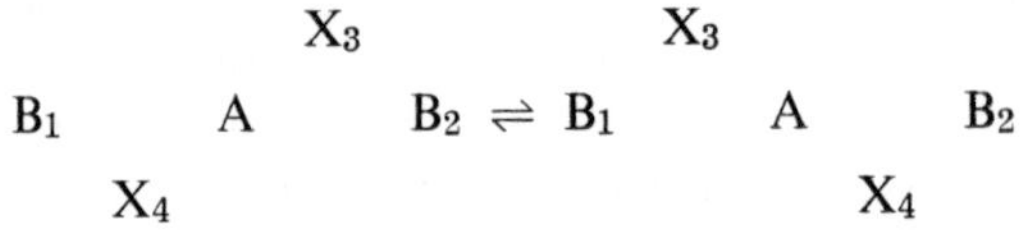

$$\begin{matrix} X_3 & & X_3 \\ B_1 \qquad A \qquad B_2 & \rightleftharpoons & B_1 \qquad A \qquad B_2 \\ X_4 & & X_4 \end{matrix}$$

FIG. 6. Model for the non-rigid AB_2X_2 system.

They are E, (12), (34) and (12)(34). The multiplication table for the group is shown in Table 9.

TABLE 9. THE MULTIPLICATION TABLE FOR THE ELEMENTS OF THE SYMMETRY GROUP OF THE NON-RIGID AB_2X_2 SYSTEM

	E	(12)	(34)	(12)(34)
E	E	(12)	(34)	(12)(34)
(12)	(12)	E	(12)(34)	(34)
(34)	(34)	(12)(34)	E	(12)
(12)(34)	(12)(34)	(34)	(12)	E

Every element of the group is conjugate with itself but no other element and therefore each element belongs to a unique class.[19] There are four classes and therefore there are four irreducible representations.

It is possible to derive the characters of these four irreducible representations on the basis of the vector properties of the representations and four important rules (see reference 19, p. 64). A basis for a representation of a group may be any set of algebraic functions or vectors. Let the vectors $\overrightarrow{12}$ and $\overrightarrow{34}$ be the basis, at least in part, for the group under consideration. These vectors transform as follows:

$$
\begin{array}{cccc}
\text{E} & (12) & (34) & (12)(34) \\
\overrightarrow{12} & -\overrightarrow{12} & \overrightarrow{12} & -\overrightarrow{12} \\
\overrightarrow{34} & \overrightarrow{34} & -\overrightarrow{34} & -\overrightarrow{34}
\end{array}
$$

The matrix representations are

$$
\begin{array}{cccc}
\text{E} & (12) & (34) & (12)(34) \\
\begin{bmatrix} 1, & 0 \\ 0, & 1 \end{bmatrix} &
\begin{bmatrix} -1, & 0 \\ 0, & 1 \end{bmatrix} &
\begin{bmatrix} 1, & 0 \\ 0, & -1 \end{bmatrix} &
\begin{bmatrix} -1, & 0 \\ 0, & -1 \end{bmatrix}
\end{array}
$$

The vectors $\overrightarrow{12}$ and $\overrightarrow{34}$ are not mixed by the elements E, (12), (34), (12)(34) and therefore each forms an independent representation of the group.

$$
\begin{array}{ccccc}
 & \text{E} & (12) & (34) & (12)(34) \\
\Gamma_{\overrightarrow{12}} & 1 & -1 & 1 & -1 & \overrightarrow{12} \\
\Gamma_{\overrightarrow{34}} & 1 & 1 & -1 & -1 & \overrightarrow{34}
\end{array}
$$

These representations are orthogonal and the sums of the squares of the characters in these representations equal the order of the group (four elements). Thus they satisfy two of the four rules. The product of their characters within each class produces a new representation,

$$
\begin{array}{ccccc}
\Gamma = \Gamma_{\overrightarrow{12}} \times \Gamma_{\overrightarrow{34}} & \text{E} & (12) & (34) & (12)(34) \\
 & 1 & -1 & -1 & 1
\end{array}
$$

which is orthogonal to both of them and therefore independent. The remaining independent representation which must be orthogonal to those defined above has the dimension $\chi(\text{E}) = 1$ (third rule in reference 19).

$$
\begin{array}{cccc}
\text{E} & (12) & (34) & (12)(34) \\
1 & 1 & 1 & 1
\end{array}
$$

The character table for the point group of the non-rigid AB_2X_2 system is therefore

	E	(12)(34)	(34)	(12)	
Γ_1	1	1	1	1	
Γ_2	1	1	-1	-1	$\overrightarrow{34}$
Γ_3	1	-1	1	-1	$\overrightarrow{12}$
Γ_4	1	-1	-1	1	

This point group is isomorphic with C_{2v}, the point group of the rigid AB_2X_2 system. It is obvious therefore that the factorization of the secular determinant will be identical both for the rigid and non-rigid AB_2X_2 system. The AB_2 part of the spectrum is composed of three superimposed ab_2 sub-spectra corresponding to $m_X = 1, 0, -1$ with statistical weight $1 : 2 : 1$ respectively as predicted by the effective frequency method.[13, 14] Table 10 gives the schematic factorization of the secular matrix. The X_2 part of the spectrum can be isolated in Table 10 by conserving $m(AB_2)$ $(\frac{3}{2}, \frac{1}{2}, -\frac{1}{2}, -\frac{3}{2})$.

TABLE 10. SCHEMATIC DIAGRAM OF THE FACTORIZATION OF THE SECULAR MATRIX OF THE AB_2X_2 SYSTEM

m_T	A_1			A_2	B_1			B_2
$\frac{5}{2}$	1							
$\frac{3}{2}$	2	1			1			1
$\frac{1}{2}$	2	2	1	1	1	1		2
$-\frac{1}{2}$	1	2	2	1		1	1	2
$-\frac{3}{2}$		1	2				1	1
$-\frac{5}{2}$			1					
m_X	1	0	-1	0	1	0	-1	0

The $(1, 1, 1 + 1)$ sub-patterns give rise to lines at ν_X. The $(2, 2, 2)$ sub-patterns give rise to explicit transition energies but cannot be identified with a well-known independent system.

6. NUCLEI WITH SPIN $>\frac{1}{2}$

6.1. *Groups of Equivalent Nuclei*†

An interesting situation (of theoretical interest alone) arises in the case of magnetically equivalent particles of spin $>\frac{1}{2}$. It appears there is no longer a one-to-one correspondence between the full symmetry and the composite particle methods, and that classification of spin states by the two methods is different for groups of three or more particles. This is illustrated in Table 11 for three particles of spin 1. From this table it can be seen that the Sp functions belong to A_1, S to A_2, and Qt to E. The three independent T functions for each eigenvalue of $\mathbf{F}_z$ transform as $A_1 + E$. It would seem, then, that classification of the functions by symmetry, in addition to eigenvalues of $\mathbf{F}^2$ brings further simplification for the spin 1 case. But as the Hamiltonian can be written entirely in terms of $\mathbf{F}$ operators, the three functions *are*

† This sub-section is taken from unpublished work by Lynden-Bell.

TABLE 11. CLASSIFICATION OF SPIN STATES FOR THREE EQUIVALENT SPIN 1 PARTICLES

m_T	No. of basis functions	Typical function (m values)	Symmetry approach (S_3)	Composite particle approach†
3	1	111	A_1	Sp(1)
2	2	110	$A_1 + E$	Sp(1) + Qt(2)
1	$\begin{cases} 3 \\ 3 \end{cases}$	100 11–1	$\left.\begin{array}{l} A_1 + E \\ A_1 + E \end{array}\right\}$	Sp(1) + Qt(2) + T(3)
0	$\begin{cases} 6 \\ 1 \end{cases}$	1–10 000	$\left.\begin{array}{l} A_1 + 2E + A_2 \\ A_1 \end{array}\right\}$	Sp(1) + Qt(2) + T(3) + S(1)

† Number of independent functions in brackets.

degenerate, and the composite particle method always gives the maximum factorization. If, on the other hand, permutation symmetry alone were used, mixing could still in principle occur—for example, between the two A, or E functions for $m_T = 0$.

6.2. Correlations with Spectra from Spin ½ Nuclei

As is apparent from the discussion of Section 4, the composite particle approach for groups of magnetically equivalent spin ½ nuclei uses basis spin wave functions that are essentially those appropriate to single spins which may have $I > \frac{1}{2}$. Thus each composite particle sub-spectrum is the *complete* spectrum for a spin system which may contain spins $>\frac{1}{2}$. For instance, the $AA'X_2X_2'$ system (see Section 4.3.3) may be factorized as the sum of DDSS, DDST, DDTS and DDTT states. The DDTT sub-spectrum is identical to the whole spectrum of a molecule of type AA'XX' with $I(X) = I(X') = 1$. An instance of a molecule falling into this class has been reported[47] for the case of N_2F_2. In fact the ^{19}F NMR spectra were used to show that both isomers of N_2F_2 (i) and (ii) have the 1,2-difluorodiazine structure.

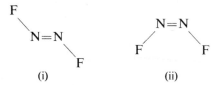

(i)   (ii)

In these particular cases, the $J_{XX'} = 0$ approximation was eventually used,[47] and the equations became identical to those derived for $AA'X_nX_n'$ systems with $J_{XX'} = 0$ and all spins ½ (Section 10.1). Due to the algebraic simplifications (Section 7.4) lines occur in the N_2F_2 spectra at *all* of the positions of the A resonances of the $AA'X_2X_2'$ spin ½ system, although the relative intensities are different. In general, however, it is dangerous to apply equations obtained for spin $\frac{1}{2}$ cases to those of spin 1 systems since only *some* of the transitions are common to both. In particular, it is not in general advisable to construct model systems containing only spin $\frac{1}{2}$ nuclei to simulate spin $>\frac{1}{2}$ cases (e.g. a hypothetical H_2C—CH_2 of D_{2d} symmetry to simulate D—C≡C—D). At the very least this increases the amount of work rather than reducing it; at the worst, since the choice of an appropriate model symmetry is difficult, spurious results may be obtained unless great care is taken. However, a more general treatment may be of great value if computation is employed. One particular computer programme,[43] written in FORTRAN, calculates the spectra of up to ten composite particles (subject only to the size of the computer store, but normally limited by a maximum size of sub-matrix that can be accommodated for diagonalization). The results are presented independently for each sub-spectrum; thus the proramme may be used for spin $>\frac{1}{2}$ nuclei by selection of an appropriate sub-spectrum of a spin $\frac{1}{2}$ composite particle calculation, the redundant results being thrown away.

7. ALGEBRAIC SIMPLIFICATIONS

7.1. *Deceptively Simple Spectra*[1, 28]

Deceptively simple spectra are spectra in which one, several, or all the existing sub-spectra are degenerate in that they approach collapse into a single line. The conditions for the existence of deceptively simple spectra have been derived[1] for the specific cases ABX, AA'XX' and ABXY. These conditions have been generalized for NMR spectra of systems of the type AB_nX_q using the principles of effective Larmor frequencies.[28] The discussion of deceptively simple spectra can be further extended and simplified by applying the method of sub-spectral analysis as for example in the AA'XX' system. The AA' part of this spectrum is a superposition of two a_2 and two ab sub-spectra. It appears deceptively simple if one or both ab sub-spectra degenerate into single lines. The conditions for degeneracy of AB spectra are

(a) for the shift $\delta^2/2J < \Delta_{1/2}$, and
(b) for the amplitude $\delta^2/2J^2 < i$.

Substitution of the effective shifts and effective coupling constants of the AA'XX' ab sub-spectra (relations (6) and (7)) immediately gives the following conditions for the degeneration of one sub-spectrum;

either	or
$\frac{1}{8}(J_{AX} - J_{AX'})^2/|J_{AA'} + J_{XX'}| < \Delta_{1/2};$	$\frac{1}{8}(J_{AX} - J_{AX'})^2/|J_{AA'} - J_{XX'}| < \Delta_{1/2}$
$\frac{1}{8}(J_{AX} - J_{AX'})^2/(J_{AA'} + J_{XX'})^2 < i;$	$\frac{1}{8}(J_{AX} - J_{AX'})^2/(J_{AA'} - J_{XX'})^2 < i.$

Degeneration of both ab sub-spectra at once is possible for

$$\tfrac{1}{8}(J_{AX} - J_{AX'})^2/|(|J_{AA'}| - |J_{XX'}|)| < \Delta_{1/2}$$

and

$$\tfrac{1}{8}(J_{AX} - J_{AX'})^2/(|J_{AA'}| - |J_{XX'}|)^2 < i.$$

The first of these two conditions turns out to be more critical since:

(a) the use of computers of average transients allows detection of weak lines, i.e. reduces i, and

(b) it is possible to boost the observable intensities of such lines relative to the strong lines, without appreciable broadening, by employing high radiofrequency powers, which saturate the stronger lines of the spectrum. This is because the saturation factor depends on the probability per unit time of an induced transition between two levels, and this in turn is proportional to the intensity of the appropriate spectral line.

The general condition for degeneracy of sub-spectra of the type ab_n ($n > 1$) into a single line is more stringent. It is necessary that the effective chemical shift is less than the line width.[28]

7.2. "Virtual Coupling"[51]

The application of sub-spectral methods to the interpretation of deceptively simple spectra leads to deeper insight and a more general definition of the phenomenon. "Virtual coupling," however, when looked at in the same light can be seen to be a misleading interpretation of sub-spectral phenomena.

The observation that the X part of the ABX spectrum displays a four-line spectrum, even if J_{AX} is zero, has been attributed to the fact that the coupling between the nuclei A and B is so strong that nucleus X "sees" nucleus A even though it is not coupled to it (virtual coupling).[51] Similarly, it seemed at first surprising that the AB part of the spectrum consists of eight lines when $J_{AX} = 0$. The arguments of sub-spectral analysis, however, show that the two ab sub-spectra comprising the AB part of the ABX spectrum can be characterized by the same coupling constant and the following differences of effective Larmor frequencies as effective chemical shifts,

$$\delta_{ab} = \delta_{AB} \pm \tfrac{1}{2}(J_{BX} - J_{AX})$$

It is therefore neither J_{AX} nor J_{BX} but their difference ($J_{BX} - J_{AX}$) which describes the appearance of the spectrum. Even in cases where one of the two coupling constants is zero there will still be two different ab sub-spectra observed in the AB part and the X part of the spectrum must consist of four lines. The condition for the "disappearance" of "virtual coupling" with

$J_{AX} = 0$ can be derived from the equations given in Section 7.1 if corresponding ab transitions overlap. The condition is

$$[J_{AB}^2 + (\delta_{AB} + \tfrac{1}{2}J_{BX})^2]^{1/2} - [J_{AB}^2 + (\delta_{AB} - \tfrac{1}{2}J_{BX})^2]^{1/2} < \Delta_{1/2}$$

Since this difference depends upon the magnitude of the relatively complex expression

$$\frac{\delta_{AB}J_{BX}}{(J_{AB}^2 + \delta_{AB}^2 + J_{BX}^2)^{1/2}}$$

the existence of "virtual coupling" may obviously not be explained uniquely as due to J_{AB} being strong. The phraseology of "virtual coupling" is, furthermore, misleading because it implies that splitting in the spectrum should only occur when there is direct coupling present and therefore encourages first-order interpretation of spectra when this is not valid.

7.3. *Degeneracy of Sub-spectra*

The spectra of complex systems may appear to be quite simple when conditions prevail which result in the exact superposition (degeneracy) of two or more sub-spectra. The AA'XX' system again provides a good illustration of such a possibility. The expressions (6) and (7) give the sub-spectral transformations for the symmetric and antisymmetric ab sub-spectra of the AA'(AA'XX') part. The chemical shifts are identical for the two sub-spectra $(\delta_{ab} = J_{AX} - J_{AX'})$ and therefore these two sub-spectra will be identical when the effective coupling constants are equal. This condition is fulfilled when either of the coupling constants $J_{AA'}$ or $J_{XX'}$ is zero. The AA'(AA'XX') spectrum then has only six lines, two from the a_2 sub-spectra and four from the doubly degenerate ab spectrum.

Alternatively, the two ab sub-spectra degenerate to exactly superimposed single lines when $J_{AX} = J_{AX'}$, the condition necessary for magnetic equivalence (see Section 2.2).

7.4. *Simplifications Arising when Some Coupling Constants are Negligibly Small*

The spectrum may be simplified when some coupling constants are approximately zero due to overlap of lines in ways which cannot be attributed to ab sub-spectra degenerating into a_2 (as in Section 7.1) nor to degeneracy between sub-spectra (as in Section 7.3). This may occur, for example, when mixing between some wave functions is zero in the absence of coupling but when this gives rise to no further breakdown of the energy level pattern. A specific example should make this clear. The AA'X$_2$X$_2'$ system, when $J_{XX'} = 0$, is discussed below.

The breakdown of the energy level diagram due to symmetry has already been noted for the $AA'B_2B_2'$ system (Section 4.3.3). The X approximation leads to further factorization in the case considered here. The energy level patterns for the DDTT(s) and DDTT(a) sub-systems, with $m_A + m_{A'} = 0$, are indicated schematically in Table 12.

TABLE 12. SCHEMATIC ENERGY LEVEL PATTERNS FOR THE DDTT(s) AND DDTT(a) SUB-SYSTEMS OF THE $AA'X_2X_2'$ SYSTEM, WITH $m_A + m_{A'} = 0$

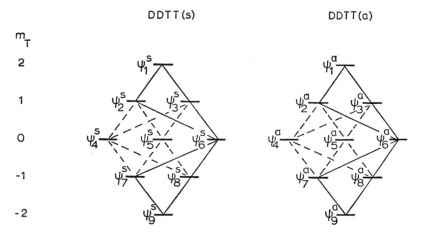

The wave functions referred to are those of Table 13.

TABLE 13. WAVE FUNCTIONS (COMPOSITE PARTICLE NOTATION) AND DIAGONAL ELEMENTS FOR THE DDTT(s) AND DDTT(a) SUB-SYSTEMS OF THE $AA'X_2X_2'$ SYSTEM ($J_{XX'} = 0$)

m_T		s	Diagonal element†	a	Diagonal element†
2	ψ_1	$(T^1T^1)_+$	$\frac{1}{4}J_{AA'}$	$(T^1T^1)_-$	$-\frac{3}{4}J_{AA'}$
1	ψ_2	$2^{-1/2}(T^1T^0 + T^0T^1)_+$	$\frac{1}{4}J_{AA'}$	$2^{-1/2}(T^1T^0 + T^0T^1)_-$	$-\frac{3}{4}J_{AA'}$
1	ψ_3	$2^{-1/2}(T^1T^0 - T^0T^1)_-$	$-\frac{3}{4}J_{AA'}$	$2^{-1/2}(T^1T^0 - T^0T^1)_+$	$\frac{1}{4}J_{AA'}$
0	ψ_4	$2^{-1/2}(T^1T^{-1} + T^{-1}T^1)_+$	$\frac{1}{4}J_{AA'}$	$2^{-1/2}(T^1T^{-1} + T^{-1}T^1)_-$	$-\frac{3}{4}J_{AA'}$
0	ψ_5	$2^{-1/2}(T^1T^{-1} - T^{-1}T^1)_-$	$-\frac{3}{4}J_{AA'}$	$2^{-1/2}(T^1T^{-1} - T^{-1}T^1)_+$	$\frac{1}{4}J_{AA'}$
0	ψ_6	$(T^0T^0)_+$	$\frac{1}{4}J_{AA'}$	$(T^0T^0)_-$	$-\frac{3}{4}J_{AA'}$
-1	ψ_7	$2^{-1/2}(T^0T^{-1} + T^{-1}T^0)_+$	$\frac{1}{4}J_{AA'}$	$2^{-1/2}(T^0T^{-1} + T^{-1}T^0)_-$	$-\frac{3}{4}J_{AA'}$
-1	ψ_8	$2^{-1/2}(T^0T^{-1} - T^{-1}T^0)_-$	$\frac{3}{4}J_{AA'}$	$2^{-1/2}(T^0T^{-1} - T^{-1}T^0)_+$	$\frac{1}{4}J_{AA'}$
-2	ψ_9	$(T^{-1}T^{-1})_+$	$\frac{1}{4}J_{AA'}$	$(T^{-1}T^{-1})_-$	$-\frac{3}{4}J_{AA'}$

† Omitting terms in ν_X.

The wave functions of the AA′ nuclei are referred to by the subscript plus and minus signs for $2^{-1/2}(\alpha\beta + \beta\alpha)$ and $2^{-1/2}(\alpha\beta - \beta\alpha)$, respectively. It can be shown readily that ψ_6^s no longer mixes with ψ_4^s and ψ_5^s, nor ψ_6^a with ψ_4^a

and ψ_5^a, if $J_{XX'} = 0$. The remaining off-diagonal elements are $\frac{1}{2}$L for $m_T = \pm 1$ and L for $m_T = 0$ (L $= J_{AX} - J_{AX'}$). The diagonal elements are noted in Table 13, omitting the chemical shift contributions. Two important features emerge from a study of the Hamiltonian. Firstly, the eigenvalues depend only on L and $J_{AA'}$. Now it can be shown by use of spin inversion operators that in general for AA'A'', ..., XX'X'' ... spin systems the signs of $J_{AA'}$ and $J_{XX'}$ relative to J_{AX} and $J_{AX'}$ do not affect the spectrum. In the present instance, with $J_{XX'} = 0$, this means that the spectrum is invariant to the sign of $J_{AA'}$. Now the DDTT(s) and DDTT(a) sub-Hamiltonians for $m_A + m_{A'} = 0$ differ only by (i) a constant difference $J_{AA'}$ between corresponding levels and (ii) the sign of $J_{AA'}$. It follows that the X transitions are therefore identical for the two sub-spectra, i.e. there is degeneracy in the sense of Section 7.3. However, the second feature is that there is degeneracy between transitions *within* each sub-Hamiltonian, that is degeneracy of transitions in the upper half of the energy level diagram with those in the lower half. In fact only eight distinct lines are possible. Those four involving T^1T^1, T^0T^0 and $T^{-1}T^{-1}$ (the solid lines in Table 12) form a true xy sub-spectrum with $\delta_{xy} = L$ and $J_{xy} = J_{AA'}$ centred about ν_X. It should be emphasized that this arises in spite of the fact that there is no *isolated* (1, 2, 1) sub-pattern of energy levels. The remaining transitions indicated by broken lines (Table 12) also form a quartet of lines symmetrical about ν_X. This quartet may be legitimately called a sub-spectrum but it does not have the correct xy intensities. The two types of transitions (solid and broken lines) have been referred to as the $\chi = 1$ and $\chi = 2$ quartets respectively.[12] Consideration of the DDST, DDTS and DDSS states indicates that all the other A transitions for $m_A + m_{A'} = 0$ fall at the $\chi = 1$ or $\chi = 2$ positions, thus giving a very simple spectrum since the effective Larmor frequency approach shows that the X transitions for $m_A + m_{A'} = \pm 1$ are single lines.

Such considerations give analogous results for the X transitions of all $AA'X_nX_n'$ spin systems with $J_{XX'} = 0$. However, although the approach gives a clear insight into the principles involved, it can be rather tedious and for cases of algebraic simplifications due to zero coupling constants it is often quicker to ignore the symmetry aspects and to deal with non-symmetrized basis functions. In this way it has been shown[29] for the general $AA'X_nX_n'$ spin system with $J_{XX'} = 0$ that the X resonance consists of a strong doublet of separation N, and n "quartets" with line positions given by closed form expressions.

It is necessary to emphasize a warning about splitting up a spin system into part systems when some coupling constants are known to be negligibly small. For instance, it might be argued that the X resonances of the $A_3A_3'M_2M_2'X$ system with $J_{AA'} = J_{MM'} = J_{AM'} = 0$ (common enough when ethyl groups are present in a molecule) could be obtained from a model

$M_2M_2'X$ system. Such "approximations" should always be shown to be valid by appropriate arguments based on sub-spectra. The whole of this Section 7 should show the dangers of relying on intuition alone in this context.

8. METHODS OF IDENTIFICATION OF SUB-SPECTRA

8.1. *Identification by Inspection*

This method is limited to the simplest sub-spectra of type a_2 or ab. A typical case is the AA' part of an $AA'XX'MR$ spectrum[30] which contains eight a_2 and eight ab sub-spectra.

$$AA'(AA'XX'MR) = (a_2)_{\pm 1,\ \pm 1/2,\ \pm 1/2} \qquad \text{symmetric}$$
$$+ (ab)_{0,\ \pm 1/2,\ \pm 1/2} \quad \text{symmetric}$$
$$+ (ab)_{0,\ \pm 1/2,\ \pm 1/2} \quad \text{antisymmetric}$$

Figure 7 provides an example of the AA' part of an $AA'XX'MR$ system divided into four constituent $aa'(aa'xx')$ sub-spectra which can be further divided into the a_2 and ab sub-spectra (see Fig. 1.)

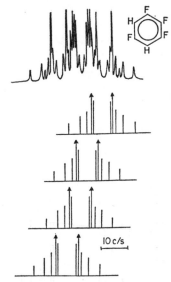

F$_{IG}$. 7. The AA' part of an $AA'XX'MR$ spectrum (^1H spectrum at 100 Mc/s of 1,2,3,5-tetrafluorobenzene) and its composite sub-spectra.

$$AA'(AA'XX'MR) = AA'(AA'XX')_{+1/2,+1/2} + AA'(AA'XX')_{+1/2,-1/2}$$
$$+ AA'(AA'XX')_{-1/2,+1/2} + AA'(AA'XX')_{-1/2,-1/2}$$

Each of these sub-spectra can be further broken down as shown in Fig. 1. Arrows indicate that lines have twice the intensity shown.

8.2. *Identification by Graphical Methods*

Graphical methods are applicable where sub-spectra have line positions which can be presented in a two-dimensional plot $v = f(J/\delta)$, for example $a_m b_n$ sub-spectra. This method has been used to pick out the ab_2 sub-spectra of ABB′XX′ systems[15, 31] The relative line positions of the complete spectrum are transferred as long parallel lines onto a transparent sheet of paper. This sheet is then put on top of a graph which shows the relative line positions of the relevant sub-spectrum as a function of (J/δ). The upper sheet of paper is rotated and moved across the graph until the intersections of the required number of lines with the curves of the graph (there are eight for ab_2 sub-spectra) fall on a straight line which is parallel to the chemical shift axis of the graph.

Some doubt may arise as to the uniqueness of the sets of lines picked out to represent the sub-spectra. However, there are a number of useful criteria which can be used to help erase ambiguity. For example, within the ABB′XX′

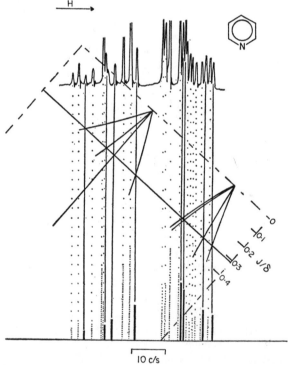

FIG. 8. The graphical identification of an ab_2 sub-spectrum in the ABB′(ABB′XX′) spectrum (^{1}H spectrum at 60 Mc/s of pyridine). (See text for an explanation of the method.)

system both ab_2 sub-spectra should give values of J_{AB} which agree within the experimental error of measuring the relative line positions. Where lines are sufficiently well resolved the total intensity within each ab_2 sub-spectrum should be identical and each sub-spectrum has the characteristic intensity distribution associated with AB_2 spectra (with varying J/δ). The values of other molecular parameters which can be derived from the analysis of the two ab_2 sub-spectra, J_{AX} and $J_{BX} + J_{BX'}$, can usually be estimated approximately from comparisons with similar molecules. Marked deviations from expected values should be regarded with suspicion (except where there are reasonable grounds to expect deviations).

The method is illustrated in Fig. 8 where one of the ab_2 sub-spectra of an ABB'XX' spectrum is pinpointed.

8.3. *Identification by Relaxation Behaviour*

The lines of NMR spectra of molecules containing several magnetic nuclei may be differentially broadened by characteristic relaxation processes. In general it may be possible to pick out the sub-spectra and in addition to assign the sub-spectra to different symmetry classifications on this basis. One possible method would be to induce additional specific relaxation processes by the addition of a paramagnetic compound, for example, DPPH. The ab sub-spectra of an AA'XX' system have been assigned to their respective symmetry classifications using this general method.[32] In addition the sub-spectral transformations ((6) and (7)) reveal that the relative signs of the coupling constants $J_{AA'}$ and $J_{XX'}$ may be determined by this assignment.

The natural line widths observed in NMR spectra can assist the assignment without recourse to artificial perturbation such as the use of DPPH as suggested above. For example in AB_2 spectra the single DS line is sharper than DT lines.[33] This is illustrated in Fig. 9. An additional possibility is to make use of the nuclear Overhauser effect. This has been used in the case of *cis*-1,2-difluoroethylene.[27]

8.4. *Identification by Tickling Techniques*

This method is applicable to all cases of sub-spectra and is superior to all the other methods mentioned above.[52]

9. SPECIAL USES OF SUB-SPECTRAL ANALYSIS

9.1. *Simplification of Perturbation Treatment*

The introduction of sub-spectral principles may considerably simplify the perturbation treatment of spectra just as it clarified the concept of deceptive simplicity.

H

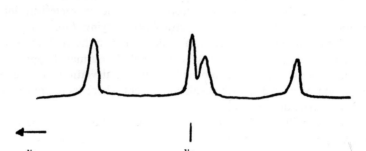

ν_B ν_A

FIG. 9. Differential broadening of the DS line at ν_A of an AB_2 spectrum; an illustration of specific relaxation effects which may be used to identify sub-spectra. The ^1H spectrum at 30·5 Mc/s of the isolated proton in $CHCl_2CH_2Cl$.

The transition from $AA'XX'MR$ to $AA'XX'MY$ provides the basis for a discussion of a typical case.[29] The XX' part of the $AA'XX'MR$ system is a superposition of x_2 and xy sub-spectra. The $XX'Y$ part of the $AA'XX'MY$ system is, however, a superposition of x_2y and xyz sub-spectra. In the case where the deviation from $AA'XX'MR$ towards $AA'XX'MY$ is not large, that is where x_2y and xyz sub-spectra are relatively weakly coupled, straightforward perturbation theory may be used. The Y spectrum, for example, then contains the following transition energies to second order (J^2/δ), (x_2y sub-spectrum)

$$\nu_y + J_{xy} + \tfrac{1}{2}J^2_{xy}/\delta_{xy}$$
$$\nu_y + \tfrac{1}{2}J^2_{xy}/\delta_{xy}$$
$$\nu_y$$
$$\nu_y - J_{xy} + \tfrac{1}{2}J^2_{xy}/\delta_{xy}$$

Substituting the sub-spectral transformations,

$$\nu_y = \nu_R \pm \tfrac{1}{2}J_{MR} \pm J_{AR}$$
$$\nu_x = \nu_X \pm \tfrac{1}{2}J_{MX} \pm \tfrac{1}{2}(J_{AX} + J_{AX'})$$
$$J_{xy} = J_{RX},$$

it can be seen that a characteristic splitting will be observed in the central part of R(Y) spectrum which is due to the perturbation and has the magnitude

$$\tfrac{1}{2}J^2_{xy}/\delta_{xy} = \tfrac{1}{2}J^2_{RX}/\{\nu_R - \nu_X \pm \tfrac{1}{2}(J_{MR} - J_{MX}) \pm [J_{AR} - \tfrac{1}{2}(J_{AX} - J_{AX'})]\}$$

The general procedure in applying the principles of sub-spectral analysis in this way is first to derive the sub-spectral transformations, then to use ordinary perturbation theory for the simple sub-spectra and finally to substitute the sub-spectral transformations into the perturbation formula.

9.2. *Spectra of Oriented Molecules*

The principles of sub-spectral analysis are not limited in their application to NMR spectra in the normally accepted sense. They can be applied to the analysis of NMR spectra of oriented molecules too. The Hamiltonians of such systems contain additional terms which are due to direct dipole–dipole interaction involving the constants A_{ik}.[34]

$$\mathscr{H} = \sum_i \nu_i F_z(i) + \sum_{i<k} (J_{ik} + 2A_{ik}) F_z(i) \cdot F_z(k)$$
$$+ \tfrac{1}{2} \sum_{i<k} (J_{ik} - A_{ik})(I_i^+ I_k^- + I_i^- I_k^+)$$

The Hamiltonian of an oriented AB system may be compared with the Hamiltonian of the ab sub-spectrum of an oriented AB_3 system as an example. The trivial transformations

$$\nu_a = \nu_A \qquad J_{ab} = J_{AB}$$
$$\nu_b = \nu_B \qquad A_{ab} = A_{AB}$$

are obtained.

10. ILLUSTRATIVE APPLICATIONS

10.1. $AA'X_3X_3'$ *Spin System*

10.1.1. *General*

This system provides a good example for the variety of approaches that may be made, depending on the particular circumstances involved.

The concept of the full NMR symmetry group has been applied to the $AA'X_3X_3'$ spin system and the results correlated[7] with those from the composite particle approach (Section 5). Table 14 shows the resultant energy level pattern with the symmetry and composite particle notations. The symmetry group used is a permutation group of order 72. It can be seen that if composite particle theory is used, the only additional symmetry that need be considered is that relating the two "halves" of the spin system, that is a C_2 element in this case. This is equivalent to using a symmetry group comprising $P_3 \times P_3$ plus a C_2 element. It is clear that the simple rules of the composite particle approach are in practice more convenient than the use of the full symmetry classification, particularly as the evaluation of transition

TABLE 14. SCHEMATIC ENERGY LEVEL DIAGRAM FOR THE GENERAL $AA'X_3X_3'$ SYSTEM

m_T	$A_1 \equiv DDQQ(s)$		$A_2 \equiv DDQQ(a)$		$G_1 \equiv DDDD(s)$ $(g=4)$	$G_2 \equiv DDDD(a)$ $(g=4)$	$G_3 \equiv DDDQ$† $(g=2)$	
4	1		1		1		1	1
3	1 1	1	1 1	1	1 1	1	1 1	2
2	1 1	2	1 1	2	1 1	1	1 1	4
1	1 1	3	1 1	3	1 1	2	1 1	4
0		4	1	4	1	1	1	2
−1	1 1	3	1 1	3	1 1	2	1 1	
−2	1 1	2	1 1	2	1 1	1	1 1	
−3	1	1	1	1	1		1	
−4					1	1	1 1 1	
$m(AA')$	1	0 −1	1	0 −1	1 0 −1	1 0 −1	1 0 −1	0 −1

† The DDQD functions form a set of levels, not given in the figure, that are degenerate with those of the DDDQ state.

energies and intensities may be performed without a detailed expansion of the wave functions in terms of the basic product functions.

It may be noted from Table 14 that there are several independent sub-systems that can be described in terms of known simple systems. The G_1 and G_2 functions give rise to an aa'xx' sub-system (which itself may be described in terms of ab and a_2 sub-spectra for the aa' part and xy and x_2 sub-spectra for the xx' part). In addition, functions characterized by $m(AA') = \pm 1$ give rise respectively to two x_3x_3' sub-systems where the effective chemical shift is given directly by effective Larmor frequency principles. Finally sub-systems such as DDQQ describe sub-spectral systems, involving particles with spin I not necessarily $\frac{1}{2}$, which are not well known.

10.1.2. $J_{XX'}$ Negligible

In this instance it is most convenient to work from first principles and to use neither the symmetry nor the composite particle approaches explicitly (see references 11, 12, 35 and Section 7.4).

The special case when only three coupling constants are important occurs frequently for a symmetrical $CH_3CHCHCH_3$ group, and several such spectra have been described.[11, 36, 35] In those examples $J_{AX'}$ is a long-range coupling ($|J_{AX'}| < 1$ c/s) and consequently $L \simeq N$. By taking $J_{XX'} = 0$ the AA' part of the spectrum may be seen to consist of a superposition of a_2 and ab sub-spectra, using the method of effective Larmor frequencies. Details of these sub-spectra are given in Table 15.

TABLE 15. CONSTITUENT SUB-SPECTRA OF THE AA' PART OF THE $AA'X_3X_3$ SYSTEM WITH $J_{XX'} = 0$

$m_X + m_{X'}$	m_X	$m_{X'}$	Sub-spectrum type	Sub-spectrum number*	Degeneracy	Sub-spectrum centre‡	J_{ab}	δ_{ab}
3	$\frac{3}{2}$	$\frac{3}{2}$	a_2	(i)	1	$\frac{3}{2}N$	—	—
2	$\frac{3}{2}$	$\frac{1}{2}$	2ab†	(ii)	6	N	$J_{AA'}$	L
2	$\frac{1}{2}$	$\frac{3}{2}$						
1	$\frac{1}{2}$	$\frac{1}{2}$	a_2	(iii)	9	$\frac{1}{2}N$	—	—
1	$\frac{3}{2}$	$-\frac{1}{2}$	2ab†	(iv)	6	$\frac{1}{2}N$	$J_{AA'}$	$2L$
1	$-\frac{1}{2}$	$\frac{3}{2}$						
0	$\frac{3}{2}$	$-\frac{3}{2}$	2ab†	(v)	2	0	$J_{AA'}$	$3L$
0	$-\frac{3}{2}$	$\frac{3}{2}$						
0	$\frac{1}{2}$	$-\frac{1}{2}$	2ab†	(vi)	18	0	$J_{AA'}$	L
0	$-\frac{1}{2}$	$\frac{1}{2}$						

* Sub-spectra (vii), (viii), (ix) and (x) are identical to (iv), (ii) and (i), respectively, but inverted through ν_A.

† Degenerate pair of ab sub-spectra.

‡ Relative to ν_A.

It is clear that the strongest two lines (comprising together $\frac{9}{32}$ of the total AA′ intensity) have a separation of N, since they arise from x_2 sub-spectra with $m_X = m_{X'} = \pm\frac{1}{2}$. The second most intense pair of lines, arising from ab sub-spectra with $J_{ab} = J_{AA'}$, $\delta_{ab} = L$, have separation $(L^2 + J_{AA'}^2)^{1/2} - J_{AA'}$, so if it is assumed that $L \simeq N$ an approximate value of $J_{AA'}$ may be obtained. Figure 10 shows the calculated and observed[36] spectra for ϵ-2,3,5,6-tetramethylpiperazine dihydrochloride (molecular diagram shown below); the positions of ring methyl groups are indicated; hydrogen atoms occupy the remaining ring positions.

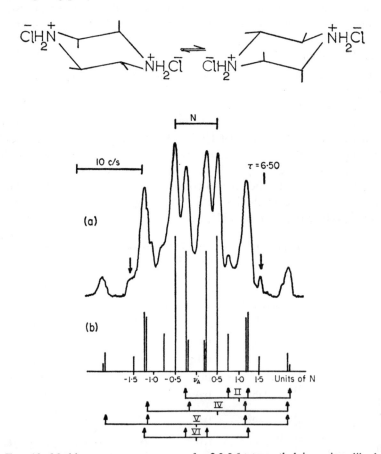

FIG. 10. Methine proton resonances of ϵ-2,3,5,6-tetramethylpiperazine dihydrochloride, i.e. the AA′ part of an AA′X_3X_3' spin system with $J_{XX'} = 0$. (a) Observed 40 Mc/s spectrum of D_2O solution. (b) Calculated spectrum. In this case there is considerable overlapping of lines due to the fact that $J_{AA'} \simeq N$. Some of the ab sub-spectra are indicated. The a_2 sub-spectra give lines at $\pm N/2$, $\pm 3N/2$ (the arrows in (a) indicate the latter). Coupling constants are N = 6·6, L = 7·2, $J_{AA'} = 6·5$ c/s.

The X_3X_3' spectrum is also rather simple when $J_{XX'} = 0$, and closed-form expressions for the lines may be obtained (Section 7.4 and reference 12). In the $AA'X_3X_3'$ case there is a strong pair of lines (totalling half the XX' intensity) of separation N. These arise from x_3x_3' sub-spectra for which $m_A + m_{A'} = \pm 1$. The remaining lines may be described in terms of three sub-spectra of four lines each. The sub-spectra may be denoted by a parameter χ (see Fig. 14). The sets of lines are symmetrical about ν_X with separations given by equations (16) (i)–(iii).

$$
\begin{aligned}
\chi = 1 \text{ lines:} \ & (L^2 + J_{AA'}^2)^{1/2} \pm J_{AA'} && \text{(i)} \\
\chi = 2 \text{ lines:} \ & (4L^2 + J_{AA'}^2)^{1/2} \pm (L^2 + J_{AA'}^2)^{1/2} && \text{(ii)} \\
\chi = 3 \text{ lines:} \ & (9L^2 + J_{AA'}^2)^{1/2} \pm (4L^2 + J_{AA'}^2)^{1/2} && \text{(iii)}
\end{aligned}
\qquad (16)
$$

The lines described by (i) form a true xy sub-spectrum with $\delta_{xy} = L$ and $J_{xy} = J_{AA'}$ although there is no isolated 1, 2, 1, pattern within the DDQQ states of Table 14 even for $J_{XX'} = 0$. The other two sub-spectra do not conform to the intensities of the xy pattern. The intensities are in fact given by equation (17a)

$$
\tfrac{1}{2}(1 \pm q)\, \frac{5!}{(2 + \chi)!\,(3 - \chi)!} \qquad (17a)
$$

where

$$
q = \frac{\chi(\chi - 1)L^2 + J_{AA'}^2}{\{[\chi^2 L^2 + J_{AA'}^2][(\chi - 1)^2 L^2 + J_{AA'}^2]\}^{1/2}} \qquad (17b)
$$

where $\chi = 1, 2$ and 3 for expressions (16) (i), (ii) and (iii) respectively. Two consequences of the analysis stand out in importance. Firstly, it follows that $L > N$ and thus J_{AX} and $J_{AX'}$ are of opposite sign if any of the *inner* lines[16] (those with the minus sign) fall outside the N doublet. Secondly, $J_{AA'}$ can be measured directly from the $\chi = 1$ xy sub-spectrum if all four lines can be observed. Figure 11 shows[36] this method of measuring $J_{AA'}$ for α-2,3,5,6-tetramethylpiperazine dihydrochloride (molecular diagram shown below); the positions of the ring methyl groups are indicated; hydrogen atoms occupy the remaining ring positions.

10.1.3. $J_{XX'}$ Small (Perturbation Approach)

When $J_{XX'}$ is much less than N, L or $J_{AA'}$ it is possible to proceed from the above analysis, appropriate for $J_{XX'} = 0$, by allowing mixing (due to finite $J_{XX'}$) only between spin wave functions that are degenerate otherwise. This

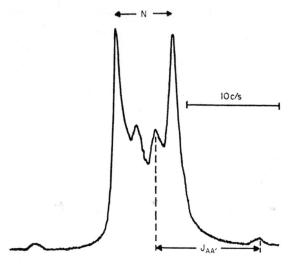

FIG. 11. Methyl proton resonances of α-2,3,5,6-tetramethyl piperazine dihydro-chloride, $X_3X_3'(AA'X_3X_3')$, in D_2O at 60 Mc/s showing direct measurement of $J_{AA'}$.

approach has been used[40] to solve the fluorine spectrum of fluoro-N,N'-dimethyl-1,3,2,4-diazadiphosphetidine:

$$CH_3-N \diagdown \overset{PF_3}{\diagup} \diagdown N-CH_3$$

The spectrum is shown in Fig. 12. Approximate analysis for the ${}^{19}F$ part of the F_3PPF_3 spins on the basis of Section 10.1.2 yields $J_{PF} = \pm922$ c/s, $J_{PF'} = \mp32$ c/s, $|J_{PP'}| = 210$ c/s. The effect of $J_{XX'}$ on the diagonal Hamiltonian elements of a typical wave function that is a solution for $J_{XX'} = 0$ is to add $pmJ_{XX'}$ where $p = m_X$ and $m = m_{X'}$. This causes a splitting of each transition of equations (16) (i), (ii) and (iii) into four lines with spacing $J_{XX'}$ and relative intensities within each group of $1:9:9:1$ for (i), $2:3:3:2$ for (ii) and $1:0:0:1$ for (iii). Figure 13 shows how this correction alone leads to qualitative agreement with observation for the outer bands of (i) for the diazadiphosphetidine, although the spacings are actually not exactly $J_{XX'}$.

The off-diagonal corrections are more complicated but are readily obtained using composite particle theory.[40] The results of the calculations are shown in Fig. 13(b) for the "outer" band given by equations (16) (i). The value of

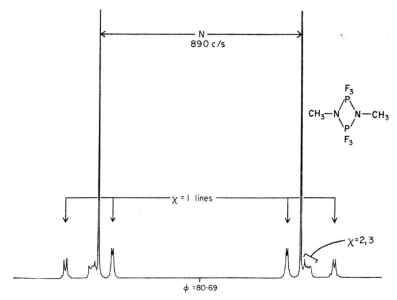

FIG. 12. Observed ^{19}F spectrum at 56·4 Mc/s of fluoro-N,N'-dimethyl-1,3,2,4-diazadiphosphetidine $[X_3X'_3(AA'X_3X'_3)]$ showing sub-spectra.

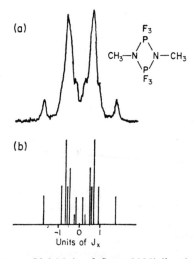

FIG. 13. ^{19}F spectrum at 56·4 Mc/s of fluoro-N,N'-dimethyl-1,3,2,4-diazadiphosphetidine $[X_3X'_3(AA'X_3X'_3)]$; $\chi = 1$ sub-spectrum, one of the outer bands. (a) Observed (b) Calculated by perturbation approach of Section 10.1.3.

$3J_{XX'}$ is given by the average of the spacings between the outermost lines in the inner and outer $\chi = 1$ bands. Moreover, the splitting of the $\chi = 3$ lines due to the additions to the diagonal elements ($pmJ_{XX'}$) is not further complicated by off-diagonal terms and therefore if the lines of this sub-spectrum can be discovered, the splitting as above gives $3J_{XX'}$ directly from the spectrum (measurement for the diazadiphosphetidine gives $J_{FF'} = 8\cdot7$ c/s).

A spectrum similar to that of the diazadiphosphetidine has been recorded[37] but the analysis has not been carried out in this case, although it is clear that J_{AX} and $J_{AX'}$ are of opposite sign.

10.1.4. *All Four Coupling Constants Important*

The equivalence of symmetry and composite particle approaches has already been discussed for this spin system in Section 10.1.1. The classification of the energy levels is shown in Table 14. The effective Larmor frequency method shows that there is a pair of lines in the AA' spectrum at $\nu_A \pm \frac{3}{2}N$ (a$_2$ sub-spectra with $m_X = m_{X'} = \pm\frac{3}{2}$) and an intense pair in the XX' spectrum at $\nu_X \pm \frac{1}{2}N$ (x$_3$x$_3'$ sub-spectra with $m_A = m_{A'} = \pm\frac{1}{2}$). Closed-form expressions may be written for all the transitions of the DDDD spin state, which forms an aa'xx' sub-spectrum, giving a pattern of lines which is the same in the aa' and xx' parts of the spectrum (see Section 2.2). Closed-form expressions can also be obtained for some of the other lines, but those remaining involve the diagonalization of fourth-order and third-order sub-matrices.[25] The spectrum of ethane-$^{13}C_2$ has been analyzed in this way[38] and it has been made clear that the relative signs of J_{AX} and $J_{AX'}$ are readily obtained but the relative signs of $J_{AA'}$ and $J_{XX'}$ can only be determined from the A spectrum, not the X spectrum. Moreover, the signs of $J_{AA'}$ or $J_{XX'}$ relative to J_{AX} or $J_{AX'}$ cannot be obtained at all from considerations of transition energies and intensities in the single resonance spectrum.[25]†

10.1.5. *Deceptively Simple Spectra* (see Section 7.4)

When $J_{XX'} = 0$, a deceptively simple spectrum is obtained if $L^2 < 2$ $|J_{AA'}|\Delta_{\frac{1}{2}}$ where $\Delta_{\frac{1}{2}}$ is the full line width of a single line at half-height. The X$_3$X$_3'$ part consists of a triplet under these conditions[1, 12] approximating to the X$_6$(A$_2$X$_6$) spectrum. In general, however, the relative heights of the triplet lines will not be exactly 1 : 2 : 1 (unless $L^2/|J_{AA'}|$ is negligible) and, conversely, the width of the central line will be greater than that of the outer lines (which are true single lines). Figure 14 shows the changes in the X$_3$X$_3'$(AA'X$_3$X$_3'$) spectrum as $|L/J_{AA'}|$ varies, case (d) being the deceptively simple one. At first sight it would appear that little information can be obtained from the deceptively simple spectrum except the value of N. Figure

---

† The definition of L in reference 25 is not the same as that used here; moreover, the spin system is designated A$_3$A$_3'$X$_2$ rather than AA'X$_3$X$_3'$.

15 illustrates such a case (actually for an $AA'X_6X_6'$ system), that of tetra-methyldiphosphine.[39, 29] However, it is possible to measure $L^2/|J_{AA'}|$ by taking account of the anticipated shape of the central band. The equations developed for the $AA'X_6X_6'$ system, with Lorentzian line-shapes (using values of $\Delta_{1/2}$ from the outer, single, lines) have been used to compute[39] a band shape for the central band for various values of $L^2/|J_{AA'}|$. Comparison with the experimental band shape gives a value for this parameter; the best criterion to use in such an analysis is actually the ratio of line heights of the

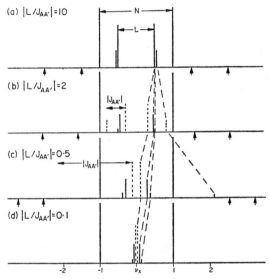

FIG. 14. The X_3X_3' part of an $AA'X_3X_3'$ spectrum ($J_{XX'} = 0$) for various values of $|L/J_{AA'}|$, showing the sub-spectra. The intense lines of separation N are those of the x_3x_3' sub-spectra. The dotted lines in (b) (c) and (d) are those of the $\chi = 1$ ab type sub-spectrum. The outer lines of the $\chi = 2$ and 3 sub-spectra are very weak and are indicated by arrows in (a), (b) and (c). Case (d) is "deceptively simple" if the lines of the central band cannot be resolved. The frequency scale is in units of L with an arbitrary ratio N/L = 2.

central and outer bands under conditions of very slow sweep rates but avoiding saturation. Alternatively, the band width or general band shape of the central band may be used. By this method the value $L^2/|J_{AA'}| = 0.38$ has been obtained for tetramethyldiphosphine.[39] In this case the value of $|J_{AA'}|$ is 179·7 c/s (from measurements of very weak "satellite" peaks—see Section 7.1). Thus $L = 8.25 \pm 0.2$ c/s, and since $N = 14.15 \pm 0.15$ c/s, this result gives $J_{AX} = 11.2$, $J_{AX'} = 2.95$ c/s. It should be noted that in such deceptively simple cases use of ^{13}C satellites does not remove the "deception" unless the A and A' are made considerably non-equivalent. For this reason

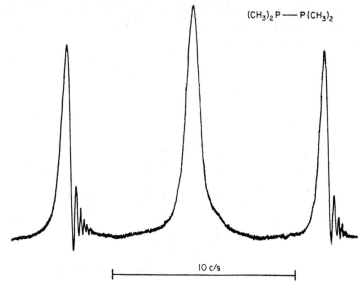

FIG. 15. Observed 100 Mc/s proton spectrum of tetramethyldiphosphine, showing
deceptive simplicity.

[13]C satellites are of little use for tetramethyldiphosphine, and the values of
J_{AX} and $J_{AX'}$ estimated from them previously[29] are incorrect.

A similar band shape computation has been carried out for the $AA'X_6X_6'$
spin-system of metal complexes involving the 4-methyl-2,6,7-trioxa-1-
phosphabicyclo (2.2.2) octane ligand[41]:

In these compounds $J_{AA'}$ is the parameter to be found from the spectrum
and $L \rightleftharpoons N$ since $J_{AX'}$ (through five bonds) is undoubtedly very small.
This type of treatment may be applied to many systems, since usually either
$J_{AA'}$ or L may be estimated fairly accurately. It should be noted that in the
deceptively simple cases under discussion the immediate sub-spectral breakdown is into the two outer lines (x_{12} sub-spectra) and the overlapping
sub-spectra which compose the central band.

10.2. *The Spectrum of Bis-pentafluorosulphur Peroxide*, SF_5OOSF_5

Due to the stereochemistry of the SF_5 group this molecule constitutes[42] a spin system of type $AA'B_4B_4'$ the axial fluorine nuclei being denoted A and A'. It can be shown using composite particle principles that there are nine spin states. These are listed in Table 16 with appropriate degeneracies g, and total relative intensity contributions.

TABLE 16. DEGENERACY AND RELATIVE INTENSITY CONTRIBU-
TIONS OF THE SPIN STATES OF THE $AA'B_4B_4'$ SYSTEM

State	g	Total relative intensity	Intensity
DDSS	4	16	0·3
DDST⎫† DDTS⎭	6	336	6·6
DDSQt⎫† DDQtS⎭	2	400	7·8
DDTT	9	1188	23·2
DDQtT⎫† DDTQt⎭	3	2280	44·5
DDQtQt	1	90	17·6
		5120	

† Doubly degenerate (the factor g does not include this degeneracy but the total intensity does). As explained in Section 4.3.3, there is no point in constructing symmetrical and anti-symmetrical combinations.

Figure 16(a) shows the observed 56·4 Mc/s spectrum. It bears a close resemblance to a simple AB_4 case with $R = J_{AB}/\delta_{AB} = 2·0$, indicating that coupling between the two SF_5 groups is weak. The sharp single line to low field is, on this basis, at ν_A. Its intensity is considerably greater than can be attributed to the DDSS state. This is undoubtedly due to the fact that the long-range coupling constants $J_{AA'}$ and $J_{AB'}$ are negligibly small (i.e. less than the line width, 0·9 c/s, of the line at ν_A). In such a case the DDST state, for instance, may be regarded as comprising isolated DS and DT states (relative intensities have to be considered rather carefully, using equation (14), and in this case DDST = 6 DS + 6 DT). The fact that singlet states may be ignored means that there are no terms dependent on $J_{BB'}$ in the sub-Hamiltonian for the DDST state, nor for any of the other states of Table 16 containing S particles. Reconsideration[44] of the spin states on this basis gives the intensities of Table 17.

The intensity of the single line at ν_A is consistent with this breakdown, and it may confidently be asserted that $|J_{AA'}| < 0·9$ c/s, $|J_{AB'}| < 0·9$ c/s. The

TABLE 17. TOTAL RELATIVE INTENSITIES OF THE MODIFIED SPIN STATES OF THE $AA'B_4B_4'$ SYSTEM WHEN $J_{AB'}$ AND $J_{BB'}$ ARE LESS THAN $\Delta_{1/2}$

State	Total relative intensity	% Intensity
DDQtQt	900	17·6
DDQtT DDTQt	2280	44·5
DDTT	1188	23·2
DQt[a]	360	7·0
DT[b]	264	5·2
DS[c]	128	2·5

[a] From DDQtS and DDSQt.
[b] From DDTS and DDST.
[c] From DDSS, DDQtS, DDSQt, DDTS and DDST.

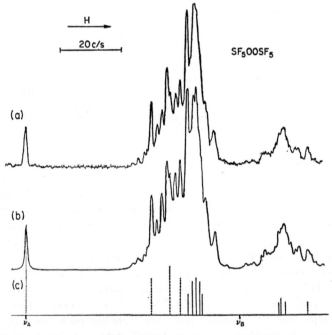

FIG. 16. ^{19}F spectrum (56·4 Mc/s) of SF_5OOSF_5, an $AA'B_4B_4'$ spin system. (a) Observed. (b) Computed spectrum assuming $J_{AB'} = J_{AA'} = 0$. (c) Partial calculation of ab_4 lines which form a true sub-spectrum (when intensities are adjusted as in Section 10.2) under the conditions of algebraic simplification. The full lines are those of the DQt state, the dotted line is due to the DS state, and the dashes indicate DT lines.

overall pattern is much more complex than the simple AB_4 case, and so $J_{BB'}$ is not negligible, although it is undoubtedly small. It is possible, from an AB_4 calculation, to estimate values of J_{AB} and δ_{AB} and to make a partial analysis based on the sub-spectra not involving $J_{BB'}$. This is shown in Fig. 16(c). It might be thought that such a calculation could be used as a basis for a perturbation approach (with $J_{BB'}$ as the perturbation), i.e. treating the system as $(AB_4)_2$. However, some of the energy levels are accidentally nearly degenerate and this perturbation approach is therefore not valid. However, it is quite feasible to write a computer program, based on the composite particle method, to calculate spectra, and since J_{AB} and δ_{AB} are known approximately from the AB_4 approximation, it is a simple matter to fit the observed spectrum by varying the remaining parameter $J_{BB'}$. One such computer program, UEANMR 2, presents the results[43] separately for each sub-spectrum (spin state), and has been extended[44] so that lines are given Lorentzian shapes and the final spectrum obtained from overlapping lines is plotted automatically by the computer. The result[44] for $S_2F_{10}O_2$ is given in Fig. 16(b) for $J_{BB'} = -4\cdot6$ c/s (the spectrum is sensitive to a change of $\pm 0\cdot1$ c/s in $J_{BB'}$ but it is probable that a similar fit could be obtained[44] using a different value of J_{AB}/δ_{AB} with a modified $J_{BB'}$).

ACKNOWLEDGEMENTS

The authors wish to thank Dr. J. G. Verkade and Mr. C. M. Woodman for work seen in advance of publication. They are also very grateful to Dr. Ruth Lynden-Bell for supplying unpublished work and to both Dr. Lynden-Bell and Mr. Woodman for valuable discussions. The editor of *Molecular Physics* is thanked for permission to reproduce Figs. 12 and 13 and the editor of *Physical Review* is thanked for permission to reproduce Fig. 9.

APPENDIX

Sub-spectral Patterns for Systems of up to Four Nuclei Involving Magnetically Equivalent Groups

Two nuclei

m	A_2 T	S	AB DD		AX DD	
1	1		1	1		
0	1	1	2	1		1
−1	1		1			1
m_X				$\frac{1}{2}$		$-\frac{1}{2}$

Three nuclei

m	A₃		AB₂		ABC
	Q	D	DT	DS	DDD
$\frac{3}{2}$	1		1		1
$\frac{1}{2}$	1	1	2	1	3
$-\frac{1}{2}$	1	1	2	1	3
$-\frac{3}{2}$	1		1		1

AX₂

m	DT		DS	
$\frac{3}{2}$	1			
$\frac{1}{2}$	1	1	1	
$-\frac{1}{2}$	1	1		1
$-\frac{3}{2}$	1			
m_X	$\frac{1}{2}$	$-\frac{1}{2}$	$\frac{1}{2}$	$-\frac{1}{2}$

ABX

m	DDD	
$\frac{3}{2}$	1	
$\frac{1}{2}$	2	1
$-\frac{1}{2}$	1	2
$-\frac{3}{2}$		1

AMX

m	DDD			
$\frac{3}{2}$	1			
$\frac{1}{2}$	1	1	1	
$-\frac{1}{2}$		1	1	1
$-\frac{3}{2}$				1
m_X	$\frac{1}{2}$	$\frac{1}{2}$	$-\frac{1}{2}$	$-\frac{1}{2}$
m_M	$\frac{1}{2}$	$-\frac{1}{2}$	$\frac{1}{2}$	$-\frac{1}{2}$

Four nuclei

m_T	A₄			AB₃		A₂B₂			
	Qt	T	S	DQ	DD	TT	TS	ST	SS
2	1			1		1			
1	1	1		2	1	2	1	1	
0	1	1	1	2	2	3	1	1	1
-1	1	1		2	1	2	1	1	
-2	1			1		1			

m_T	AB₂C		ABCD
	DTD	DSD	DDDD
2	1		1
1	3	1	4
0	4	2	6
-1	3	1	4
-2	1		1

m_T	AX₃				A₂X₂							
	DQ		DD		TT			TS			ST	SS
2	1				1							
1	1	1	1		1	1		1			1	
0	1	1	1	1	1	1	1		1		1	1
-1	1	1		1		1	1			1	1	
-2		1					1					
m_X	$\frac{1}{2}$	$-\frac{1}{2}$	$\frac{1}{2}$	$-\frac{1}{2}$	1	0	-1	1	0	-1	0	0

AB_2X

m_T	DTD		DSD	
2	1			
1	2	1	1	
0	2	2	1	1
-1	1	2		1
-2		1		
m_X	$\frac{1}{2}$	$-\frac{1}{2}$	$\frac{1}{2}$	$-\frac{1}{2}$

ABX_2

m_T	DDT			DDS
2	1			
1	2	1		1
0	1	2	1	2
-1		1	2	1
-2			1	
m_X	1	0	-1	0

ABCX

m_T	DDDD	
2	1	
1	3	1
0	3	3
-1	1	3
-2		1
m_X	$\frac{1}{2}$	$-\frac{1}{2}$

A_2MX

m_T	TDD				SDD			
2	1							
1	1	1	1		1			
0	1	1	1	1		1	1	
-1		1	1	1				1
-2				1				
m_X	$\frac{1}{2}$	$\frac{1}{2}$	$-\frac{1}{2}$	$-\frac{1}{2}$	$\frac{1}{2}$	$\frac{1}{2}$	$-\frac{1}{2}$	$-\frac{1}{2}$
m_M	$\frac{1}{2}$	$-\frac{1}{2}$	$\frac{1}{2}$	$-\frac{1}{2}$	$\frac{1}{2}$	$-\frac{1}{2}$	$\frac{1}{2}$	$-\frac{1}{2}$

ABMX

m_T	DDDD			
2	1			
1	2	1	1	
0	1	2	2	1
-1		1	1	2
-2				1
m_X	$\frac{1}{2}$	$\frac{1}{2}$	$-\frac{1}{2}$	$-\frac{1}{2}$
m_M	$\frac{1}{2}$	$-\frac{1}{2}$	$\frac{1}{2}$	$-\frac{1}{2}$

ABXY

m_T	DDDD		
2	1		
1	2	2	
0	1	4	1
-1		2	2
-2			1
m_X	1	0	1

AMRX

m_T	DDDD							
2	1							
1	1	1	1		1			
0		1	1	1	1	1	1	
-1				1		1	1	1
-2								1
m_X	$\frac{1}{2}$	$\frac{1}{2}$	$\frac{1}{2}$	$\frac{1}{2}$	$-\frac{1}{2}$	$-\frac{1}{2}$	$-\frac{1}{2}$	$-\frac{1}{2}$
m_R	$\frac{1}{2}$	$\frac{1}{2}$	$-\frac{1}{2}$	$-\frac{1}{2}$	$\frac{1}{2}$	$\frac{1}{2}$	$-\frac{1}{2}$	$-\frac{1}{2}$
m_M	$\frac{1}{2}$	$-\frac{1}{2}$	$\frac{1}{2}$	$-\frac{1}{2}$	$\frac{1}{2}$	$-\frac{1}{2}$	$\frac{1}{2}$	$-\frac{1}{2}$

The composite particle notation S, D, T, etc., has been used to characterize the various spin states. The notation DD, DDD, etc., is trivial but has been included for completeness.

Where relevant the conservation of the spin quantum numbers m_X, m_M, m_R has been used to pick out the transitions of the strongly coupled part, A_aB_b. Alternatively, where applicable, $m(A_aB_b)$ can be conserved to pick out X_xY_y transitions.

c

REFERENCES

1. R. J. ABRAHAM and H. J. BERNSTEIN, *Can. J. Chem.* **39**, 216 (1961).
2. H. M. McCONNELL, A. D. McLEAN and C. A. REILLY, *J. Chem. Phys.* **23**, 1152 (1955).
3. J. A. POPLE, W. G. SCHNEIDER and H. J. BERNSTEIN, *High-Resolution Nuclear Magnetic Resonance*, McGraw-Hill Book Co., Inc. (1959).
4. P. L. CORIO, *Chem. Rev.* **60**, 363 (1960).
5. E. BRIGHT WILSON, *J. Chem. Phys.* **27**, 60 (1957).
6. H. C. LONGUET-HIGGINS, *Mol. Phys.* **6**, 445 (1963).
7. C. M. WOODMAN, *Mol. Phys.* **11**, 109 (1966).
8. N. F. BANNERJEE, T. P. DAS, A. E. SAHA, *Proc. Roy. Soc.* A **226**, 490 (1954).
9. J. S. WAUGH and F. W. DOBBS, *J. Chem. Phys.* **31**, 1235 (1959).
10. D. R. WHITMAN, L. ONSAGER, M. SAUNDERS and H. T. DUBB, *J. Amer. Chem. Soc.* **82**, 67 (1960).
11. F. A. L. ANET, *J. Amer. Chem. Soc.* **84**, 747 (1962).
12. R. K. HARRIS, *Can. J. Chem.* **42**, 2275 ((1964).
13. J. A. POPLE and T. SCHAEFER, *Mol. Phys.* **3**, 547 (1961).
14. P. DIEHL and J. A. POPLE, *Mol. Phys.* **3**, 557 (1961).
15. P. DIEHL, R. G. JONES and H. J. BERNSTEIN, *Can. J. Chem.* **43**, 81 (1965).
16. R. G. JONES, R. C. HIRST and H. J. BERNSTEIN, *Can. J. Chem.* **43**, 683 (1965).
17. P. DIEHL, *Helv. Chim. Acta* **48**, 567 (1965).
18. R. G. JONES and S. M. WALKER, *Mol. Phys.* **10**, 363 (1966).
19. F. A. COTTON, *Chemical Applications of Group Theory*, Interscience Publishers (Wiley & Sons, Inc.).
20. R. M. LYNDEN-BELL (unpublished research).
21. T. B. GRIMLEY, *Mol. Phys.* **6**, 329 (1963).
22. P. DIEHL and D. TRAUTMANN, *Mol. Phys.* **11**, 531 (1966).
23. D. CHAPMAN and R. K. HARRIS, *J. Chem. Soc.* 237 (1963).
24. J. S. WAUGH and E. L. WEI, *J. Chem. Phys.* **43**, 2308 (1965).
25. R. M. LYNDEN-BELL, *Mol. Phys.* **6**, 601 (1963).
26. W. G. SCHNEIDER, H. J. BERNSTEIN and J. A. POPLE, *Can. J. Chem.* **35**, 1487 (1957).
27. K. KUHLMANN and J. D. BALDESCHWIELER, *J. Amer. Chem. Soc.* **85**, 1010 (1963).
28. P. DIEHL, *Helv. Chim. Acta* **47**, 1 (1964).
29. R. K. HARRIS and R. G. HAYTER, *Can. J. Chem.* **42**, 2282 (1964).
30. E. LUSTIG and P. DIEHL, *J. Chem. Phys.* **44**, 2974 (1965).
31. J. A. LADD and R. G. JONES, *Spectrochim. Acta* **22**, 1964 (1966).
32. R. M. LYNDEN-BELL, *Proc. Roy. Soc.* A **286**, 337 (1964).
33. W. A. ANDERSON, *Phys. Rev.* **102**, 151 (1956).
34. G. ENGLERT, A. SAUPE and A. POVH, *Advances in Chemistry Theories*, American Chemical Society, Washington (in press).
35. A. A. BOTHNER-BY and C. NAAR-COLIN, *J. Amer. Chem. Soc.* **84**, 743 (1962).
36. R. K. HARRIS and N. SHEPPARD, *J. Chem. Soc.* Sec, B, 200 (1966).
37. R. J. CLARK and E. O. BRIMM, *Inorg. Chem.* **4**, 651 (1965).
38. R. M. LYNDEN-BELL and N. SHEPPARD, *Proc. Roy. Soc.* A **269**, 385 (1962).
39. E. G. FINER and R. K. HARRIS, *Mol. Phys.* (in press) (1967).
40. R. K. HARRIS and C. M. WOODMAN, *Mol. Phys.* **10**, 437 (1966).
41. D. G. HENDRICKER, R. E. McCARLEY, R. W. KING and J. G. VERKADE, *Inorg. Chem.* **5**, 639 (1966).
42. R. K. HARRIS and K. J. PACKER, *J. Chem. Soc.* 3077 (1962).
43. C. M. WOODMAN (unpublished work).
44. E. G. FINER and R. K. HARRIS (unpublished work).
45. J. W. EMSLEY, J. FEENEY and L. H. SUTCLIFFE, *High Resolution Nuclear Magnetic Resonance Spectroscopy*, Vol. 1. Pergamon Press (1965).
46. P. L. CORIO, *J. Mol. Spectroscopy* **8**, 193 (1962).

47. J. H. Noggle, J. D. Baldeschwieler and C. B. Colburn, *J. Chem. Phys.* **37**, 182 (1962).
48. R. K. Harris and K. J. Packer, *J. Chem. Soc.* 4736 (1961).
49. C. I. Merrill, S. M. Williamson, G. H. Cady and D. F. Eggers, *Inorg. Chem.* **1**, 215 (1962).
50. See, for example, L. I. Schiff, *Quantum Mechanics*, McGraw-Hill Book Co., Inc.
51. J. I. Musher and E. J. Corey, *Tetrahedron* **18**, 791 (1962).
52. P. Diehl, (unpublished work).

CHAPTER 2

THE ISOTOPE SHIFT

H. Batiz-Hernandez† and R. A. Bernheim

Department of Chemistry, Whitmore Laboratory,
The Pennsylvania State University,
University Park, Pennsylvania

CONTENTS

ABSTRACT

Shifts in nuclear magnetic resonance line positions brought about by isotopic sub-stitution have been known for some time. In this review, the origin of the shift is first examined empirically by making correlations of experimental values of the isotope shift with molecular geometry and with nuclear magnetic interaction properties such as the spin–spin coupling. From a large amount of tabulated experimental data it is apparent that for geminal heavy isotopic substitution the isotope shift is larger for molecules with smaller H—X—H bond angles, where X is a first-row element. In addition, a definite empirical relationship exists between the ^{13}C—F spin–spin splitting and the isotope shift in the ^{19}F resonance when H is substituted by D in a H—C—F group. Since the isotope shift must be related to changes in the dynamic state of the molecule upon isotopic substitution, the theoretical approaches to the explanation of the shift have been based on calculations of the changes in magnetic shielding that accompany changes in average bond angles and internuclear distances resulting from the existence of anharmonic components of the vibrational potential. The isotope shift in the hydrogen molecule can be treated in a satisfactory way on this basis, but the situation is much more complex in the case of polyatomic molecules.

† Present address: Agricultural Experiment Station, University of Puerto Rico, Rio Piedras, Puerto Rico.

This review discusses the contributions from intramolecular electric fields, changes in bond hybridization, and the thermal distribution of molecules over the excited vibrational states to the isotope shift for polyatomic molecules. The outlook for future progress in relating the isotope shift to an individual molecular structure parameter is not very optimistic due to the variety of effects that can contribute to the observed shift.

1. INTRODUCTION

The effects of isotopic substitution on the magnetic shielding of nuclei have been known for some time and are commonly observed in high resolution nuclear magnetic resonance spectra. It is not surprising that such effects should occur since isotopic substitution changes the dynamic state of a molecule. The modified vibrational and rotational motion of the molecule results in a change in the electronic structure and, hence, the magnetic shielding of the various nuclei. This isotope shift first concerned those workers who were interested in precise measurements of the magnitude of nuclear magnetic moments. Subsequently, the effect was observed in the conventional high resolution NMR spectra of many isotopically substituted compounds.

The isotope shift is quite common and leads one to ask whether it is ever responsible for complications in the interpretation of NMR spectra and whether it can be used as a tool for the investigation of molecular structure. Accordingly, this report reviews the experimental observation of isotope shifts and their correlation with other experimentally obtained quantities such as bond angles and distances and indirect spin coupling constants. The various theoretical approaches to the explanation of the isotope shift will also be discussed, and the consequences with respect to molecular structure considered. It will become evident that the factors which result in the small experimentally observed shifts are numerous; perhaps too numerous to make the isotope shift a really valuable molecular structure tool. However, some progress has been made on understanding the isotope shift, and a discussion of the present status of this effect seems timely.

One possible importance of the isotope chemical shift lies in its use for evaluating the variation in molecular dimensions upon isotopic substitution. Recent experimental findings[1-4] have led to the interesting conclusion that heavy isotopic substitution has the effect of shortening the average length of the bond holding the isotope (primary isotope effect) as well as the remaining bonds (secondary isotope effects) in the molecule. If these data can be effectively related to the isotope shift the results might be used to predict the isotope effect on the molecular dimensions for other molecules, provided the shifts can be measured.

The first experimental observation of the isotope shift was obtained for hydrogen by Wimett[5] in 1953. His values for the differences in the proton magnetic shielding constants were $\sigma(H_2) - \sigma(D_2) = 0 \cdot 065 \pm 0 \cdot 059 \times 10^{-6}$ and $\sigma(HD) - \sigma(H_2) = 0 \cdot 048 \pm 0 \cdot 032 \times 10^{-6}$ for the proton magnetic resonance. Previously, Ramsey[6] had predicted the existence of an isotope effect in molecular NMR spectra and had proposed a relationship which would allow one to make quantitative estimates of the shift resulting from the changes in the dynamic state of the isotopic molecules, i.e. different zero-point energies. More interest was generated in 1957 and 1958 when Tiers[7, 8] observed upfield isotope shifts in the proton and fluorine resonances of deuterated molecules and with the semi-quantitative explanation of the isotope shift by Gutowsky[9] based on the differences in electrostatic deformations of the electron distribution caused by a neighbour atom in the same molecule. A *gem*-substituted deuterium neighbour with a smaller zero-point vibrational amplitude than a hydrogen neighbour produces less electrostatic deformation of the electronic distribution, resulting in an increase in shielding or an upfield shift in the proton resonance of a CHD group compared with the proton resonance of a CH_2 group.

It is not surprising that the most satisfactory explanation of the isotope shift is that which is applied to the hydrogen molecule. Starting with the early work of Ramsey,[6] there have been several theoretical treatments of the phenomenon in hydrogen.[10-13] In the study by Saika and Narumi[10] a calculation of the differences in shielding constant among H_2, HD and D_2 is given for discrete vibration-rotational states of hydrogen. In the study by Marshall[11] the isotope shift for hydrogen was obtained from a knowledge of the variation of the shielding constant as a function of internuclear distance $\sigma(R)$ and subsequent averaging of $\sigma(R)$ over the zero-point vibrations for H_2, HD and D_2. These approaches to the problem will be discussed later.

Isotope shifts in polyatomic molecules have been considered by a number of workers. Besides the electrostatic explanation originally offered by Gutowsky,[9] Lauterbur interpreted the large shifts of the ^{59}Co resonance of $K_3Co(CN)_6$, when ^{13}C and ^{15}N are substituted into the molecule,[14] in terms of the theory developed for the temperature dependence of the shielding[15] which makes use of the fifteen normal mode frequencies of the complex in the ground and excited electronic states. Loewenstein and Shporer[16] also examined the ^{13}C isotope shift on the ^{59}Co resonance in $K_3Co(CN)_6$ and, interpreting it in terms of the pressure and temperature dependence of the ^{59}Co shielding,[15] found that substitution of one ^{12}C by ^{13}C decreases the mean cobalt–ligand distance ($R \simeq 1 \cdot 8$ Å) by $1 \cdot 2 \times 10^{-3}$ per cent. While isotope effects on the ^{59}Co resonance lend themselves to quantitative calculation, the isotope effects on the shielding of protons is quite small and a quantitative treatment is difficult. The changes in the diamagnetic part of the shielding arising from the slight variations in bond angles and distances upon deuterium substitution were considered by the authors of this review[17] and were found to be large enough to account for the observed shifts. Finally, the effect of thermal population of excited vibrational modes on the magnetic shielding has been used by Petrakis and Sederholm[18] to discuss the isotope shift. In this review the different explanations mentioned above are considered. While no attempt is made to tabulate all the isotope shift measurements that have been performed, rather extensive lists of data are presented, mainly for the purpose of illustrating the various correlations that can be made with other molecular structure parameters.

2. EXPERIMENTAL MEASUREMENTS OF THE ISOTOPE SHIFT

Isotope effects on magnetic shielding have been observed in the magnetic resonance signals of several nuclei for a large number of isotopically substituted molecules. For the case in which protons are substituted by deuterium atoms the effect has been measured on the proton and fluorine resonances. In addition, a substantial amount of data is available for fluorine resonances in molecules where ^{12}C has been substituted by ^{13}C. These molecules include cases where the substitution is both at a neighbour carbon atom $\delta(F—^{12}C; F—^{13}C)$ and at the next nearest neighbours $\delta(F—^{12}C—^{12}C; F—^{12}C—^{13}C)$.[19-28] Other systems that have been studied, although to a lesser extent, include $\delta(H—^{12}C; H—^{13}C)$,[28-30] $\delta(F—^{28}Si; F—^{29}Si)$,[31,32] $\delta(F—^{32}S; F—^{33}S)$,[32] $\delta(F—^{32}S; F—^{34}S)$,[32] $\delta(F—^{80}Se; F—^nSe$, where n indicates the 76, 77, 78 and 82 isotopes of Se),[33] in addition to the $K_3Co(CN)_6$ compound mentioned above[14, 16] where the effects of ^{13}C and ^{15}N substitution were observed on the ^{59}Co resonance.

For almost all cases, heavy isotopic substitution shifts the NMR signal of a nearby nucleus toward a higher magnetic field which means that the substitution results in an increase in shielding of neighbouring nuclei. The only exceptions to this general trend that have been noted are cis-CHF=CDF[34] and the deuterated ammonium ions[35] where the proton resonance is shifted downfield from the non-deuterated species. The effect in the latter species is thought to be due to solvent interactions[35] and will be discussed briefly in the next section.

The magnitude of the isotope shift is generally dependent on how remote the isotopic substitution is from the nucleus under observation. As might be expected, the largest shifts occur when the substitution is next to the nucleus under observation, and the effect becomes smaller as the number of bonds separating the substituted isotope and the resonant nucleus increases. For example, when a deuterium atom is substituted for a hydrogen atom to give trans-CHF=CDF the isotope shift in the proton resonance is 0·005 ppm,[34] while the shift is 0·035 ppm for HTeD and HSeD[36] and 0·040 ppm for

HD.[37] The deuterium isotope effect on the fluorine magnetic resonance is 0·162 ppm for the *cis*-fluorine in $CF_2=CFD$ and 0·530 ppm for the fluorine *geminal* to the deuterium substituents in *cis*-CFD=CFD.[34] The magnitude of the shift is also a function of the resonant nucleus. The typical value of a *geminal* deuterium isotope effect on ^{19}F resonances is 0·60 ppm, as observed in n-C_3F_7D,[8] while the *geminal* deuterium isotope effect on proton resonances is typically 0·019 ppm as observed in CH_3D.[38] This difference reflects the nearly two orders of magnitude range in chemical shifts observed for ^{19}F as compared with the total range of proton chemical shifts.

As might be expected, the isotope shift is largest where the fractional change in mass upon isotopic substitution is largest. The effect of substituting ^{13}C for ^{12}C produces changes up to 0·194 ppm in the ^{19}F resonance and changes up to 0·007 ppm in the proton resonance where the observed nucleus is directly bonded to ^{13}C. As can be noted from the above examples, deuterium substitution results in much larger shifts even when it is an extra bond away. A smaller fractional change in mass results in smaller changes in the dynamic state of the molecule.

In general, the isotope chemical shift is approximately proportional to the number of atoms in the molecule that have been substituted by isotopes. This fact was observed in the experiments of Bernheim and Lavery,[38] Bernheim and Batiz-Hernandez,[39] Kanazawa, Baldeswieler and Craig,[34] and Barfield and Grant.[40] Lauterbur,[14] in his measurements on ^{15}N and ^{13}C substituted $K_3Co(CN)_6$ complexes, obtained an almost linear increase in the isotope shift of the ^{59}Co resonance as the number of ^{13}C and ^{15}N substituents increased. The data from these various sources have been listed in Table 1. Such an additive relationship may very well be directly related to the additivity of zero-point vibrational energies which result in the so-called "law of the mean" for the zero-point vibrational energies of isotopically substituted molecules.[41-4] Moreover, it is interesting to note that the small deviations from additivity of the isotope shift have their counterparts in deviations from the law of the mean for the zero-point energies as has been observed for the deuterated methanes.[43] The additivity of the isotope shift is also exhibited by *vicinal* isotopic substitution as shown in Table 1 for the fluorine resonance of the deuterated difluoroethylenes.

3. CORRELATIONS

A reasonable step in understanding the isotope shift is the correlation of isotope shift data with quantities that might also be related to the small changes in molecular electronic structure upon isotopic substitution. An empirical linear relationship between the ^{13}C isotope effect on the fluorine

TABLE 1. ISOTOPE SHIFTS FOR VARIOUS AMOUNTS OF ISOTOPIC SUBSTITUTION
The bold letter indicates the nucleus for which the shift is reported.

	Nuclear spin I	δ(ppm)	δ(ppm)/I	Ref.
$^{15}NH_3$	0	0		39
$^{15}NH_2D$	1	$0\cdot029 \pm 0\cdot002$	$0\cdot029 \pm 0\cdot001$	
$^{15}NHD_2$	2	$0\cdot053 \pm 0\cdot003$	$0\cdot027 \pm 0\cdot001$	
CH_4	0	0		38
CH_3D	1	$0\cdot019 \pm 0\cdot001$	$0\cdot019 \pm 0\cdot001$	
CH_2D_2	2	$0\cdot027 \pm 0\cdot003$	$0\cdot014 \pm 0\cdot001$	
CHD_3	3	$0\cdot045 \pm 0\cdot004$	$0\cdot015 \pm 0\cdot001$	
CH_3CN	0	0		40
CH_2DCN	1	$0\cdot012 \pm 0\cdot001$	$0\cdot012 \pm 0\cdot001$	
CHD_2CN	2	$0\cdot023 \pm 0\cdot001$	$0\cdot012 \pm 0\cdot001$	
CH_3CO_2H	0	0		40
CH_2DCO_2H	1	$0\cdot012 \pm 0\cdot001$	$0\cdot012 \pm 0\cdot001$	
CHD_2CO_2H	2	$0\cdot025 \pm 0\cdot001$	$0\cdot013 \pm 0\cdot001$	
CH_3NO_2	0	0		40
CH_2DNO_2	1	$0\cdot015 \pm 0\cdot001$	$0\cdot015 \pm 0\cdot001$	
CHD_2NO_2	2	$0\cdot029 \pm 0\cdot001$	$0\cdot014 \pm 0\cdot001$	
$K_3{}^{59}Co(^{13}C^{14}N)i(^{12}C^{14}N)_{6-i}$	0	0		14
	1	$0\cdot914 \pm 0\cdot004\dagger$	$0\cdot914 \pm 0\cdot004$	
	2	$1\cdot832 \pm 0\cdot026\dagger$	$0\cdot916 \pm 0\cdot013$	
	3	$2\cdot743 \pm 0\cdot013\dagger$	$0\cdot914 \pm 0\cdot004$	14
	4	$3\cdot659 \pm 0\cdot020\dagger$	$0\cdot915 \pm 0\cdot005$	
	5	$4\cdot580 \pm 0\cdot014\dagger$	$0\cdot916 \pm 0\cdot003$	
	6	$5\cdot463 \pm 0\cdot065\dagger$	$0\cdot910 \pm 0\cdot011$	
$K_3{}^{59}Co(^{12}C^{15}N)i(^{12}C^{14}N)_{6-i}$	0	0		14
	6	$1\cdot18 \pm 0\cdot004\dagger$	$0\cdot197 \pm 0\cdot007$	
$K_3{}^{59}Co(^{13}C^{14}N)i(^{12}C^{14}N)_{6-i}$	0	0		16
	1	$0\cdot851 \pm 0\cdot050\ddagger$	$0\cdot851 \pm 0\cdot050$	
	1	$0\cdot814 \pm 0\cdot232**$	$0\cdot814 \pm 0\cdot232$	
	2	$1\cdot709 \pm 0\cdot071\ddagger$	$0\cdot854 \pm 0\cdot035$	
	2	$1\cdot441 \pm 0\cdot348**$	$0\cdot720 \pm 0\cdot174$	
	3	$2\cdot560 \pm 0\cdot106\ddagger$	$0\cdot853 \pm 0\cdot035$	
$^{14}NH_4{}^+$	0	0		35, 45
$^{14}NH_3D^+$	1	$-0\cdot013 \pm 0\cdot001$	$-0\cdot013 \pm 0\cdot001$	
$^{14}NH_3D^+$	1	$-0\cdot015$	$-0\cdot015$	
$^{14}NH_2D_2{}^+$	2	$-0\cdot030$	$-0\cdot015$	
$^{14}NHD_3{}^+$	3	$-0\cdot045$	$-0\cdot015$	
$BH_4{}^-$	0	0		85
BH_3D^-	1	$0\cdot020$	$0\cdot020$	
$BH_2D_2{}^-$	2	$0\cdot040$	$0\cdot020$	
 F $\quad$ F $\;\;\searrow\;\;\swarrow$ $\quad$C=C $\;\;\swarrow\;\;\searrow$ H $\quad$ H	0	0		34
 F $\quad$ F $\;\;\searrow\;\;\swarrow$ $\quad$C=C $\;\;\swarrow\;\;\searrow$ H $\quad$ D	1	$0\cdot418$	$0\cdot418$ (*trans*)$\dagger\dagger$	
 F $\quad$ F $\;\;\searrow\;\;\swarrow$ $\quad$C=C $\;\;\swarrow\;\;\searrow$ H $\quad$ D	1	$0\cdot502$	$0\cdot502$ (*gem*)$\dagger\dagger$	

Table 1 (Continued)

Structure	Nuclear spin I	δ(ppm)	δ(ppm)/I	Ref.
F, F (top) / D, D (bottom), C=C	2	0·920	0·418 (trans)†† 0·502 (gem)††	
H, F (top) / F, H (bottom), C=C	0	0		34
H, F (top) / F, D (bottom), C=C	1	0·293	0·293 (cis)††	
H, F (top) / F, D (bottom), C=C	1	0·498	0·498 (gem)††	
D, F (top) / F, D (bottom), C=C	2	0·783	0·289 (cis)†† 0·494 (trans)††	
F, H (top) / F, H (bottom), C=C	0	0		34
F, H (top) / F, D (bottom), C=C	1	0·260	0·260 (trans)††	
F, H (top) / F, D (bottom), C=C	1	0·210	0·210 (cis)††	
F, D (top) / F, D (bottom), C=C	2	0·468	0·260 (trans)†† 0·208 (cis)††	

†, ‡, **. Measurements of the NMR ^{59}Co signals at 15·351, 14·1, and 4·3 Mc/s, respectively.

†† Position of deuterium atom with respect to resonant fluorine nucleus.

resonance, $\delta(F—^{13}C; F—^{12}C)$, with the $^{13}C—F$ spin–spin coupling interaction has been found by Frankiss.[28] Meaningful correlations have been found for the deuterium isotope effect on the proton resonance, $\delta(H—C—D; H—C—H)$, $\delta(H—N—D; H—N—H)$, with bond angles and distances[17, 45] in an attempt to relate the isotope shift in these systems to the slight bond rehybridization accompanying isotopic substitution. The success of this latter approach is limited by the lack of enough microwave and electron diffraction measurements, but the magnitude of the shift arising from these effects is quite reasonable. The effects of solvent on the isotope shift have been explored in only a few cases but can be of sufficient magnitude to confuse attempts at correlating isotope shifts with spin–spin coupling or molecular geometry.

In this section experimental data is presented which illustrates the correlation of the ^{13}C isotope shift of the ^{19}F resonance with the $^{13}C—F$ spin coupling. In addition, the available molecular geometry information for those molecules where isotope shift data are available are collected.

3.1. The ^{13}C Isotope Shift of the ^{19}F Resonance Correlated with the $^{13}C—F$ Spin–Spin Coupling

In the paper published by Frankiss,[28] several relationships between the isotope shifts of the fluorine resonance $\delta(F—^{13}C; F—^{12}C)$ and the $^{13}C—F$ spin–spin coupling constant, J_{CF}, were found for various saturated and unsaturated ^{13}C substituted fluorine compounds. The isotope shifts for the various fluoromethanes behaved according to the empirical equation.

$$\delta(F—^{13}C; F—^{12}C) = 0\cdot001 + 4\cdot62 \times 10^{-4}J_{CF}. \tag{1}$$

For molecules containing CF groups bonded to C, N, P, As, O, S, Se, F, Cl, Br, I and Hg, a different relationship was obtained. This is represented by

$$\delta(F—^{13}C; F—^{12}C) = 0\cdot007 + 4\cdot36 \times 10^{-4}J_{CF}. \tag{2}$$

and was found to hold very well for molecules containing only first row atoms and hydrogen, or less than three second-row elements bonded directly to the ^{13}C atom. Still another relationship was obtained for fluorine bound to an unsaturated (sp^2) ^{13}C atom and is given by

$$\delta(F—^{13}C; F—^{12}C) = -0\cdot039 + 5\cdot04 \times 10^{-4}J_{CF} \tag{3}$$

These equations were arrived at by a regression analysis of the actual data using the method of least squares where the maximum deviation of most of the data was within a reasonable estimate of the experimental error. It is interesting to note that the term dependent on J_{CF} is nearly the same in each equation. In Tables 3, 4 and 5 the results of the applications of equations (1),

(2) and (3) to the experimental data are summarized. These tables include values taken from the papers of Frankiss,[28] Harris,[25] Muller and Carr[24] and Bacon and Gillespie.[23]

TABLE 2. EFFECT OF MASS NUMBER ON THE ISOTOPE SHIFT OF THE
^{19}F RESONANCE

	Isotopic mass n	δ(ppm)	Ref.
$^nSeO_2F_2$	76	-0.026 ± 0.003	33
	77	-0.020 ± 0.005	
	78	-0.013 ± 0.002	
	80	0	
	82	0.014 ± 0.002	
$^nSF_6$	32	0	32
	33	0.027 ± 0.002	
	34	0.053 ± 0.004	

TABLE 3. THE CORRELATION OF ^{13}C ISOTOPE SHIFTS OF THE FLUORINE MAGNETIC RESONANCE, $\delta(F—^{13}C; F—^{12}C)$, WITH $^{13}C—F$ SPIN–SPIN COUPLING CONSTANTS J_{CF} USING EQUATION (1). The bold letter indicates the resonant nucleus and the J_{CF} reported is for the attached carbon atom.

	J_{CF}(c/s)	$\delta(F—^{13}C; F—^{12}C)$ ppm(exp)	$\delta(F—^{13}C; F—^{12}C)$ ppm(calc)	Ref.
CF_4	259	0.118	0.121	28
	257	0.105	0.120	24
CF_3H	274	0.126	0.128	28
	272	0.133	0.127	24
CF_2H_2	235	0.115	0.109	28
	232	0.143	0.108	24
CFH_3	158	0.072	0.076	28
	158	0.067	0.076	24
$C_6H_5CH_2F$	165	0.107	0.077	24

Although no explanation for these correlations has been offered it seems quite reasonable that the amount of change in hydridization of the C—F bond upon ^{13}C isotopic substitution would depend upon the original electron distribution in the C—F bond. The magnitude of the spin coupling reflects this original electron distribution and is not very sensitive to the isotopic substitution as has been shown in some examples of the deuterium isotope effect on the proton resonance.[38, 39] The facts that the fluorine chemical shift is dominated by the second-order, paramagnetic term and that perturbation theory must be carried to second order to describe the spin coupling, may also contribute to the significance of the correlation.

TABLE 4. THE CORRELATION OF ^{13}C ISOTOPE SHIFTS OF THE FLUORINE MAGNETIC RESONANCE, $\delta(F—^{13}C; F—^{12}C)$, WITH $^{13}C—F$ SPIN–SPIN COUPLING CONSTANTS J_{CF} USING EQUATION (2). The bold letter indicates the resonant nucleus and the J_{CF} reported is for the attached carbon atom.

	J_{CF}(c/s)	$\delta(F—^{13}C; F—^{12}C)$ ppm(exp)	$\delta(F—^{13}C; F—^{12}C)$ ppm(calc)	Ref.
CF_3Cl	299	0·152	0·137	24
CF_3Br	324	0·152	0·148	24
CF_3I	345	0·149	0·157	25, 26
	344	0·132	0·157	24
CF_3CCl_3	283	0·131	0·130	19
$(CF_3)_2CF_2$	265	0·115	0·123	24
$(CF_3O)_2$	268	0·122	0·124	25
$(CF_3)_2O$	265	0·116	0·123	25
$(CF_3)_2PCl$	320	0·138	0·147	25
$(CF_3)_2N$	269	0·125	0·124	24
CF_3CO_2H	283	0·129	0·130	19, 24
CF_3CH_2OH	278	0·116	0·128	24
CF_3CO_2Et	284	0·130	0·131	24
CF_3COCH_3	289	0·120	0·133	24
CF_3CH_3	271	0·128	0·125	24
$m-(CF_3)_2C_6H_4$	272	0·128	0·126	24
$p-(CF_3)_2C_6H_4$	271	0·128	0·125	24
$(CF_3)_2N—NO_2$	274	0·133	0·126	25
CF_3CH_2Cl	274	0·108	0·126	24
CF_3CH_2Br	272	0·129	0·125	24
$CF_3CCl=CCl_2$	274	0·140	0·126	24
$(CF_3)_2S$	309	0·148	0·142	25
$(CF_3S)_2$	314	0·142	0·144	25
CF_3SCFS	312	0·141	0·143	25
$(CF_3)_2Se$	331	0·145	0·151	25
$(CF_3Se)_2$	337	0·142	0·154	25
$cis-CF_3CClCClCF_3$	275	0·137	0·127	21
$trans-CF_3CClCClCF_3$	276	0·132	0·127	21
CF_3CCCF_3	256	0·142	0·119	24
$(CF_3)_2NSF_2$	263	0·147	0·122	24
$(CF_3)_2CF_2$	285	0·106	0·104	24
$(CF_3S)_2Hg$	308	0·153	0·141	25
CF_3SH	304	0·146	0·140	25
CF_3SNCO	309	0·138	0·142	25
$(CF_3)_2AsCl$	344	0·139	0·157	25
$(CF_3Se)_2Hg$	333	0·151	0·152	25
$(CF_3)_2Hg$	356	0·136	0·162	25
CF_2ClCH_3	288	0·147	0·133	24
CF_2Br_2	357	0·168	0·163	25
	358	0·168	0·163	24
CF_2Cl_2	324	0·170	0·148	25
	325	0·164		24
CF_2BrCF_2Br	312	0·139	0·143	25
CF_2ClCF_2Cl	299	0·140	0·137	25
$CHCl_2F$	294	0·156	0·135	25
CCl_3F	337	0·194	0·155	25
	337	0·192	0·155	24
CBr_3F	372	0·191	0·169	25
C_6H_5COF	344	0·118	0·157	24
$(CH_3)_3CF$	167	0·114	0·080	24

TABLE 5. THE CORRELATION OF ^{13}C ISOTOPE SHIFTS OF THE FLUORINE MAGNETIC RESONANCE, $\delta(F—^{13}C; F—^{12}C)$, WITH ^{13}C—F SPIN–SPIN COUPLING CONSTANTS J_{CF} USING EQUATION (3). The bold letter indicates the resonant nucleus and the J_{CF} reported is for the attached carbon atom.

	$J_{CF}(c/s)$	$\delta(F—^{13}C; F—^{12}C)$ ppm(exp)	$\delta(F—^{13}C; F—^{12}C)$ ppm(calc)	Ref.
C**S**F₂	366	0·143	0·145	28
C**O**F₂	308	0·121	0·116	28
C**F**₂CCl₂	289	0·103	0·107	19, 28
CF**Cl**CCl₂	303	0·112	0·114	19, 28
CF₃S**C**FS	395	0·16	0·160	28
cis-C**F**ClCFCl	299	0·113	0·112	22
trans-C**F**ClCFCl	290	0·107	0·107	22
(C**O**F)₂	366	0·133	0·145	23
C**O**FCOCl	377	0·132	0·151	23
CH₃C**O**F	353	0·125	0·139	24
HC**O**F	369	0·128	0·147	24

3.2. Isotope Chemical Shift and Molecular Geometry

When the available isotope chemical shift data is examined, it is seen that the magnitude of the shift decreases with an increase in s character of the bond holding the isotope.[17] A strong inverse dependence with the average linear distance between the substituted site and the resonant nucleus is also noticeable. In a simple covalent molecule the degree of hybridization depends upon the average coordinates of its various constituent atoms, and, therefore, it may be possible to correlate the isotope shift with the geometry of the molecule.

With this purpose Bernheim and Batiz-Hernandez have investigated the *geminal* deuterium isotope effect in the proton resonance of some molecules containing first row elements.[17] The data is sparse, but for monodeuteration a general decrease in the isotope shift with an increase in the average proton–proton distance, $D(H—H)$ was noted in addition to a decrease in isotope shift with an increase in the geminal angle, $H\hat{X}H$, where X is a first-row atom. These results in addition to some other data found to follow a similar trend are presented in Tables 6 and 7. In a later section the connection between this data and the dependence of the isotope shift on the bond hybridization parameters[17] is reviewed.

It is evident from Table 6, however, that the molecular dimensions are not the only quantities involved. One of the deviations from the general trend is for formaldehyde. The proton resonance of CHDO exhibits a solvent dependence[54] of the isotope shift. Another abnormality is noted in the case of the deuterated ammonium ions,[35] where the isotope shift is *downfield*

TABLE 6. EXPERIMENTAL DATA FOR THE DEUTERIUM ISOTOPE SHIFT OF THE PROTON RESONANCE AND MOLECULAR GEOMETRY.

The bold letter proton is that for which the resonance is observed. Isotope shifts are reported in parts per million, a positive shift being upfield with respect to the unsubstituted compound. The internuclear distance, $D(H—H)$, is the distance between the resonant nucleus and the site of substitution and is listed in units of a_0, the Bohr radius.

	Solvent	Isotope shift δ(ppm)	Refs.	$H\hat{X}H$	$D(H—H)$	Refs.
H_2	gas	0	37		1·3983	46
HD		0·040				
H_2Te	CS_2	0	36	89° 30′	4·4018	47
HDTe		0·035				
H_2Se	CS_2	0	36	91°	3·9286	48
HDSe		0·035				
H_2S	CS_2	0	49	92° 10′	3·6345	50, 51
HDS		0·033				
H_2O	aq. ac.	0	52	104° 56′	2·8708	2, 51, 53
HDO		0·030 ± 0·003				
CH_2O		0	54	123° 26′	3·5644	51
CHDO	TMS	0·029				
	THF	0·023				
	CH_3CN	0·026				
$^{15}NH_3$	neat	0	39	106° 47′	3·0760	51
$^{15}NH_2D$		0·029 ± 0·002				
$^{15}NHD_2$		0·053 ± 0·003				
PH_3	neat	0	55	93° 30′	3·9061	56
PHD_2		0·004 ± 0·002				
CH_4	CCl_4	0	39	109° 28′	3·3772	57
CH_3D		0·019 ± 0·001				
CH_2D_2		0·027 ± 0·003				
CHD_3		0·045 ± 0·004				
BH_4^-	H_2O	0	85	109°28′	3·872	86
BH_3D^-		0·020				
$BH_2D_2^-$		0·040				
$^{14}NH_4^+$	H_2SO_4	0	38	109° 28′	3·2020	58, 59
$^{14}NH_3D^+$		−0·015				
$^{14}NH_2D_2^+$		−0·030				
$^{14}NHD_3^+$		−0·045				
SiH_4	C_5H_{12}	0	60	109° 28′	4·569	61
$SiHD_3$		0·008				
CH_3COCH_3	neat	0	9	108° 30′	3·3310	62
CD_3COCH_2D		0·034 ± 0·001				
CH_3CN	neat	0	40	109° 16′	3·4270	63
CH_2DCN		0·012 ± 0·001				
CH_3COOH	neat	0	40			
CH_2DCOOH		0·012 ± 0·001				
$C_6H_5CH_3$	neat	0	7			
$C_6H_5CH_2D$		0·015 ± 0·002				
CH_3NO_2	neat	0	40			
CH_2DNO_2		0·015 ± 0·001				
$EtCH_2CN$	neat	0	40			
$EtCHDCN$		0·014 ± 0·001				
CH_2Cl_2	CCl_4	0	16, 17	112°	3·3462	64

Table 6   (*Continued*)

	Solvent	Isotope shift δ(ppm)	Refs.	HX̂H	D(H—H)	Refs.
CHDCl₂		0·0125 ± 0·0003				
CH₂Br₂	CCl₄	0	16, 17	109° 28′	3·3772	65
CHDBr₂		0·0137 ± 0·0002				
CH₂I₂	CCl₄	0	16, 17	109° 28′	3·3772	66
CHDI₂		0·0135 ± 0·0002				
H₂C=CH₂	TMS	0	67	116°	3·4422	68
H₂C=CDH		0·0085				
C₆H₅CH=CH₂	neat	0	69			
C₆H₅CD=CH₂		{0·0090 − (H *cis* to D)				
		{0·0048 − (H *trans* to D)				
cis-CHF=CHF	CCl₃F	0	34			
cis-CHF=CDF		−0·002				
trans-CHF=CHF	CCl₃F	0	34			
trans-CHF=CDF		0·005				
CH₂=CF₂	CCl₃F	0	34			
CHD=CF₂		0·010				

TABLE 7. EXPERIMENTAL DATA FOR THE ^{13}C ISOTOPE SHIFT OF THE ^{19}F RESONANCE AND MOLECULAR GEOMETRY.

The resonant fluorine nucleus is in bold letter. Isotope shifts are in parts per million, a positive shift being upfield with respect to the unsubstituted compound. CF bond distances are listed in units of a_0, the Bohr radius.

	Solvent	Isotope shift δ(ppm)	Refs.	FĈH	D(C—F)	Ref.
^{12}**CF**₄	neat	0		109° 28′	2·5057	84
^{13}**CF**₄		0·118	28			
		0·105	24			
^{12}**CF**₃H	neat	0		108° 48′	2·5170	84
^{13}**CF**₃H		0·126	28			
^{12}**CClF**₃	neat	0	28	108° 36′	2·5094	84
^{13}**CClF**₃		0·152				
^{12}**ClF**₃	neat	0			2·5094	84
^{13}**ClF**₃		0·149	25, 26			
		0·132	24			
^{12}**CF**₃CC^{12}**CF**₃	neat	0		107° 30′		84
^{13}**CF**₃CCC^{13}**CF**₃		0·142	24			
^{12}**CH**₂**F**₂	neat	0		108° 18′	2·5661	84
		0·115	28			
^{13}**CH**₂**F**₂		0·143	24			
^{12}**CCl**₂**F**₂	neat	0			2·5510	84
^{13}**CCl**₂**F**₂		0·170	25			
		0·164	24			
^{12}**CH**₃**F**		0			2·6171	84
^{13}**CH**₃**F**		0·072	28			

instead of upfield. This reversal in sign is very likely due to the electrostatic effects arising from hydrogen bonding with the sulphuric acid solvent which is very polar. Hydrogen bond formation is known to result in downfield chemical shifts.[71-3] In addition, it is known that "hydrogen bonding" is stronger for deuterated species than for non-deuterated species,[74] due partly to the lower zero-point vibrational frequencies of the $—X \cdots D \cdots Y—$ bond compared with the $—X \cdots H \cdots Y—$ bond. The downfield shift would be expected to increase with the number of deuterium substituents since the degree of solvation of the ion is expected to increase with the number of N—D— solvent bonds. This is indeed what is observed.[35]

4. THEORETICAL EXPLANATIONS OF THE ISOTOPE SHIFT

The various theories that have been proposed in order to obtain numerical estimates of the isotope shift are discussed below. First, the general formulism for H_2, HD and D_2, derived by Ramsey,[6] for the paramagnetic contribution, and by Saika and Narumi[10] for the diamagnetic contribution to the total shielding constant together with Marshall's treatment[11] are examined. For molecules other than hydrogen no complete theory exists, but the studies that have been made offer at least a semiquantitative estimate of the effect. Those that will be considered here are the electrostatic deformation treatment[9] with modifications[52] and the bond rehybridization considerations.[17] In addition, the various suggestions[14, 16] advanced for the explanation of the effect in ^{13}C and ^{15}N substituted $K_3Co(CN)_6$ and the consideration of excited vibrational states[18] will be discussed.

4.1. *The Hydrogen Molecule*

The theory of vibrational and centrifugal effects in molecules derived by Ramsey,[6] and later extended by Saika and Narumi[10] for treating the isotope effect, gives a satisfactory explanation of the occurrence of an upfield isotope chemical shift arising from heavy isotopic substitution. In the formulation the total shielding constant, σ, was represented by a power series in (R/R_e), where R is the internuclear distance and R_e is the equilibrium value of R. An isotope chemical shift is estimated by noting the differences in the average value of (R/R_e) among the isotopic molecular species. The value of the equilibrium internuclear distance is invariant upon isotopic substitution if the Born–Oppenheimer approximation holds. Consequently, the major contribution to the variation in the shielding comes from the small differences in the average distance due to different zero-point energies and the anharmonic parts of the potential function.

In his treatment, Ramsey considered only the (R/R_e) variations of the high frequency or paramagnetic contribution, σ^p, while Saika and Narumi included similar variations in the diamagnetic contribution, σ^d. For an isotopic molecule x in a vibrational state v and rotational state J, the expectation value of the shielding constant, $_v^x\langle\sigma\rangle_J$ can be written as a sum of the diamagnetic and paramagnetic contributions:

$$_v^x\langle\sigma\rangle_J = {}_v^x\langle\sigma^d\rangle_J + {}_v^x\langle\sigma^p\rangle_J$$
$$= (e^2/3mc^2)_v^x\langle\Psi_{0\lambda} \left| \sum_k r_k^{-1} \right| \Psi_{0\lambda}\rangle_J$$
$$+ \tfrac{2}{3}\mu_0 \sum_{n\lambda'}{}' \{1/_{(E_n-E_0)}\}\{H_{0\lambda n\lambda'}L_{n\lambda'0\lambda} - L_{0\lambda n\lambda'}H_{n\lambda'0\lambda}\}, \tag{4}$$

where

$$-\mu_0 L_{0\lambda n\lambda'} = \sum_k {}_v^x\langle\Psi_{0\lambda} \left| m_{zk} \right| \Psi_{n\lambda'}\rangle_J \tag{5}$$

and

$$\mu_0 H_{n\lambda'0\lambda} = {}_v^x\langle\Psi_{n\lambda'} \left| \sum_k m_{zk}/r_k^3 \right| \Psi_{0\lambda}\rangle_J \tag{6}$$

and where the origin of r_k is the nucleus for which the shielding is desired, and $\sum_k$ is over all the electrons. The symbols λ and λ' indicate the orientation of the molecule with respect to the applied magnetic field in the ground and excited electronic states of the molecule, Ψ_0 and Ψ_n, respectively.

For diatomic molecules the paramagnetic term can be assumed[6] to vary with average internuclear spacing as $(R/R_e)^m$, in which case the paramagnetic contribution becomes

$$_v^x\langle\sigma^p\rangle_J = \tfrac{2}{3}\mu_0 G_e[_v^x\langle(R/R_e)^m\rangle_J]. \tag{7}$$

Ramsey has discussed[6] how the spin-rotation constant measured in two different rotational states could be used to determine G_e and m. The diamagnetic term also will vary with average internuclear spacing as has been discussed by Saika and Narumi.[10] The latter workers arrived at two possible expansions of $1/r_k$ in terms of R/R_e which can be used to obtain two possible values of the diamagnetic contribution. The two expansions are

$$1/r_{k(+)} = 2\cdot8960 - 4\cdot9240(R/R_e) + 4\cdot3172(R/R_e)^2 - 1\cdot3745(R/R_e)^3, \tag{8}$$

and

$$1/r_{k(-)} = -0\cdot6400 + 3\cdot5997(R/R_e)^{-1} - 2\cdot9507(R/R_e)^{-2} + 0\cdot9057(R/R_e)^{-3}. \tag{9}$$

The corresponding diamagnetic contribution resulting from the $1/r_{k(+)}$ and $1/r_{k(-)}$ expansions are $_0^x\langle\sigma^d\rangle_J^{(+)}$ and $_0^x\langle\sigma^d\rangle_J^{(-)}$, respectively, where only the ground vibrational state, $v = 0$, has been considered.

In order to evaluate the total shielding for each isotopic molecule, x (and hence, the isotope shift), the important quantity to be obtained is $\langle(R/R_e)^m\rangle_J$. Ramsey[6] pointed out how this could be done by expressing R/R_e in terms of $(R - R_e)/R_e$:

$$_v^x\langle(R/R_e)^m\rangle_J = \sum_{q=0}^{\infty} [n!/(n - q)! \, q!]_v^x\langle(R - R_e)^q/R_c^q\rangle_J \tag{10}$$

and obtaining the expectation value of the latter quantity by assuming a Morse potential for the nuclear vibrational function. The expectation values of $(R - R_e)^q/R_c^q$ can be calculated for different rotational states, and the ratio

$$_v^x\langle(R/R_e)^m\rangle_{J'} \quad \text{to} \quad _v^x\langle(R/R_e)^m\rangle_J$$

found for two rotational states J' and J. The same ratio can be found experimentally by measuring a quantity such as the rotational magnetic moment in the two states, and the value of m is inferred by comparing the experimental value with the theoretical value.

From molecular beam studies Ramsey determined m to be either $-1\cdot9$ or $+1\cdot7$. However, the limits of error on these quantities and the corresponding error in the diamagnetic contribution to the shielding were large. Saika and Narumi found the value $m = -1\cdot9$ more consistent with a comparison of their calculated total shielding for H_2, HD, and D_2 with the available experimental isotope shifts at that time, although Ramsey considered the choice of $m = +1\cdot7$ more probable. The recent more accurate measurements by Dayan et al.[37] for $\sigma(HD) - \sigma(H_2) = 0\cdot040 \pm 0\cdot01$ ppm further supports the $m = -1\cdot9$ choice.

The problem of the isotope shift in the NMR spectra of H_2, HD and D_2 was further examined by Marshall.[11] In addition to the nuclear vibrational wave functions for the isotopic species, the magnetic shielding constant as a function of internuclear separation $\sigma(R)$ had been calculated[70] assuming that the paramagnetic part of the shielding σ^p could be neglected. Marshall obtained a quadratic fit of $\sigma(R)$ in the neighbourhood of the equilibrium separation R_e as a function of $\sigma(R_e)$:

$$\sigma(R) = \sigma(R_e) + (R - R_e)\sigma'(R_e) + \tfrac{1}{2}(R - R_e)^2\sigma''(R_e). \tag{11}$$

The observed magnetic shielding, σ_{obs}, is

$$\sigma_{obs} = \sigma(R_e) + \overline{(R - R_e)}\sigma'(R_e) + \tfrac{1}{2}\overline{(R - R_e)^2}\sigma''(R_e) \tag{12}$$

where the functions of deviation from the equilibrium internuclear separation are averaged over the vibrational wave functions:

$$\overline{R - R_e} = \int \Psi^2(R - R_e)\,dR, \tag{13}$$

and

$$\overline{(R - R_e)^2} = \int \Psi^2(R - R_e)^2\,dR. \tag{14}$$

Marshall's results are

$$\sigma(D_2) - \sigma(H_2) = 0\cdot055 \times 10^{-6},$$
$$\sigma(D_2) - \sigma(HD) = 0\cdot030 \times 10^{-6},$$
$$\sigma(HD) - \sigma(H_2) = 0\cdot025 \times 10^{-6}.$$

The value of $\sigma(HD) - \sigma(H_2) = 0\cdot025$ ppm is to be compared with the experimental value of $0\cdot040$ ppm,[37] and with the theoretical values of Saika and Narumi[10] in the rotational state $J = 0$ which are:

$$\sigma(HD) - \sigma(H_2) = 0\cdot035 \text{ ppm}$$

for $m = -1\cdot9$ and the above expansion $1/r_{(-)}$, and

$$\sigma(HD) - \sigma(H_2) = 0\cdot057 \text{ ppm}$$

for $m = -1\cdot9$ and the above expansion $1/r_{(+)}$. If the average of the two expansions is taken the resulting value is

$$\sigma(HD) - \sigma(H_2) = 0\cdot047 \text{ ppm.}$$

We conclude that the deuterium isotope shift for the hydrogen molecule is fairly well understood.

4.2. Polyatomic Molecules

The theoretical calculation of the isotope shift for polyatomic molecules is much more complicated than that for the hydrogen molecule. A quantitative treatment would require accurate wave functions and a knowledge of how they are affected by the various vibrational degrees of freedom of the molecule before and after isotopic substitution. Nevertheless, several semi-quantitative approaches to the consideration of isotope shifts in polyatomic molecules have been made.

The first suggestion how the problem might be approached was made by Gutowsky who considered an electrostatic deformation model.[9] In this treatment the contribution to the magnetic shielding of a proton by intra-molecular electric fields is examined. In the presence of an electrostatic field the magnetic shielding is decreased by an amount proportional to the square of the electric field.[75, 77] If the electric field, E, is due to an electric charge on an atom in the molecule, the decrease in shielding will be directly proportional to the fourth power of the separation between the charge and the resonant nucleus. A charge of q electron units on an atom at a distance b Å from the proton whose resonance is observed will decrease the shielding at the proton by an amount $2(q^2/b^4) \times 10^{-5}$. The fact that a deuterium atom has smaller zero-point vibrational amplitudes than a hydrogen atom means that it will be less effective in decreasing the shielding than would a proton. Estimating the decrease in the average value of $1/b^4$ to be 10^{-2} it is found that a point charge of $q = 0\cdot65$ must be placed on a geminal deuterium atom in a CHD group to account for the observed shifts of the order of 10^{-8}. This quantity of charge, which is high, could be decreased by allowing for a spatial distribution of charge rather than a point charge at the nuclear site.

The electrostatic deformation model has also been used by Holmes, Kivelson and Drinkard[52] to discuss their measurement of the deuterium isotope effect in water. They suggest an exponential form for the expression of the electric field induced shift to take account of the fact that the effect of an electric field on the magnetic shielding can be expressed as a power series in E^2.

An alternative approach to the explanation of the isotope shift in poly-atomic molecules is through a consideration of the changes in hybridization parameters that would accompany the small changes in average molecular geometry upon isotopic substitution. The changes in average atomic co-ordinates will occur if the vibrational potentials have anharmonic components. An estimate of the magnitude of this contribution has been made by Bernheim and Batiz-Hernandez[17] in which the change in the diamagnetic part of the shielding, σ^d, was calculated as a function of bond hybridization parameters. It was found that the change in shielding accompanying a reasonable change in average bond angle was sufficient to explain the observed effects. An example of the changes that are expected are given in Table 8 where the diamagnetic shielding constant calculated with Slater type localized equivalent orbitals is tabulated as a function of the H—C—H bond angle in a molecule such as CH_3X where X = H or D. The observed shift of 0·019 ppm in the proton resonance of CH_3D as compared with CH_4 could be explained on this basis if the H—C—H angle decreased by about 10 minutes in going from CH_4 to CH_3D. However, it was also found that the same change could be accounted for if the C—H bond length shortened by about 0·0012 atomic units. Similar calculations have been performed for NH_3.

TABLE 8. THE DIAMAGNETIC SHIELD-
ING CONSTANT FOR PROTONS IN
$CH_3X(X=H, D)$ AS A FUNCTION OF
H—C—H BOND ANGLE.

The C—H bond distance is constant and assumed to be completely covalent.

HĈH	$\sigma^d(\theta)$
105°	30·774
106°	30·645
107°	30·529
108°	30·424
109°	30·331
109° 28′	30·285
110°	30·246
111°	30·170
112°	30·100
113°	30·037
114°	29·979
115°	29·927

The paramagnetic contribution to the magnetic shielding, σ^p, will also be affected by changes in geometry accompanying isotopic substitution. However, for molecules such as H_2 or CH_4, σ^p is small compared with σ^d, and the contribution of changes in σ^p to the isotope shift can be argued[17] to be less than the changes in σ^d.

The ionic character of the bond holding the nucleus whose shift is being observed will affect the calculated isotope shift slightly.[17] In NH_3 the change in σ^d for a given change in the H—N—H angle is slightly greater if the N—H bond is considered to have ionic character. This effect is demonstrated in Table 9 where σ^d is calculated as a function of H—N—H angle for

TABLE 9. THE DIAMAGNETIC SHIELDING CONSTANT FOR PROTONS IN NH_3 AS A FUNCTION OF THE H—N—H BOND ANGLE FOR A COVALENT N—H BOND, $(\lambda/\mu) = 1$, AND FOR A PARTIALLY IONIC BOND, $(\lambda/\mu) = 0.84$.

	$\sigma^d(\theta)$	
HNH	$(\lambda/\mu) = 1$	$(\lambda/\mu) = 0.84$
100°	31·937	33·868
101°	31·756	33·684
102°	31·593	33·517
103°	31·444	33·366
104°	31·308	33·226
105°	31·183	33·097
106°	31·066	32·977
106° 47′	30·980	32·889
107°	30·957	32·865
108°	30·856	32·760
109°	30·782	32·661
110°	30·671	32·566

(a) a completely covalent N—H bond and (b) a N—H bond with enough ionic character to obtain agreement between the calculated and experimental electric dipole moments of the ammonia molecule.[78] This is described by the ratio of coefficients (λ/μ) of the nitrogen hybrid and hydrogen atomic orbitals which are used to form the localized bond orbital:

$$\Psi_{N-H} = \lambda\Psi_N + \mu\Psi_H$$

There is considerable evidence to show that average bond distances are shortened upon heavy isotope substitution, but there is not very much information concerning the effect on changes in bond angles. In order to state unambiguously that a given bond angle decreases upon heavy isotope substitution, the geometry of the molecule must be over-determined by some experiment, such as a study of the microwave spectra of a large number of isotopically substituted species. The differences in bond distances between CD_4 and CH_4 has been investigated by electron diffraction techniques[79] and with the Raman effect[57] where it is found that the C—D distance is shorter than the C—H distance. There is also evidence[57] that the average C—H distance is shorter in CH_3D than in CH_4 if CH_3D is assumed to retain a tetrahedral geometry. The calculated diamagnetic contribution to the proton magnetic shielding is $\sigma^d = 30\cdot28522$ for CH_4 and $\sigma^d = 30\cdot29784$ for tetrahedral CH_3D where the C—H bond distance was $2\cdot06733$ atomic units in CH_4 and $2\cdot06670$ atomic units in CH_3D.[17, 80] If, indeed, methane actually remained tetrahedral upon mono-deuterium substitution, the calculated change in shielding is about two-thirds the experimental value.

While the magnitude of the effect can be accounted for on the basis of either a small change in bond distance or angle, it must be realized that the shift is very likely a result of both these effects in addition to any contribution from intramolecular electric fields. The possibility of relating the isotope shift to any single one of these parameters is remote at this time because of the problem of sorting out all of the individual contributions.

There have been other suggestions regarding the details of the source of the isotope shift. Petrakis and Sederholm,[18] in a study of the temperature variation of the chemical shift, considered the effect of a Boltzmann distribution of molecules over the available excited vibrational energy states of the molecule. The chemical shift of the magnetic resonance of a nucleus in the molecule will depend upon a properly weighted average of the chemical shift contributions from the available vibrational states. For example, it is concluded[18] that excitation of a CH_2 rock, wag, or twist gives rise to a chemical shift of 0·47 ppm. Substitution of H by D should produce a shift of less than 0·003 ppm which is less than, but of the order of magnitude of the observed shifts.

The very large isotope shifts of the ^{59}Co magnetic resonance in $K_3Co(CN)_6$ caused by selective substitution of ^{12}C by ^{13}C and ^{14}N by ^{15}N have been discussed by Lauterbur[14] and Loewenstein and Shporer.[16] This is a case where the chemical shift of the resonant nucleus is dominated by the para-magnetic contribution to the magnetic shielding.[81–3] Again, the shift can be interpreted[17] in terms of the theory advanced to explain the temperature and pressure dependence of the Co chemical shift[15] where the contribution from excited vibrational states is invoked. For purposes of actual computation of an isotope shift, the average geometry of the molecule would have to be calculated using the thermal distribution over excited vibrational states rather than just the ground, or zero-point, state.

5. CONCLUSIONS

In the preceding discussion we have tried to indicate the present status of the problem of trying to interpret the isotope shift of a nuclear magnetic resonance frequency. Tables of experimental data indicate that there is a relation between the isotope shift and molecular geometry and spin–spin coupling. Theoretically, the effect is fairly well understood for the hydrogen molecule in terms of an average over the different zero-point vibration for the isotopically substituted species. However, for polyatomic molecules the situation is much more complex due to the variety of interactions that can result in changes in magnetic shielding accompanying isotopic substitution. The present outlook for the ultimate use of the isotope shift as a molecular structure tool is not very optimistic for this reason. In order to relate the experimental shift to a single parameter, such as a change in average bond angle, requires a knowledge of all the other changes taking place, such as changes in average bond distances.

There are, however, several individual situations that possibly might yield useful information. For the heavier atoms, such as cobalt, the magnetic shielding is dominated by the paramagnetic, second-order, term. The large shifts found for ^{59}Co and expected to occur for other heavy nuclei might very easily turn out to be directly related to structural changes accompanying isotopic substitution. The cases where solvent effects seem to dominate the magnetic shielding, such as in the deuterated ammonias, may be another area in which isotopic shifts may be useful as well as the investigations that yield definite correlations with spin–spin coupling constants.

In order for the applications of the isotope effect to progress much beyond the present stage it will be necessary to have some independent studies of the precise details of the effects of isotopic substitution on molecular geometry, as well as separate investigations of the contribution of the excited vibrational states to the magnetic shielding.

REFERENCES

1. L. S. BARTELL, K. KUCHITSU and R. J. DeNEUI, *J. Chem. Phys.* **35,** 1211 (1961).
2. V. W. LAURIE and D. R. HERSCHBACH, *J. Chem. Phys.* **37,** 1687 (1962).
3. L. S. BARTELL and H. K. HIGGINBOTHAM, *J. Chem. Phys.* **42,** 851 (1965).
4. R. H. SHWENDEMAN and J. D. KELLY, *J. Chem. Phys.* **42,** 1132 (1965).
5. T. F. WIMETT, *Phys. Rev.* **91,** 476 (1953).
6. N. F. RAMSEY, *Phys. Rev.* **87,** 1075 (1952).
7. G. V. D. TIERS, *J. Amer. Chem. Soc.* **79,** 5585 (1957).
8. G. V. D. TIERS, *J. Chem. Phys.* **29,** 963 (1958).
9. H. S. GUTOWSKY, *J. Chem. Phys.* **31,** 1683 (1959).
10. A. SAIKA and H. NARUMI, *Can. J. Phys.* **42,** 1481 (1964).
11. T. W. MARSHALL, *Mol. Phys.* **4,** 61 (1961).
12. N. J. HARRIS, R. G. BARNES, P. J. BRAY and N. F. RAMSEY, *Phys. Rev.* **90,** 260 (1953).
13. W. E. QUIN, J. M. BAKER, J. T. LA TOURRETTE and N. F. RAMSEY, *Phys. Rev.* **112,** 1929 (1958).
14. P. C. LAUTERBUR, *J. Chem. Phys.* **42,** 799 (1965).
15. G. B. BENEDEK, R. ENGLMAN and J. A. ARMSTRONG, *J. Chem. Phys.* **39,** 3349 (1963).

16. A. LOEWENSTEIN and M. SHPORER, *Mol. Phys.* **9**, 293 (1965).
17. R. A. BERNHEIM and H. BATIZ-HERNANDEZ, *J. Chem. Phys.* **45**, 2261 (1966).
18. L. PETRAKIS and C. H. SEDERHOLM, *J. Chem. Phys.* **35**, 1174 (1961).
19. G. V. D. TIERS, *J. Phys. Soc. Japan* **15**, 354 (1960).
20. G. V. D. TIERS, *J. Phys. Chem.* **66**, 945 (1962).
21. G. V. D. TIERS, *J. Chem. Phys.* **35**, 2263 (1961).
22. G. V. D. TIERS and P. C. LAUTERBUR, *J. Chem. Phys.* **36**, 1110 (1962).
23. J. BACON and R. J. GILLESPIE, *J. Chem. Phys.* **38**, 781 (1963).
24. N. MULLER and D. T. CARR, *J. Phys. Chem.* **67**, 112 (1963).
25. R. K. HARRIS, *J. Mol. Spectroscopy* **10**, 309 (1963).
26. H. SPIESECKE and W. G. SCHNEIDER, *J. Chem. Phys.* **35**, 722 (1961).
27. R. K. HARRIS, *J. Phys. Chem.* **66**, 768 (1962).
28. S. G. FRANKISS, *J. Phys. Chem.* **67**, 752 (1963).
29. J. N. SHOOLERY, L. F. JOHNSON and W. A. ANDERSON, *J. Mol. Spectroscopy* **5**, 100 (1960).
30. H. DREESKAMP and E. SACKMAN, *Z. Phyzik. Chem.* (Frankfurt) **27**, 136 (1961).
31. G. V. D. TIERS, *J. Inorg. Nucl. Chem.* **16**, 363 (1961).
32. R. J. GILLESPIE and J. W. QUAIL, *J. Chem. Phys.* **39**, 2555 (1963).
33. T. BIRCHALL, S. L. CROSSLEY and R. J. GILLESPIE, *J. Chem. Phys.* **41**, 2760 (1964).
34. Y. KANAZAWA, J. D. BALDESWIELER and N. C. CRAIG, *J. Mol. Spectroscopy* **16**, 325 (1965).
35. G. FRAENKEL, Y. ASAHI, H. BATIZ-HERNANDEZ and R. A. BERNHEIM, *J. Chem. Phys* **44**, 4647 (1966).
36. H. SCHMIDBAUR and W. SIEBERT, *Z. Naturforsch.* **20b**, 596 (1965).
37. E. DAYAN, G. WIDENLOCHER and M. CHAIGNEAU, *Compt. Rend.* **257**, 2455 (1963).
38. R. A. BERNHEIM and B. J. LAVERY, *J. Chem. Phys.* **42**, 1464 (1965).
39. R. A. BERNHEIM and H. BATIZ-HERNANDEZ, *J. Chem. Phys.* **40**, 3446 (1964).
40. M. BARFIELD and D. M. GRANT, *J. Amer. Chem. Soc.* **83**, 4726 (1961).
41. H. J. BERNSTEIN and A. D. E. PULLIN, *J. Chem. Phys.* **21**, 2188 (1953).
42. J. BIGELEISEN and P. GOLDSTEIN, *Z. Naturforsch.* **18a**, 205 (1963).
43. J. BIGELEISEN, R. E. WESTON and M. WOLFSBERG, *Z. Naturforsch.* **18a**, 210 (1963).
44. M. WOLFSBERG, *Z. Naturforsch.* **18a**, 216 (1963).
45. H. BATIZ-HERNANDEZ, Ph.D. THESIS, The Pennsylvania State University (1965).
46. G. HERZBERG, *Spectra of Diatomic Molecules*, D. Van Nostrand Company, Inc., Princeton, New Jersey (1945).
47. A. W. JACHE, P. W. MOSER and W. GORDY, *J. Chem. Phys.* **25**, 209 (1956).
48. K. ROSSMAN and J. W. STRALEY, *J. Chem. Phys.* **24**, 1276 (1956).
49. H. SCHMIDBAUR and W. SIEBERT, *Chem. Ber.* **97**, 2090 (1964).
50. M. T. EMERSON and D. F. EGGERS, *J. Chem. Phys.* **37**, 251 (1962); H. C. ALLEN, Jr., *The Infrared Band Systems of Hydrogen Sulfide*, Ph.D. Thesis, University of Washington, Seattle (1951).
51. G. HERZBERG, *Infrared and Raman Spectra of Polyatomic Molecules*, D. Van Nostrand Company, Inc., New Jersey (1945).
52. J. R. HOLMES, D. KIVELSON and W. C. DRINKARD, *J. Chem. Phys.* **37**, 150 (1962).
53. D. W. POSENER and M. W. P. STRANDBERG, *Phys. Rev.* **95**, 374 (1954).
54. B. L. SHAPIRO, R. M. KOPCHIK, and S. J. EBERSOLE, *J. Chem. Phys.* **39**, 3154 (1963).
55. R. M. LYNDEN-BELL, *Trans. Faraday Soc.* **57**, 888 (1961).
56. M. H. SIRVETZ and R. E. WESTON, Jr., *J. Chem. Phys.* **21**, 898 (1953).
57. E. H. RICHARDSON, S. BRODERSEN, L. KRAUSE, and H. L. WELSH, *J. Mol. Spectroscopy* **8**, 406 (1962).
58. H. S. GUTOWSKY, G. E. PAKE and R. BERSOHN, *J. Chem. Phys.* **22**, 643 (1954).
59. D. P. STEVENSON and J. A. IBERS, *Ann. Rev. Phys. Chem* **9**, 359 (1958).
60. J. J. TURNER, *Mol. Phys.* **3**, 417 (1960).
61. S. R. POLO and M. K. WILSON, *J. Chem. Phys.* **22**, 1559 (1954).
62. J. D. SWALEN and C. C. COSTAIN, *J. Chem. Phys.* **31**, 1562 (1959).
63. L. F. THOMAS, E. I. SHERRARD and J. SHERIDAN, *Trans. Faraday Soc.*, **51**, 619 (1955).

64. R. J. MYERS and W. D. GWINN, *J. Chem. Phys.* **20**, 1420 (1952).
65. F. L. VOELZ, F. F. CLEVELAND, A. G. MEISTER and R. B. BERNSTEIN, *J. Opt. Soc. Am.* **43**, 1061 (1953).
66. J. M. DOWLING and A. G. MEISTER, *J. Chem. Phys.* **22**, 1042 (1954).
67. G. S. REDDY and J. H. GOLDSTEIN, *J. Mol. Spectroscopy* **8**, 475 (1962).
68. L. S. BARTELL and R. A. BONHAM, *J. Chem. Phys.* **27**, 1414 (1957).
69. E. I. SNYDER, *J. Phys. Chem.* **67**, 2873 (1963).
70. T. W. MARSHALL and J. A. POPLE, *Mol. Phys.* **3**, 339 (1960).
71. R. A. OGG, *J. Chem Phys.* **22**, 560 (1954).
72. R. A. OGG, *Helv. Phys. Acta* **30**, 89 (1957).
73. W. G. SCHNEIDER, H. J. BERNSTEIN and J. A. POPLE, *J. Chem. Phys.* **28**, 601 (1958).
74. C. G. SWAIN and R. F. BADER, *Tetrahedron* **10**, 182; C. G. SWAIN, R. F. BADER and E. R. THORNTON, *Tetrahedron* **10**, 200 (1960).
75. T. W. MARSHALL and J. A. POPLE, *Mol. Phys.* **1**, 199 (1958).
76. H. S. GUTOWSKY, *Ann. N.Y. Acad. Sci.* **70**, 786 (1958).
77. P. J. FRANK and H. S. GUTOWSKY, *Arch. Sci. (Geneva)* **11**, 215 (1958).
78. A. B. F. DUNCAN and J. A. POPLE, *Trans. Faraday Soc.* **49**, 217 (1953).
79. L. S. BARTELL, K. KUCHITSU and R. J. DENEUI, *J. Chem. Phys.* **35**, 1211 (1961).
80. H. BATIZ-HERNANDEZ, Ph.D. Thesis, The Pennsylvania State University, 1965.
81. R. FREEMAN, G. R. MURRAY and R. E. RICHARDS, *Proc. Roy. Soc.* (London) A **242**, 455 (1957).
82. J. S. GRIFFITH and L. E. ORGEL, *Trans. Faraday Soc.* **53**, 601 (1957).
83. S. S. DHARMATTI and C. R. KANEKAR, *J. Chem. Phys.* **31**, 1436 (1959).
84. L. PAULING, *The Nature of the Chemical Bond*, 3rd ed., Cornell University Press, Ithaca, N.Y., 1960.
85. R. E. MESMER and W. L. JOLLY, *J. Amer. Chem. Soc.* **84**, 2039 (1962).
86. P. T. FORD and R. E. RICHARDS, *Disc. Faraday Soc.* **19**, 230 (1955).

CHAPTER 3

NUCLEAR SPIN RELAXATION STUDIES OF MOLECULES ADSORBED ON SURFACES

K. J. PACKER

School of Chemical Sciences, University of East Anglia, Norwich

CONTENTS

GLOSSARY OF SYMBOLS

T_1	nuclear magnetic spin-lattice or longitudinal relaxation time.
T_2	nuclear magnetic spin–spin or transverse relaxation time.
T_a	relaxation time (longitudinal or transverse) of phase having the longer relaxation time in a two-phase system.
T_b	relaxation time (longitudinal or transverse) of phase having the shorter relaxation time in a two-phase system.
T_a'	apparent value of T_a in presence of exchange of nuclei.
T_b'	as for T_a'.

T_M relaxation time (longitudinal or transverse) of nuclei in a molecule forming part of an adsorbed monolayer.

T_b'' apparent relaxation time of phase b in a multiphase system when the magnetization obeys a Gaussian decay law.

τ any time interval.

τ_c correlation time for nuclear magnetic relaxation interaction.

τ_i jump time of a molecule undergoing a quantized diffusion process.

τ_k critical correlation time in the apparent phase change theory.

τ_M correlation time for molecular motion in an adsorbed monolayer.

τ^* median correlation time in a distribution of correlation times.

τ_0 pre-exponential factor in transition state theory expression for molecular jump time.

$\mathbf{I}$ nuclear spin angular momentum operator with components $\mathbf{I}_x$, $\mathbf{I}_y$ and $\mathbf{I}_z$.

I nuclear spin quantum number associated with operator $\mathbf{I}^2$; is also the maximum eigenvalue of $\mathbf{I}_z$ in units of $\hbar$.

$\hbar$ Planck's constant divided by 2π.

γ nuclear magnetogyric ratio.

m_J quantum number associated with a component of the angular momentum of a molecule.

H_0 static, magnetic field.

H_1 radiofrequency magnetic field, perpendicular to H_0.

H any value of magnetic field in direction of H_0.

$H(0)$ value of H at which resonance occurs.

ΔH distance from resonance in units of magnetic field.

$\langle\langle \Delta H^2 \rangle\rangle$ second moment of a resonance absorption.

$\mathscr{H}$ Hamiltonian operator.

$\mathscr{H}_0$ time independent Hamiltonian operator.

$\mathscr{H}(t)$ time dependent Hamiltonian operator.

$\mathscr{H}_{DD}$ Hamiltonian operator for nuclear dipole–dipole coupling.

ω angular frequency, radians/sec.

ω_0 Larmor precession frequency of nucleus.

J_{ij} spin–spin coupling constant for scalar interaction between nuclei i and j.

$J(\omega)$ spectral density of a relaxation interaction at frequency ω.

$g(\tau)$ correlation function, for relaxation interaction, at time τ.

ω_a, ω_b resonance frequencies of nuclei in phases a and b in a multiphase system.

$(m \,|\mathscr{H}(t)|\, n)$ matrix element of Hamiltonian operator $\mathscr{H}(t)$ between states m and n of the system under consideration.

$(\;)(\;)$	bar indicates an ensemble average.
$\mathbf{r}_{ij}$	radius vector between nuclei i and j.
θ_{ij}	angle between $\mathbf{r}_{ij}$ and H_0.
ϕ_{ij}	azimuthal angle of $\mathbf{r}_{ij}$ relative to H_0.
$a_{(q)}$	time independent function of nuclear spin angular momentum operators.
(f_q)	time dependent function of lattice parameters.
$f^*_{(q)}$	complex conjugate of $f_{(q)}$
P_a	fractional population of phase having the longer relaxation time in a two-phase system.
P_b	fractional population of phase having the shorter relaxation time in a two-phase system.
P'_a, P'_b	apparent values of P_a and P_b in presence of exchange of nuclei.
C_a, C_b	specific rates of exchange of nuclei from phase a to b and vice versa.
$M(t)$	time dependent component of resultant nuclear magnetization.
$M_\perp(t)$	transverse magnetization at time t.
$M_z(t)$	longitudinal magnetization at time t.
M_0	equilibrium longitudinal nuclear magnetization.
μ	maximum observable component of nuclear magnetic moment.
θ	surface coverage of adsorbate on adsorbent.
p	number of nearest neighbours of an adsorbed molecule.
ν	vibration frequency characteristic of an adsorbed molecule in its potential well.
λ	lifetime of a nucleus in a phase.
m	number of jumps the molecule containing a nucleus makes before that nucleus transfers from one phase to another.
σ^2_0	rigid lattice second moment for a system of nuclear spins.
f_m	fraction of nuclei in the mobile phase in the apparent phase change theory.
$\langle \Delta H^* \rangle$	average activation enthalpy for a distribution of activation enthalpies.
σ_H	width parameter for a distribution of activation enthalpies.
ΔH^*	activation enthalpy.
ΔS^*	activation entropy.
ΔG^*	activation free energy.

1. INTRODUCTION

Since their discovery in the late 1940's, nuclear spin resonance phenomena in bulk matter have found application in an increasing number of areas, covering the biological, chemical and physical sciences. The reason for this wide scope of application lies primarily in the ability of magnetic nuclei to reflect, in their resonance characteristics, relatively small changes in the structure and motion of the molecules or atoms containing them without significantly affecting these properties. In other words, the magnetic nucleus is an ideal probe for relatively subtle effects occurring on the molecular level. Of the various parameters used to describe nuclear spin resonance phenomena the relaxation times T_1 and T_2 are perhaps the most sensitive to changes in molecular environment. Jardetzky has illustrated this in an account[1] of nuclear magnetic relaxation studies in systems of biological significance.

The purpose of this article is to discuss the use of nuclear magnetic resonance, with particular emphasis on relaxation-time measurements, in the study of surface phenomena. Since the subject of surface chemistry is of interest to many scientists who are not specialists in the field of nuclear magnetic resonance, and since the application of this technique to surface effects requires the use of those parts of the subject least familiar to chemists, only a minimal knowledge of nuclear magnetic resonance has been assumed. Those already familiar with the basic concepts of nuclear spin relaxation will find it unnecessary to read Sections 2 and 3 which are devoted to developing these ideas and to discussing briefly certain of the experimental methods involved.

Section 4 outlines some important theoretical ideas which have direct relevance to the interpretation of relaxation times of nuclei in adsorbed molecules. Among these are the effects of the rate of nuclear transfer between different adsorbed states on the observed state populations and relaxation times;[2, 3] the effect of anisotropic molecular motion on relaxation times[4] and the consequence of the existence of a broad distribution of correlation times for molecular motion on the observable relaxation behaviour.[5]

Section 5 describes several particular studies on systems of fairly diverse nature. The selection of material for this section was made on the basis of the ability of particular studies to illustrate the points made earlier in the article and is not intended to be a comprehensive bibliography of work in this area. Any shortcomings in this selection are solely the fault of the author.

Winkler[6] has published an account of work in this area, covering the period up to 1960–1. This has formed a starting-point for this article. Another, shorter, review article by Winkler covering some aspects of spin relaxation and dielectric studies of adsorbed molecules has been published recently.[80]

2. NUCLEAR MAGNETIC RESONANCE AND RELAXATION

A nucleus which has a maximum observable component of spin angular momentum $I\hbar$ also has a magnetic moment, μ, given by the relationship:

$$\mu = \gamma I\hbar \tag{1}$$

where γ, the magnetogyric ratio, is a constant for a given nuclear species. When placed in a strong static magnetic field, H_0, such a nucleus has $(2I + 1)$ energy states corresponding to the $(2I + 1)$ allowed orientations of its magnetic moment vector with respect to the applied field. N such nuclei will be distributed among the $(2I + 1)$ energy levels according to the Boltzmann distribution provided that the spin system is in thermal contact with its environment and that the nuclei are not interacting strongly with each other. Thus when a spin system is placed in a static magnetic field a mechanism must exist for establishing the equilibrium population differences required by the Boltzmann distribution. This mechanism is called spin–lattice relaxation, where the word lattice is taken to represent the thermal bath in which the nuclei are situated, i.e. the translational, rotational, etc., degrees of freedom of the system. A spin–lattice relaxation time, T_1, is defined which characterizes the time scale of this energy transfer, between spin system and lattice. In practice the spin–lattice relaxation is often exponential in its time dependence.

Any system, such as a nucleus, which has a magnetic moment and angular momentum carries out a precessional motion when placed in a static magnetic field. If two nuclei are precessing in phase at a given time, then as time runs on they will irreversibly lose their phase coherence. This process is called spin–spin relaxation and is characterized by a second relaxation time, T_2.

An alternative description of the nature of T_1 and T_2 is found in the macroscopic description of nuclear magnetic resonance by means of Bloch's famous equations.[7] In this treatment the behaviour of the resultant nuclear magnetization of the sample is described by a set of phenomenological equations. In these equations the return of the longitudinal component of the magnetization (the component in the direction of H_0) towards its equilibrium value is the spin–lattice relaxation process. (T_1 is often called the longitudinal relaxation time.) This process necessitates energy transfer between the spin system and the lattice, whereas the decay of any component of magnetization in the plane perpendicular to the static field H_0, that is the transverse or spin–spin relaxation process, conserves the energy of the spin system.

The nuclear magnetic resonance experiment consists of applying to the spin system, polarized by H_0, a second and much smaller magnetic field, H_1, rotating at an angular frequency ω, in the plane at right angles to the direction of H_0. Resonance is achieved by varying the frequency ω or the field H_0 until the precession frequency of the spins $\omega_0(-\gamma H_0)$ and ω are equal. When this condition is satisfied, H_1 and the nuclear spins are in phase and significant absorption of energy can take place into the spin system from the rotating field H_1. As is well known, for values of H_0 in the kilogauss range, ω_0 lies in the radiofrequency region of the electromagnetic spectrum.

For a Boltzmann distribution to be produced when a spin system is placed in a static magnetic field, a coupling must exist between the spins and their

D

environment which has the characteristics of H_1. That is, the coupling must have a component of the correct polarization and frequency to enable it to induce transitions between the energy levels of the spin system. The types of time dependent coupling which a magnetic nucleus may experience can be summarized as follows:

(i) A dipole–dipole coupling with other magnetic moments (electronic or nuclear) modulated by molecular translation and rotation or, in the case of some electronic magnetic moments, by rapid relaxation of the electronic spin. This type of coupling is the most widespread and in the types of system to be discussed in this article is usually assumed to be the dominant mechanism. As such it will be discussed in more detail at a later stage.

(ii) A scalar coupling with other magnetic moments of the form $J_{ij}\mathbf{I}_i \cdot \mathbf{I}_j$, modulated by chemical exchange of one or more of the nuclei involved or by rapid relaxation of one or more of the spins.

(iii) A coupling of the rotational magnetic moment of the molecule containing the nucleus under study to the magnetic moment of that nucleus (the spin–rotation interaction), the coupling being modulated by random changes in m_J, the rotational magnetic quantum number of the molecule, brought about by frequent molecular collisions. This mechanism is of most importance in gases or in liquids at high temperatures.

(iv) A coupling of the nucleus with the static magnetic field H_0, when the screening tensor of the nucleus is anisotropic. This coupling is modulated by the rotational motion of the molecules.

(v) Nuclei which have $I > \frac{1}{2}$ have an electric quadrupole moment which may couple to inhomogeneous electric fields at the nucleus. These may be intramolecular or intermolecular in origin and may be modulated by molecular rotation (intramolecular) or by both rotation and translation (intermolecular). This coupling is usually an order of magnitude larger than the others and when present is often the dominant cause of relaxation.

Detailed and quantitative equations describing the contribution of a particular coupling mechanism to both T_1 and T_2 for a given system of spins, are only accessible by the use of relatively lengthy theoretical procedures. These are usually based on some form of time dependent perturbation theory. The most usual methods of approach are those of the semi-classical density operator technique[8–10] and the linear response theory of Kubo and Tomita.[11, 12] In their classic paper on nuclear magnetic relaxation Bloembergen, Purcell and Pound[13] used ordinary time dependent perturbation theory.

Whilst it is not possible here to go into the details of these methods it is perhaps necessary to review briefly the important concepts introduced by them. We shall use as an example the dipole–dipole coupling mechanism in view of its general interest and its specific importance for the subject matter of this article.

The Hamiltonian for a magnetic nucleus in a strong static magnetic field may be separated into time independent and time dependent parts:

$$\mathscr{H} = \mathscr{H}_0 + \mathscr{H}(t) \tag{2}$$

$\mathscr{H}(t)$ represents one or more of the time dependent couplings listed above. It is this time dependent part of the Hamiltonian which is effective in producing relaxation. As an illustration, consider the Hamiltonian representing the dipole–dipole coupling between two nuclei i and j. This may be written as:

$$\mathscr{H}_{DD} = \gamma_i \gamma_j \hbar^2 \left[\frac{\mathbf{I}_i \cdot \mathbf{I}_j}{r_{ij}^3} - \frac{3(\mathbf{I}_i \cdot r_{ij})(\mathbf{I}_j \cdot r_{ij})}{r_{ij}^5} \right] \tag{3}$$

where r_{ij} is the radius vector between nuclei i and j.

This Hamiltonian is conveniently rewritten[13] as

$$\mathscr{H}_{DD} = \frac{\gamma_i \gamma_j \hbar^2}{r_{ij}^3} [A + B + C + D + E + F] \tag{4}$$

where:

$$A = I_{zi} I_{zj} (1 - 3 \cos^2 \theta_{ij})$$
$$B = -\tfrac{1}{4}[I_i^+ I_j^- + I_i^- I_j^+](1 - 3 \cos^2 \theta_{ij})$$
$$C = -\tfrac{3}{2}[I_i^+ I_{zj} + I_{zi} I_j^+] \sin \theta_{ij} \cos \theta_{ij} \, e^{-i\phi_{ij}}$$
$$D = -\tfrac{3}{2}[I_i^- I_{zj} + I_{zi} I_j^-] \sin \theta_{ij} \cos \theta_{ij} \, e^{i\phi_{ij}}$$
$$E = -\tfrac{3}{4} I_i^+ I_j^+ \sin^2 \theta_{ij} \, e^{-2i\phi_{ij}}$$
$$F = -\tfrac{3}{4} I_i^- I_j^- \sin^2 \theta_{ij} \, e^{2i\phi_{ij}} \tag{5}$$

where θ_{ij} and ϕ_{ij} are the polar and azimuthal angles made by the radius vector r_{ij} in a coordinate system in which H_0 is along the positive z direction.

If the nuclei i and j are contained in the same molecule then θ_{ij} and ϕ_{ij} are made time dependent by the rotational motion of the molecule. If i and j are in separate molecules then r_{ij} is time dependent as well and both molecular translation and rotation are sources of relaxation.

The dipole–dipole Hamiltonian as written above may be separated further into time dependent and time independent parts:

$$\mathscr{H}_{DD}(t) = \sum_q a_{(q)} f_{(q)} \tag{6}$$

where the $a_{(q)}$ terms are functions of spin operators alone and are independent of time whereas the $f_{(q)}$ terms, which are functions of the lattice variables (r_{ij}, θ_{ij} and ϕ_{ij}), carry all the time dependence.

In evaluating the contribution of a particular coupling mechanism to nuclear spin relaxation rates, expressions of the following type (or their equivalent) are encountered:

$$(m \,|\mathscr{H}(t)|\, n)(n \,|\mathscr{H}(t')|\, m) \tag{7}$$

where the bar represents an ensemble average of the product of matrix elements.

If the Hamiltonian in expression (7) is that for the dipole–dipole interaction and if it is used in the form of equation (6), then the only time dependent expressions to be evaluated are of the type:

$$\overline{f_{(q)}(t)f_{(q)}(t')} \tag{8}$$

As an example, term C in equation (5) would result in the expression:

$$\overline{[\sin\theta_{ij}(t)\cos\theta_{ij}(t)\,e^{-i\phi_{ij}(t)}][\sin\theta_{ij}(t')\cos\theta_{ij}(t')\,e^{-i\phi_{ij}(t')}]} \tag{9}$$

These expressions, (7), (8) and (9), known as auto-correlation functions, express the average correlation in time of any particular configuration of the nuclei i and j. That is, they indicate the average persistence in time of any given arrangement of nuclear moments relative to each other. These correlation functions are thus intimately concerned with the statistical nature of molecular motion in the system being studied. In general it is assumed that the motion of molecules in fluids is statistically stationary. This means that the value of a particular correlation function depends only on the difference between the times t and t' and not on the absolute value of t.

In principle it is necessary to choose a model for the molecular motion and then to evaluate the necessary correlation functions. In practice it is almost always assumed that the correlation functions may be written as:

$$g(\tau) = \overline{f_{(q)}(t)f_{(q)}(t')} = \overline{f_{(q)}(t)f(t)^{*}_{(q)}} \exp\left(-\left|\frac{\tau}{\tau_c}\right|\right) \tag{10}$$

This introduces the parameter τ_c which is called the correlation time and which characterizes in a very coarse grained fashion the time scale of molecular motion. This can be seen by considering the two extreme cases of $\tau \ll \tau_c$ and $\tau \gg \tau_c$. In the former the configuration of the system has not changed significantly since the original observation at $\tau = 0$, whereas in the latter it has completely changed: thus τ_c is the time beyond which the configuration of the system will have changed significantly. This is often given as the time for a molecule to rotate through an angle of one radian or to diffuse through a distance of one molecular diameter.

The importance of these correlation functions in calculating nuclear magnetic relaxation times is that their form determines the spectral density of the interaction producing the relaxation as a function of frequency. The theory of almost random processes[14] shows that the spectral density of such an interaction is given by the Fourier transform of its correlation function. If $J(\omega)$ is the spectral density, then using the correlation function of equation (10)

$$J(\omega) = \int_{-\infty}^{\infty} g(\tau) \exp(-i\omega\tau) \, d\tau = \overline{f_{(q)} f_{(q)}^*} \frac{2\tau_c}{1 + \omega^2 \tau_c^2} \qquad (11)$$

This indicates that the energy available for producing relaxation transitions is distributed over a frequency range given by equation (11). Figure 1 shows the spectral density as a function of correlation time. It can be seen that for long

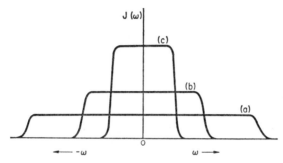

FIG. 1. The spectral density of a relaxation interaction as a function of the correlation time τ_c. (a) τ_c short; (b) τ_c intermediate; (c) τ_c long.

correlation times there is a large spectral density at low frequencies but very little at high frequencies. In general there is a cut-off frequency around τ_c^{-1} beyond which the spectral density goes rapidly to zero.

The general expressions for T_1 and T_2 for two like nuclei with $I = \frac{1}{2}$, coupled by a dipolar interaction are:[11]

$$(T_1)^{-1} = \frac{9}{4} \frac{\gamma^4 \hbar^2}{r^6} [J(\omega_0) + J(2\omega_0)]$$

and   (12)

$$(T_2)^{-1} = \frac{9}{4} \frac{\gamma^4 \hbar^2}{r^6} [\tfrac{1}{4}J(0) + \tfrac{5}{2}J(\omega_0) + \tfrac{1}{4}J(2\omega_0)]$$

which give on evaluation of the spectral densities:

$$(T_1)^{-1} = \frac{3\gamma^4 \hbar^2}{10 r^6} \left[\frac{\tau_c}{1 + \omega_0^2 \tau_c^2} + \frac{4\tau_c}{1 + 4\omega_0^2 \tau_c^2} \right] \qquad (13)$$

$$(T_2)^{-1} = \frac{3\gamma^4 \hbar^2}{20 r^6} \left[3\tau_c + \frac{5\tau_c}{1 + \omega_0^2 \tau_c^2} + \frac{2\tau_c}{1 + 4\omega_0^2 \tau_c^2} \right]$$

These expressions show up an important difference between T_1 and T_2. As τ_c increases from zero, T_1 goes through a minimum whilst T_2 continues decreasing. This is readily pictured since it is obvious that the fastest dephasing of a group of precessing nuclei due to the spread of the local magnetic field produced at the site of one of them by the others, will occur when the nuclei are in a fixed position relative to one another. Any motion reduces the spread of the effective average local field and hence reduces the rate at which the nuclei lose phase coherence. The behaviour of T_1 and T_2 with change of τ_c is illustrated in Fig. 2.

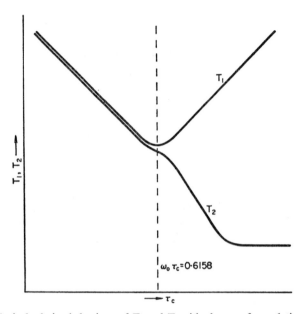

Fig. 2. Typical relative behaviour of T_1 and T_2 with change of correlation time, τ_c.

One feature of interest in equations (13) is that the minimum in T_1 occurs when $\omega_0\tau_c = 0\cdot6158$ and that the ratio of T_1/T_2 at this minimum is $1\cdot6$. Thus if this model fits the experimental data and the T_1 minimum can be located by changing the temperature (i.e. changing τ_c), or the frequency of observation (ω_0) then τ_c is known under those conditions. However, this model is based on the characterization of the molecular motion by a single correlation time. In the case of molecules adsorbed on surfaces this is usually found to be incorrect, in that a distribution of correlation times has often to be invoked and that different correlation times may determine T_1 and T_2 if the motion of a molecule is not isotropic on the average. These effects will be dealt with in more detail in Section 4.

When more than two nuclei are interacting to produce relaxation it has been shown that it is a good approximation in most instances to consider a representative nucleus and to sum separately its interactions with all the other nuclei in the system.[30] This procedure involves the assumption that there is no correlation between the various pairwise interactions of the representative nucleus. In the case of intramolecular coupling in molecules containing more than two of the nuclei being considered, this assumption is obviously incorrect, but the observed deviations from the behaviour expected if the assumption were correct are usually less than 1 per cent.

Since molecular motion plays a large part in determining T_1 and T_2, a study of these relaxation times as a function of temperature can, in principle, lead to information concerning the temperature dependence of the motion. This, of course, depends on the assumption that the temperature does not alter the mean square interaction energy of the relaxation mechanism. This is more likely to be a reasonable assumption if the relaxation is dominated by intramolecular interactions. If this is so and the motion involved is taken to be a thermally activated process showing Arrhenius type behaviour, graphical plots of $\ln (T_1)^{-1}$ or $\ln (T_2)^{-1}$ against $(T^\circ K)^{-1}$ can lead to the measurement of apparent activation energies for the process. This technique has found considerable application in the study of surface phenomena by nuclear spin relaxation.

3. EXPERIMENTAL METHODS

The two main techniques of observing nuclear magnetic resonance are those involving the use of continuous wave radiation and those involving the use of pulsed radiofrequency radiation, the latter being collectively called spin–echo techniques. Descriptions of both these methods are found in most standard texts on nuclear resonance[15–7] and will not be given here. A brief discussion will be given, however, of the use of the method of second moments in the study of the distribution and motion of nuclei since this is of direct relevance to some of the work to be described later and is not usually included in the more chemically oriented texts on nuclear resonance.

The nuclear magnetic resonance absorption of a system of spins coupled together by dipole–dipole interactions has its maximum width when the spins are stationary, except for lattice vibrations, in their equilibrium rigid lattice positions. This line width depends mainly on the number of nearest neighbours of a particular characteristic nucleus and on the geometric distribution of these about that nucleus. The width and shape of a resonance line, therefore, contains information about these properties. A parameter of particular value in dealing with situations where many nuclei are involved is the second moment of the resonance line which may be defined as:

$$\langle\langle\Delta H^2\rangle\rangle = \int_{-\infty}^{\infty} g\{H - H(0)\}\{H - H(0)\}^2\, dH \qquad (14)$$

where $g\{H - H(0)\}$ is the normalized line shape function for the resonance absorption, $H(0)$ is the field strength corresponding to the centre of the resonance line and $\Delta H = \{H - H(0)\}$. The first set of brackets indicates the quantum mechanical expectation value of ΔH while the outer brackets denote the averaging of this expectation value over a statistical ensemble of systems. Van Vleck[18] has derived an expression for the second moment of the resonance absorption of a lattice of N like spins. This may be written:

$$\langle\langle\Delta H^2\rangle\rangle = \frac{3}{2}\frac{I(I+1)}{NI^2}\mu^2 \sum_{j>k} \frac{(3\cos^2\theta_{jk} - 1)^2}{r_{jk}^6} \qquad (15)$$

where r_{jk} is the distance between spins j and k and θ_{jk} is the angle that r_{jk} makes with the external field $\mathbf{H}_0$.

For a polycrystalline sample, the angular term must be averaged over a sphere giving

$$\langle\langle\Delta H^2\rangle\rangle = \frac{6}{5}\frac{I(I+1)}{NI^2}\mu^2 \sum_{j>k} r_{jk}^{-6} \qquad (16)$$

Further terms must be added if the lattice contains spins of different species. The main difference in these terms will be a smaller numerical factor which arises because the mutual energy-conserving spin flips which can occur between two dipole–dipole coupled nuclei of the same species do not occur to a first order between unlike nuclei. Thus this method of second moments constitutes a powerful technique for investigating the distribution of nuclei in solids or in stationary adsorbed systems.

If, on raising the temperature of the system, the nuclei begin to carry out relative motion, the second moment may be diminished, i.e. so-called motional narrowing of the resonance line can occur.[19] A study of this type of effect can lead to information concerning the onset of various types of motion and to activation energies for such processes. A classic example of the use of second moments is the work of Andrew and Eades[20] on the proton resonance in benzene and deuterated benzenes, leading to an accurate value of the proton–proton distance in the benzene molecule and the information that at about 120°K and above the benzene molecules, in solid benzene, are reorienting rapidly about their hexad axes.

It is perhaps pertinent to note at this point that qualitative statements concerning the rates of various processes studied by magnetic resonance should be accompanied by some indication as to the time scale involved.

For example, the statement above, concerning the rate of reorientation of benzene molecules in solid benzene above 120°K, should be associated with the statement that it is rapid compared to the rigid lattice second moment in frequency units. In each case the terms fast or slow must be defined relative to a parameter of the nuclear magnetic resonance experiment.

4. SPECIAL CONSIDERATIONS FOR SURFACE PHENOMENA

When a molecule in a fluid interacts with a surface, two effects may occur which can change the relaxation characteristics of nuclei contained in that molecule. Firstly, the nature of the relative motion of the nuclei may be changed, both in terms of its rate and in terms of its time-averaged symmetry. Secondly, if intermolecular interactions are an appreciable source of relaxation in the bulk fluid phase, these may be changed considerably, especially at low surface coverage, leading to a change in the average interaction energy available for relaxation.

Almost all the work published in this field to date has involved the assumption of relaxation which is dominated by intramolecular dipole–dipole interactions. Because of this, the first effect mentioned above, the modification of the molecular motion, has usually been considered as the source of changes in T_1 and T_2 with parameters such as surface coverage and temperature. Out of this work, and work in related fields such as dielectric relaxation, have arisen several concepts concerning the effect, on measured relaxation times, of changes in the motional characteristics of molecules when near surfaces. Some of these will be discussed in the succeeding sections.

4.1. Multiphase Relaxation and Nuclear Exchange

Early in the study of surface interactions by nuclear spin relaxation techniques, Zimmerman and his co-workers[2, 22–3] observed two component relaxation for either or both T_1 and T_2. Two component relaxation implies that in a particular experimental determination of T_1 or T_2 two distinct nuclear magnetizations are observed that decay according to two well-defined relaxation times. These two sets of nuclei Zimmerman *et al.* ascribed to two distinct surface phases (phase in this sense not being a thermodynamic definition). It was soon realized, however, that if exchange of nuclei between these two phases could and did take place, the observed relaxation times and apparent relative populations of the phases would not be those truly characteristic of the separate phases. Zimmerman and Brittin[2] developed a stochastic theory of relaxation in multiphase systems. Using Bloch's equations as modified for exchange by McConnell,[24] Woessner[3] re-derived the results obtained by Zimmerman and Brittin and extended them to include systems with phases having different resonance frequencies. Their results may be summarized as follows. Consider a two-phase system with true fractional state populations P_a and P_b and true state relaxation times T_a and T_b (T may be either T_1 or T_2), in which exchange of nuclei from state a to b and from b to a occurs with specific rates C_a and C_b respectively. The effect of exchange is to transfer magnetization from one state to another, which means the relaxation rate of a given nucleus will be influenced by the amount of time it spends in each phase during the course of its relaxation process. The expected results are obtained for the two limits of extremely slow and extremely fast exchange. Slow and fast are defined relative to the relaxation rates $(T_a)^{-1}$ and $(T_b)^{-1}$.

For very slow exchange the observed relaxation behaviour is just the superposition of two exponentially decaying magnetizations one for each phase, the decays being characterized individually by T_a and T_b. For very fast exchange the nuclear magnetization decays as a single exponential with an effective relaxation time T_{eff}, which is the weighted average of the individual state relaxation times, i.e.

$$(T_{eff})^{-1} = \frac{P_a}{T_a} + \frac{P_b}{T_b} \tag{17}$$

The results for fast and slow exchange may be extended to any number of phases.

For intermediate rates of exchange the general result for the decay of a component of nuclear magnetization may be written:

$$M(t) = M(0)\left\{P_a' \exp\left(-\frac{t}{T_a'}\right) + P_b' \exp\left(-\frac{t}{T_b'}\right)\right\} \tag{18}$$

where for transverse relaxation $M(t) = M_\perp(t)$ and $M(0) = M_\perp(0)$; and where for longitudinal relaxation $M(t) = M_0 - M_z(t)$ and $M(0) = M_0 - M_z(0)$, M_0 being the equilibrium nuclear magnetization. P_a' and P_b' are the apparent fractional state populations and T_a' and T_b' the apparent relaxation times of states a and b. Even for a system of only two phases the general equations for P_a', P_b', T_a' and T_b' in terms of the true parameters are fairly complicated. We shall not describe them in detail here but will illustrate graphically the effects of nuclear exchanges on these parameters. Consider the two-phase system described above and arbitrarily assign the subscript b to the phase having the shorter relaxation time. This does not necessarily imply, of course, that in a given system T_{1a}' and T_{2a}' correspond to the same phase. In other words it is possible for the state with the shorter T_2 to have the longer value of T_1. Following Woessner we consider for simplicity a system in which $P_a = P_b$ (consequently $C_a = C_b$), $T_a = 100T_b$ and $\omega_a = \omega_b$.

The nuclear exchange rate is described relative to the relaxation rate of state b, the short relaxation time phase, resulting in the definition of a reduced nuclear transfer rate $C_b T_b$. Figure 3 shows the effect of nuclear transfers in this system on the apparent fractional population of the long relaxation time state, P_a'. When $C_b T_b$ is about 3, P_a' approaches unity and since $P_a' + P_b' = 1$, P_b' appears to be zero. There is, therefore, an apparent emptying of the short

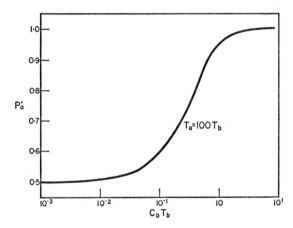

FIG. 3. The effect of reduced nuclear transfer rate, $C_b T_b$, on the apparent fractional population, P_a' of the state having the longer relaxation time, in a two state system with: $\omega_a = \omega_b$; $P_a = P_b$; $C_a = C_b$ and $T_a = 100 T_b$. (Woessner.[3])

relaxation time state. A similar diagram with $T_a = \infty$ does not differ much from that in Fig. 3. It is found in general that if $C_b T_b \geqslant 2$ it is not possible to detect two separate relaxation times T_a' and T_b' experimentally.

In Fig. 4 is shown the effect of the reduced nuclear transfer rate, $C_b T_b$, on the ratio of the apparent relaxation times to the true relaxation times for both states a and b, for a system with the same conditions as that considered above. It can be seen that the apparent relaxation time of the long relaxation time state falls more rapidly with increasing transfer rate than that of the short relaxation time state. Both apparent relaxation times do decrease, however, with increase in transfer rate. In all these considerations it is assumed that the act of transfer is not in itself a relaxation mechanism. As can be seen from Fig. 4 it is possible for P_a' and P_b' to be near their true values P_a and P_b and for T_b' to be near the value T_b while T_a' may be several times smaller than T_a.

Since T_1 is usually greater than T_2 for nuclei in adsorbed systems it is possible, as Woessner has pointed out, for two-phase behaviour to be found for T_2 and not for T_1, because it may be possible for $C_b T_{1b} \gg 1$ while $C_b T_{2b} \ll 1$.

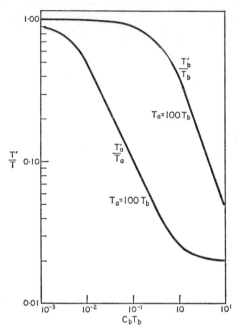

Fig. 4. The effect of nuclear exchange on the apparent relaxation times of a two state system with: $\omega_a = \omega_b$; $P_a = P_b$; $C_a = C_b$ and $T_a = 100T_b$. (Woessner.[3])

The problem of multiphase relaxation in the presence of nuclear exchange has also been treated by Beckert and Pfeifer.[81] They obtained expressions for the correlation function for the multiphase system by assuming a correlation coefficient for the exchange process. A value of zero for this coefficient implies that on exchange the lattice parameters do not show any correlation with their values previous to the exchange event and the correlation function drops instantaneously to zero. A value of unity for the coefficient implies complete retention of correlation; the system starting with the same value of the correlation function immediately after the exchange as it had immediately prior to the exchange. The theory was developed for a two-phase system in terms of the rotational diffusion correlation times and the exchange rates for each phase. The contribution of the overall correlation function to the relaxation was obtained by calculating its Fourier transform and using this in equation (12) of this article.

The considerations outlined above are obviously of great importance if such processes occur within adsorbed systems, since it is the true relaxation times and populations of the phases which are the parameters of direct interest, although, of course, the rates and activation parameters of any exchange processes are also of interest.

4.2. *Anisotropic Molecular Motion and Nuclear Spin Relaxation Times*

When a molecule interacts with a surface its thermal motion might be expected to become anisotropic to some degree. The extent of this anisotropy should perhaps depend on the strength and nature of the surface interaction and the symmetry of the adsorbed molecule. The relevance of this fact from the point of view of nuclear magnetic relaxation is that the original theory of relaxation rates envisaged isotropic motion of the nuclei, characterized by a single correlation time. The results of these treatments therefore would not be expected to hold for nuclei in adsorbed molecules.

Woessner[4] has investigated this problem using a model in which a molecule containing a pair of nuclei with $I = \frac{1}{2}$, reorients such that the internuclear vector moves rapidly about an axis perpendicular to the surface while the rotation axis itself undergoes much slower reorientation or wobbles with a fairly low frequency, with the assumption that the two types of motion are independent. The results show that the contributions of each type of motion to the relaxation phenomena are not independent but that under certain conditions it is possible for T_1 to be determined mainly by the fast reorientation while T_2 is dominated by the slower motion. This leads to the situation where $T_1 \gg T_2$ with T_1 being independent of ω_0. T_1 may also be very much greater than T_2 when the fast motion is too fast to contribute effectively to T_1 and both T_1 and T_2 are determined by the slower motion. In this case T_1 increases with increase in ω_0.

In applying the concept of anisotropic motion to the explanation of experimental results, it is found[27] that small changes in the details of the motional and orientational characteristics of a model system do not produce measureable effects on the relaxation behaviour. However, different preconceived models can be investigated and it is to be expected that the significant overall aspects of their effect on relaxation will be discernible.

Other workers have developed relaxation equations for various situations in which anisotropic motion is likely. For example, Cohen-Addad[28] has treated the problem of the relaxation of the protons in water held in zeolites.

4.3. *Distributions of Correlation Times*

Comparison of dielectric and mechanical relaxation characteristics of certain systems with their nuclear magnetic relaxation behaviour led Odajima[29] to suggest that in many systems a particular nucleus sampled many different environments during the course of its relaxation and that these environments could be characterized by a variety of different correlation times. This introduced the concept of a distribution of correlation times, the most popular being the so-called log-normal distribution in which the logarithm of the correlation time is distributed in a Gaussian fashion. The main effects of the existence of a distribution of correlation times in a system is to raise the value of T_1 at the T_1 minimum, to flatten out the change of T_1 with τ_c at the minimum and following from these results, to increase markedly the ratio T_1/T_2 at the T_1 minimum.

The effect of the existence of a broad distribution of correlation times on observable relaxation behaviour has been investigated in detail by Resing[5] for a system in which dipole–dipole coupling is assumed to be the dominant mechanism. Resing has reviewed in some detail the concepts of distribution of relaxation times, distribution of correlation times, phase and lifetime in phase, from the point of view of nuclear magnetic relaxation. For a multi-phase system, he defines the life time in phase, λ, in terms of m, the number of jump times, τ_j, a nucleus undergoes before it leaves the phase, thus:

$$\lambda = m\tau_j \tag{19}$$

In the case of relaxation dominated by intramolecular couplings, τ_j is characteristic of a given molecule-site combination and is the correlation time for the relaxation process. This is not true for intermolecular relaxation since the correlation time is then only characteristic of a group of molecules and sites (for surface effects) and reflects the jump times of all the species involved. As Resing has pointed out this may lead to a considerable distortion of the facts in the sense that the correlation times will give larger weight to the jump times of the faster moving species.

The concept of phase in this type of system is therefore directly linked to the mechanism of relaxation, since a phase is expected to "separate out," i.e. to be observable experimentally, when the ratio of the life time in phase to the relaxation time in that phase is greater than unity. The detailed conditions for this to occur are dictated, of course, by the form of the equations for T_1 and T_2 for a given mechanism of relaxation. For dipole–dipole induced relaxation Resing finds that phase separation in T_1 is not possible except for very large values of m and suggests that, if observed, it indicates the presence of another mechanism for T_1, such as interaction with paramagnetic impurities. For T_2, however, it is noted that one jump can result in phase separation. A critical jump time τ_k is defined which is that value of τ_c for which the ratio of the life time in phase to T_2 is unity. For a value of m of unity this turns out to be the condition for the onset of rigid lattice behaviour, i.e. T_2 independent of τ_c. In the presence of a broad distribution of correlation or jump times, at some given temperature, it is quite likely that a fraction of the nuclei in such a system will be experiencing correlation times shorter than τ_k and will give rise to a single exponential relaxation, the time constant of which will be the average over this fraction of nuclei. The other fraction of nuclei will all have correlation times longer than τ_k, provided that the distribution is broad enough to make the fraction with jump times near τ_k very small, and these nuclei will all relax with the rigid lattice value of T_2; the decay being Gaussian in character. The significance of this is that in the presence of a broad distribution of correlation times it is possible, by lowering the temperature of the system, to observe an apparent phase change (two-phase relaxation) which is a feature of the NMR experiment alone and is not a true thermodynamic phase change. It has been suggested that this is a new method for the study of the heterogeneity of surface adsorption sites. Examples of its application are discussed later.

The effects of a distribution of correlation times, discussed above, and those due to multiphase behaviour with exchange are very similar. It seems quite likely that in situations where several distinct adsorption phases exist that these phases will exhibit within them a distribution of correlation times. If these distributions overlap significantly then the two phases will not be detectable in the NMR experiment.[27]

4.4. *Two-dimensional Diffusion*

Kokin and Izmest'ev[82] have investigated theoretically the nuclear magnetic resonance line shape expected for a monolayer of a monatomic species undergoing two-dimensional surface diffusion. They used an ideal gas model for the diffusion process and found, to a first order, that the line shape should be independent of the correlation time. They suggest that ^{3}He could be used as a test for the theory.

5. SPIN-RELAXATION STUDIES OF SURFACE INTERACTIONS

In this section several individual studies will be discussed in detail, in an attempt to illustrate the considerations outlined above and also to show the complex nature of the results obtainable and the consequent difficulties in unambiguous interpretation.

5.1. *The Silica–Water System*

This system has been studied by several groups of workers.[2, 22–3, 31–8] Its popularity stems from the importance of silicas in industrial processes, the high surface area of many silicas and the presumed dominance of intramolecular dipole–dipole coupling in the relaxation of the protons in the water molecule.

The work which has been carried out on this system may be broadly classified in three categories. In the first category are studies of hydroxyl groups chemically bonded to the surface, these being the result of various dehydration–rehydration sequences; in the second, studies of physically adsorbed water as a function of surface coverage and in the third, studies of physisorbed water as a function of temperature at a given surface coverage. The last two categories have been combined in some cases.

O'Reilly and co-workers[34, 35] have studied the proton magnetic resonance signals from a silica gel and some silica–alumina catalysts which were dehydrated, in vacuum, at 500°C. The signals were Lorentzian in shape, this being interpreted as showing the existence of a random distribution of silanol groups, occupying only a fraction of the total possible sites. This type of distribution has been shown[39] to give rise to a Lorentzian line shape where, in the absence of motional narrowing, a Gaussian shape is usually expected.

Kvlividze *et al.*[37] have studied some silica gels by broad line nuclear magnetic resonance at low surface coverages of adsorbed water. In this paper the important point is made that surfaces must be studied by a variety of techniques if the mechanisms of adsorption, surface diffusion, etc., are to be discovered. The results, in the form of line widths, second moments and line shapes, were interpreted in terms of a non-uniform distribution of surface silanol groups together with strong adsorption of water molecules at coordinately unsaturated silicon atoms at low surface coverages. Hydrogen bonding of water molecules to silanol groups was proposed to explain the results at higher surface coverages. All the data were interpreted in the light of studies on similar systems by infrared spectroscopy and classical adsorption techniques.

Kvlividze[33] has also studied the NMR line shape, at 93°K, of the silanol protons on the surface of a silica gel, the adsorption properties of which had been studied previously. The results showed that three distinct types of surface hydroxyl groups existed, characterized by different values of the second moments of their resonance lines. The groups were as follows:

(a) Isolated hydroxyl groups, constituting 27 per cent of the total, whose nearest neighbour protons were at a distance of 5·2–5·4 Å.
(b) Isolated pairs of hydroxyl groups, either geminal or vicinal, with an inter-proton distance of 2·3–2·6 Å. This group constituted 37 per cent of the protons.
(c) Hydroxyl groups with two, three or four neighbouring protons (hydroxyl groups) at 2·3–2·6 Å; 45 per cent were in this group.

The study of the line shapes on addition of water to the sample at this low temperature showed that adsorption took place primarily on type (c) hydroxyl groups and that multilayer adsorption occurred at these centres rather than monolayer formation over the whole surface.

Measurements of the relaxation times, T_1 and T_2, of the protons in water molecules adsorbed on silica surfaces have been carried out by Zimmerman and co-workers[2, 22, 23, 31] Woessner,[32] Clifford and Lecchini[36] and

Michel.[38] The types of silica used in these studies vary considerably in their mode of preparation, the extent of knowledge as to their surface characteristics and their content of paramagnetic impurities.

Zimmerman et al.[2, 22, 23] have studied the proton relaxation times of water on a silica gel as a function of surface coverage. In these experiments they observed the multiphase behaviour which led to the development of the theories discussed earlier.[2, 3] Zimmerman and Woessner,[31] and Woessner[32] have studied the same silica gel, at various surface coverages as a function of temperature. Woessner's data cover the larger temperature range and will be discussed in greater detail.

It is perhaps an interesting commentary on the difficulties involved in this type of work to note that in one of these studies,[31] a sample which was investigated twice, with an interval of one year between the measurements, showed significantly different relaxation characteristics after the year had elapsed. In this particular study,[31] two-phase behaviour was observed in both T_1 and T_2 for samples having surface coverages of the order of one-third to three-quarters of a statistical monolayer. The results were discussed in terms of multiphase relaxation with exchange of nuclei and anisotropic rotation of the adsorbed water molecules. It was concluded that these concepts alone were sufficient to explain the relaxation data in a semi-quantitative manner and that other effects, such as those due to paramagnetic impurities, were not necessary for the interpretation of the data. It was also noted that, consistent with dielectric studies, no freezing of the adsorbed water occurred as the temperature was lowered beyond 0°C. As will be seen later this is probably due to the microporous nature of the adsorbent.

Woessner[32] studied a sample having three-quarters of a monolayer surface coverage. Figures 5, 6 and 7 show the observed temperature dependence of T_1, T_2 and P'_{2a}, the latter being the apparent fractional population of the phase having the longer T_2 value.

Although the longitudinal relaxation was apparently two-phase in character below 264°K, the two values, T'_{1a} and T'_{1b} were too close to be resolved experimentally and so the T_1 curve, below this temperature, was based on a weighted average relaxation time, which was obtained from the initial portion of the experimental relaxation curve. The T_1 data thus obtained exhibited a minimum near 255°K and above this temperature increased with an apparent activation energy of 7 kcal mole^{-1}. At higher temperatures T_1 approached a maximum.

The transverse relaxation behaviour was, as mentioned above, two-phase over the whole temperature range, the shorter value, T'_{2b} being essentially independent of temperature. The longer relaxation time, T'_{2a}, increased with increasing temperature and was a maximum at around 255°K. Below this temperature the apparent activation energy was 7 kcal mole^{-1}, the same as that

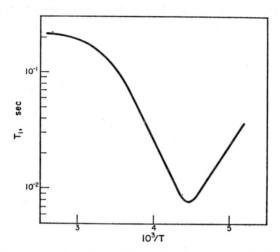

Fig. 5. The temperature dependence of the weighted average longitudinal relaxation time of protons in water adsorbed on silica gel. (Woessner.[32])

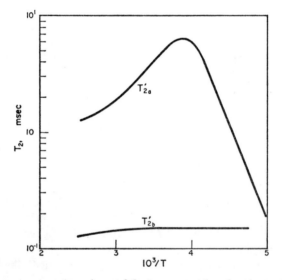

Fig. 6. The temperature dependence of the transverse relaxation times of protons in water molecules adsorbed on silica gel. (Woessner.[32])

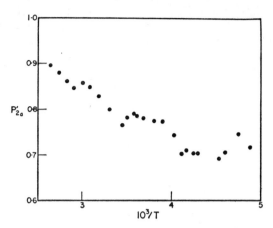

FIG. 7. The temperature dependence of the apparent population of the longer relaxation time state, P'_{2a}, of protons in water adsorbed on silica gel. (Woessner.[32])

observed for part of the T_1 data. At temperatures higher than $255°K$, T'_{2a} decreased on increasing the temperature yielding an activation enthalpy of $3·4$ kcal mole^{-1}. This behaviour is a very strong indication of the existence of nuclear exchange with a phase having a shorter value of T_2, since for normal motional relaxation processes T_2 would be expected to increase with temperature as it does in this case below $255°K$.

Using the theory of two-phase relaxation with exchange,[3] Woessner was able to calculate the nuclear exchange rates, C_a and C_b, from the observed relaxation data. This required the assumptions that, in the low temperature region well below $255°K$, the measured value of T'_{2a} was the true phase relaxation time T_{2a} unaffected by nuclear exchange; that P'_{2a} was actually P_{2a} and that because of its temperature independence, T'_{2b} was the true transverse relaxation time, T_{2b}, of the short relaxation time phase. By extrapolating the T_{2a} data into the high temperature region and by simplifying the equation giving T'_{2a} in terms of the true phase characteristics by limiting the calculations to the region in which $C_b T_{2b} \leqslant 0·3$, $T_a/T_b \gg 1$ and $P_a \sim P_b$, values of C_a were obtained over a reasonable temperature range. An Arrhenius plot for C_a was a straight line, giving an activation energy for nuclear exchange of $4·9$ kcal mole^{-1}. Woessner points out that breaking of hydrogen bonds could be involved. C_a varied from 10^2 to 10^3 sec^{-1} from 263 to $314°K$. Values of C_b were obtained with the use of another set of reasonable approximations and also by the use of the relationship $C_b = (P_a/P_b)C_a$. The values calculated by the two methods cover different temperature ranges and on an activation plot merge very well, lending support to the approximations and basic theory leading to their calculation. This plot was not a straight line, however, and hence a simple activation energy treatment could not be given.

The increase of T_1 above 220°K was interpreted as showing the existence of a rapid motion ($\omega_0^2 \tau_c^2 \ll 1$). The levelling off of T_1 at higher temperatures was taken as a strong indication that T_1 was determined by at least two motions, one of these being slow ($\omega_0^2 \tau_c^2 \gg 1$). This slow motion would contribute less to T_1 as the temperature increased thus leading to the observed trend towards a maximum value at higher temperatures. This interpretation was given more weight by the observation that at room temperature, T_1 depended on the frequency ω_0.

Winkler[6] has suggested that the physical reason for the observed multiphase relaxation behaviour reported in these systems by Zimmerman[2, 22, 23, 31] and Woessner[32] is the microporous structure of the adsorbent. A recent extensive study of the silica–water system[36] has supported this suggestion. In this work a series of well-characterized silicas were used. These were as follows:

(a) Two non-microporous silicas with similar specific surface areas but with different concentrations and distributions of surface hydroxyl groups.

(b) Two silicas with much larger specific surface areas than group (a), one being non-microporous, the other, a gel, being considerably microporous.

(c) One of the samples from (a) compressed into a disc as for infrared spectroscopy.

The relaxation times of the protons in water molecules adsorbed on these silicas were measured as a function of temperature at surface coverages of 1·4, 3·1 and 15 monolayers.

The non-microporous silicas all exhibited similar relaxation characteristics for a given surface coverage. No two-phase behaviour was observed for these samples and, as mentioned above, it is suggested that multiphase relaxation is linked closely with the microporosity of the adsorbent.

For the non-microporous samples with a surface coverage of 1·4 monolayers the relaxation data were discussed in terms of Woessner's anisotropic rotation model.[4]

The T_2 values for the same samples, with 3·1 monolayers of adsorbed water, decreased with increasing temperature. This was taken as evidence for the existence of exchange of the water protons with a group of protons having a T_2 value which was too short to be detected with the spectrometer used. It was suggested that the protons of the surface silanol groups are involved and that the exchange rate increases with increasing surface coverage (the effect not being noticeable at 1·4 monolayers).

It was also noticed that among the non-microporous adsorbents, it was the one with the lowest surface concentration of hydroxyl groups which inhibited the motion of the adsorbed water molecules the most. This was explained by suggesting that the higher surface concentration of hydroxyl groups involves a higher number of geminal arrangements leading to hydrogen bonding between the groups. This would decrease the availability of these groups for similar interactions with adsorbed water molecules.

The lack of freezing of the adsorbed water, remarked on earlier, was not observed in this work for the non-microporous silicas. By $-20°C$ almost no signal was obtained from these samples since the transverse relaxation time of ice is too short to be observed with the equipment used. No freezing of the water was observed with the silica-gel which was extremely microporous. Compressing of a non-microporous sample into a disc produced effects on the relaxation phenomena akin to the existence of microporosity, especially at the higher surface coverages of water.

The temperature dependence of the relaxation times for the silica gel, with a surface coverage of fifteen monolayers, was interpreted in terms of exchange of the water protons, at considerably different rates, with at least two distinct groups of surface hydroxyl protons.

In a recent paper, Michel[38] has thrown some doubt on the assumption usually employed in these studies that the relaxation is dominated by intra-molecular dipole–dipole coupling. He has studied the relaxation of the protons in H_2O/D_2O mixtures adsorbed on various surfaces as a function of the isotopic concentration of the mixture at constant surface coverage. For relaxation dominated by dipole–dipole coupling, the observed relaxation times would be expected to increase with increasing deuterium substitution because of the much smaller magnetic moment of the deuterium. One of the silicas studied exhibited no dependence at all, of both T_1 and T_2, on deuterium concentration and he suggests that paramagnetic impurities dominate the observed behaviour although none were detectable by electron spin resonance.

5.2. The Alumina–Water System

O'Reilly and Poole[41] have studied the distribution of transition metal oxides on γ-alumina by means of ^{27}Al NMR and by measuring the proton relaxation times of water molecules adsorbed on these surfaces. The experimental conditions under which the ^{27}Al resonance was observed were such that nuclei with T_1 values between certain limits did not contribute to the detected signal. From a consideration of the contribution of the paramagnetic transition metal ions adsorbed on the alumina surface to the spin–lattice relaxation time of the ^{27}Al nuclei and from the range of T_1 values for which

[27]Al nuclei were not observable, it was found possible to explain the observed change in intensity of the ^{27}Al NMR signal with concentration of paramagnetic ions in a manner consistent with the available electron spin resonance data. The results for the chromia–alumina system indicate the existence of magnetically isolated chromium ions at low concentrations with clusters of chromium ions having strong internal spin–spin interactions building up at higher chromium concentrations. The observed proton relaxation times for water molecules adsorbed on the surface of the chromia-impregnated alumina were interpreted in terms of proton exchange between two magnetically distinct environments, these being water molecules undergoing surface diffusion and water molecules coordinated to chromium ions. In both studies the results were in almost quantitative agreement with the proposed surface structure and relaxation mechanisms.

Winkler[6, 42] and Reball and Winkler[43] have studied the proton relaxation times of water adsorbed on γ-alumina as a function of surface coverage. They interpreted their results in terms of the microporous nature of the adsorbent. The transverse relaxation was two phase in character when the water content, W, was greater than 0.2 g H_2O/g Al_2O_3. The single phase below 0.2 g H_2O/g Al_2O_3 was ascribed to the filling of the micropores whilst the second phase, observable only above a surface coverage of 0.2 g H_2O/g Al_2O_3, was assigned to water in the spaces between the alumina granules (the so-called macropores). With the assumptions that the observed micropore relaxation was an average of that due to a monolayer of water with a relaxation time T_{2M} and that arising from the rest of the water with a relaxation time T_{2_0}, not too different from pure water, it was possible to calculate a value for T_{2M}. This was then used to calculate a value for τ_M, the correlation time of the proton–proton interaction in the monolayer. This calculation was based on the assumption of isotropic motion with a single correlation time, which as pointed out earlier is not likely to be a good model for adsorbed molecules. It was then assumed that this correlation time would determine T_{1M}, which as discussed earlier is an assumption which may not be valid. A calculation of T_{1M} on this basis gave a value an order of magnitude greater than that observed. From these results Winkler[42] suggested that T_1 was dominated by paramagnetic impurities whereas T_2 was controlled by dipole–dipole coupling. Michel,[38] using the deuterium dilution technique mentioned earlier, has claimed that both T_1 and T_2 are dominated by interaction with paramagnetic impurities in the adsorbent.

Reball and Winkler,[43] by studying γ-alumina impregnated with ferric oxide, found that in this case T_1 was two phase in character and they were able to calculate that the lifetime of a proton in the micropore phase was between 30 and 150 msec.

5.3. *Thorium Oxide*

Brey and Lawson[44] have studied the NMR line widths and relaxation times of the protons in water, methyl alcohol, ethyl alcohol and butylamine adsorbed on thorium oxide. The adsorbent was prepared by two different methods and was activated at various temperatures. The relaxation behaviour of the adsorbate depended very strongly on the mode of preparation and activation of the adsorbent. A study of the resonance lines as a function of surface coverage indicated the usual increase in width expected with decrease in surface coverage. It was noticeable that a sample of thoria produced by ignition of the oxalate exhibited weaker interactions with adsorbed water molecules than a sample obtained by hydroxide precipitation and ignition. It was also found that T_1 values were far more dependent on surface coverage for the oxide from the oxalate and it was suggested that paramagnetic centres on the surface might be the cause of this. At comparable surface coverages the organic adsorbates tended to have sharper lines despite the presence of chemically shifted nuclei. It was suggested that this could be due to either the longer organic molecules having their "tails" free of the surface thus giving longer relaxation times because of greater freedom of movement or to the larger molecules being excluded from the smaller pores in the adsorbent. This last effect is equivalent to a lowering of the available surface area and would thus lead to longer relaxation times for a given coverage provided interadsorbate effects were negligible. Some evidence for the existence of such interadsorbate effects were obtained from the spin–lattice relaxation times which in some instances decreased with increasing surface coverage. The temperature dependence of the line widths for these systems gave evidence of the successive "freezing out" of various motions of the molecules. Thus in methanol adsorbed on thoria it appeared that the maximum in the line width at about 0°C corresponded to the broadening beyond detection of the hydroxyl proton resonance, the methyl group showing essentially unhindered rotation down to temperatures well below the normal freezing point of methanol.

5.4. *Carbon*

Resing and his co-workers[5, 46, 47] have studied the proton relaxation times of water and benzene adsorbed on a high surface area charcoal (1830 m^2/g). These studies were accompanied by measurements of adsorption isotherms and other surface parameters in more or less detail. The adsorbent was porous having a narrow distribution of pore diameters centred around 27 Å. It also contained a significant concentration of paramagnetic centres (unpaired electrons) which added considerably to the difficulties of interpreting the relaxation data.

The experiments on water and benzene will be described separately but to facilitate the discussion a brief resumé of the details of the apparent phase change theory will be given.

As mentioned earlier, in the presence of a broad distribution of correlation times the spin–spin relaxation time may exhibit an apparent phase change as the temperature is lowered. In practice this means that at some temperature the observed nuclear magnetization consists of two components, one of which decays exponentially with a temperature dependent relaxation time T_{2a} and the other of which obeys a Gaussian decay law with a shorter, temperature independent relaxation time, T_{2b}. This situation arises when the distribution contains correlation times greater than a critical value τ_k, which is that correlation time which defines the onset of rigid lattice behaviour for a life time in phase of one jump assuming that dipole–dipole coupling is dominant. The fraction of the nuclei with correlation times longer than τ_k do not participate in the rapid averaging process implicit in the concept of a distribution of correlation times. Thus there is an appearance of two-phase behaviour with decreasing temperature, the fraction of nuclei with the rigid lattice relaxation time T_{2b} increasing with temperature decrease until effectively all the correlation times are longer than the critical value τ_k and only rigid lattice behaviour is observed.

In order to fit the experimental data to this theory and hence to obtain useful information as to the surface interactions, it is necessary to assume that the relaxation is dominated mainly by intramolecular dipole–dipole interactions, to postulate a model for the molecular motion and to choose a form for the distribution.

In their study of water adsorbed on charcoal Resing et al.[46] used a model of molecular motion in fluids proposed by Cohen and Turnbull.[48] In this it is the critical volume of the hole which must be opened before a molecule can jump which is the rate controlling parameter and an expression for the jump time or correlation time was obtained using this theory. This expression defines the temperature dependence of τ^*, the median correlation time of the distribution, and that of the width parameter of the distribution. The distribution was taken to be log-normal in character. The reason for choosing this model for the motion was because the water interacted relatively weakly with the surface and to a good approximation could be regarded as a capillary liquid. The value of τ^* was determined reasonably accurately from the experimental data at two points. The first of these was the T_1 minimum, at which $\omega_0\tau^* \simeq 0.6$ and the second was the half-way stage in the apparent phase transition where f_m, the fraction of nuclei in the mobile phase is one half; at this point τ^* equals τ_k. These two values and their corresponding temperatures served to establish the correct temperature dependence of τ^*. The width parameter of the distribution was established by trial and error by use of the T_1/T_2 ratio at the T_1 minimum. This enabled the temperature dependence of the width parameter to be obtained. From these parameters theoretical values of T_1, T_2, f_m and σ_0^2 were calculated. σ_0^2 is the rigid lattice value of the second moment of the system and, since intramolecular dipole–dipole coupling was assumed, had to give a value consistent with that expected for a water molecule with some slight intermolecular interactions. The values obtained fitted the observed data reasonably well and, except for noting that the width of the necessary distribution seemed rather large,

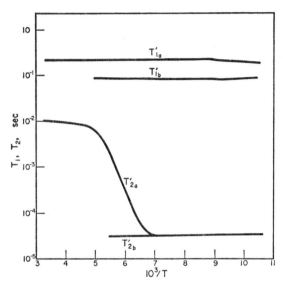

FIG. 8. The temperature dependence of the relaxation times of protons in benzene adsorbed on charcoal at a surface coverage $\theta = 1\cdot31$. (Thompson et al.[46])

Resing[5] concluded that the observed phase change is probably a facet of the NMR experiment and is not a real thermodynamic phase change. He also suggests that if the data of this type can be fitted according to the apparent phase change theory with a value of m close to unity then the phase change is indeed apparent.

In studying benzene adsorbed on the same charcoal Thompson, Krebs and Resing[47] used samples having surface coverages ranging from $\theta = 0\cdot18$ to $\theta = 1\cdot31$. The latter coverage represented saturation of the sample, i.e. all the capillaries filled. The observed relaxation data for a sample with $\theta = 1\cdot31$ are reproduced schematically in Fig. 8. The most notable features of the data are the two-phase, temperature independent, T_1 values at low temperatures, merging to a single temperature independent value at higher temperatures; the plateau in the T_2 data at the higher temperatures and the onset of two-phase T_2 behaviour at lower temperatures, the shorter value, T_{2b}, being independent of temperature. In this study both broad line and spin–echo measurements were made and the second moments determined from both techniques agreed reasonably well. The relaxation behaviour at different surface coverages was qualitatively similar to that shown in Fig. 8.

The plateau in the T_2 data at the higher temperatures was explained by postulating the existence of a small percentage of high surface energy adsorption sites, molecules adsorbed on these sites exchanging with the relatively "free" molecules adsorbed on low energy sites. This explanation is essentially that used by Woessner to interpret the temperature dependence of the T_2 values of water protons adsorbed on silica gel.[32] A calculation was performed in which 96 per cent of the benzene molecules were assumed to have an activation enthalpy for surface diffusion of 6 kcal mole^{-1} and the other 4 per cent were assumed to be characterized by a Gaussian distribution of activation enthalpies with an average value, $\langle \Delta H^* \rangle$, of 10·8 kcal mole^{-1} and a width, $\sigma_H / \langle \Delta H^* \rangle$ of 0·15. This model reproduced the plateau in T_2 at a value of about 10 msec. This corresponded to the value observed for the largest values of θ. As θ was lowered the value of T_2 at the plateau decreased, the explanation given being that there are correspondingly less "free" molecules for the tightly bound molecules to exchange with at lower surface coverages. It was suggested that these plateaux might furnish a method of detecting the presence of small amounts of high energy adsorption sites.

The low temperature value T_{2b} and the broad line data were used to calculate the second moment of the system. The value obtained was completely consistent with a benzene molecule, in an environment rather more dilute in protons than solid benzene, rotating rapidly about its hexad axis. This conclusion was reached using the data of Andrew and Eades[20] for comparison.

The T_2 data, after correction for the existence of the plateau, were interpreted in terms of the apparent phase transition theory.[5] As before, a log-normal distribution of correlation times was used. The results of the theory are shown in Fig. 9 for $\theta = 1·31$. Figure 10 shows the change of f_m with temperature for different values of θ.

The model used for surface diffusion was one in which the jump time was given by the equation

$$\tau = \tau_0 \exp\left(\Delta H^* / RT\right) \tag{20}$$

and

$$\tau_0^{-1} = p\nu \exp\left(\Delta S^* / R\right) \tag{21}$$

where ν is the vibration frequency characteristic of the adsorbed molecule in its potential well, p is the number of nearest neighbour adsorption sites, ΔS^* is the entropy of activation and ΔH^* the enthalpy of activation for surface

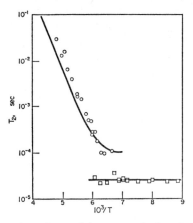

FIG. 9. The temperature dependence of the corrected relaxation times of the protons in benzene adsorbed on carbon at $\theta = 1\cdot31$ monolayer. The points are experimental data, the solid lines, the calculated dependence using the apparent phase change theory with a log-normal distribution of correlation times. (Thompson *et al.*[46])

diffusion. To obtain the best fit to the experimental data it was found necessary to put all the spread in the distribution on to the enthalpy parameter. The reason for this was that the width parameter required to produce the best fit was temperature dependent and only the enthalpy produces such an effect.

The width of the distribution of activation enthalpies was found to decrease with increasing surface coverage. It was suggested that this reflects the

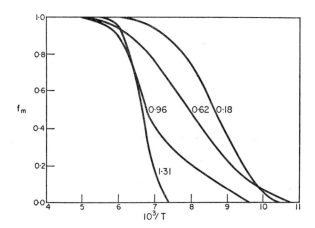

FIG. 10. The temperature dependence of f_m, the fraction of protons in the mobile phase, as a function of surface coverage, for benzene adsorbed on charcoal. The figures adjacent to each curve are the values of θ. (Thompson *et al.*[46])

decreasing effect of surface heterogeneity as more and more of the surface becomes covered with benzene molecules. The average value of the enthalpy of activation increased with surface coverage and this was explained by considering that during a jump from one site to another the benzene molecule must break off a certain number of nearest neighbour interactions and this number would increase with surface coverage. These results are reflected in the f_m data given in Fig. 10. At small values of θ the f_m versus $1/T$ curve is very broad whilst at large values of θ it is steep.

Thompson, Krebs and Resing also discuss the diffusion coefficients for surface diffusion and reasonable agreement in order of magnitude with other measurements is demonstrated.

One disquieting feature of these studies by Resing et al.[47] is the behaviour of T_1, especially in the case of the benzene system. It was suggested that the low temperature two-phase, temperature independent behaviour of T_1 is an artifact, being caused by spin–diffusion controlled relaxation arising from the paramagnetic centres. This type of mechanism has been shown[49] to give rise to non-exponential relaxation behaviour and the low temperature T_1 relaxation followed the pattern expected for this mechanism. The high temperature T_1 behaviour remained unexplained. As Resing et al.[47] pointed out T_1 should start to increase at some temperature where molecular diffusion short cuts the spin diffusion. This was not observed. Another feature which was discussed but could not be explained was the apparent lack of a contribution to σ_0^2 from the paramagnetic centres. There must then remain some doubt about the interpretations given, although the rationalization of the data in terms of surface diffusion and nuclear–dipolar coupling is impressive.

Kimmel[45] has studied the effect of carbon on the relaxation of water protons in the presence of manganous ions. These were added to bring the line widths, used to measure T_2 values, into the range measureable by the wide line spectrometer employed. T_1 values were obtained from the saturation behaviour of the line and the measured T_2 values. The results indicated that carbon reduced T_1 by about a factor of 5 whereas T_2 was reduced by a factor of 10^4. This observation was explained by postulating essentially two-phase behaviour with rapid exchange, the correlation time for the adsorbed phase being very much longer than that for the non-adsorbed phase.

There have been several studies of carbons by Uebersfeld and co-workers[51-8] employing a nuclear–electron double resonance technique. These papers will not be discussed in detail but a brief summary follows. In this work microwave power is applied at the resonance frequency of the electron spins of the sample and the transfer of polarization to the nuclei of molecules adsorbed on the surface is studied. A theoretical paper by Torrey et al.[50] has discussed the steady-state nuclear polarization produced for a system in which a fluid is bounded by a spherical surface. Two limiting cases are described. The first of these is the situation where relaxation of the nuclei takes place at the surface, in which case it is found that the enhancement of nuclear polarization is independent of the volume bounded by the spherical surface. The enhancement is found to be inversely proportional to the volume if relaxation in the bulk phase dominates. Thus a study of such systems can in principle lead to information about surface interactions and diffusion.

5.5. The Silica Gel–Benzene System

Freude et al.[59] have studied the proton relaxation times of benzene molecules adsorbed on a silica gel from 300° to 670°K. The results, corrected for change in surface coverage with temperature, were interpreted in terms of the motional processes of surface diffusion and desorption/adsorption. Activation energies for both these processes were calculated from the data with the assumption that the number of molecules adsorbed on active sites is temperature independent. These values were $\Delta G^*_{\text{diffusion}} = 2\text{--}3$ kcal mole^{-1}; $\Delta G^*_{\text{desorption}} = 10\text{--}14$ kcal mole^{-1}.

Woessner[27] has recently published an extremely detailed study of the proton relaxation times of benzene adsorbed on a silica gel. This work illustrates very effectively the substantial amount of data which is necessarily required in this type of study before a relatively unambiguous interpretation of the results can be given. Woessner had available extensive data on the surface characteristics of the silica gel (e.g. the B.E.T. surface area to N_2 and benzene and the number of hydroxyl groups per 100 Å^2 of surface.) Also available were results of heats of immersion studies and the adsorption site energy distribution for benzene adsorption. This enabled the conclusion to be drawn that the three-quarters of a statistical monolayer coverage employed in the NMR experiments consisted of one-half a monolayer on high energy adsorption sites and one-quarter of a monolayer either on low energy sites or distributed between low energy sites and multilayer formation on high energy sites.

The relaxation times were measured as a function of temperature at frequencies of 25 Mc/s and 50 Mc/s; the latter frequency was employed after one year had elapsed. Differences observed in the data at the two frequencies could not be ascribed unambiguously to either sample ageing or the improved instrumental signal-to-noise ratio at the higher frequency. The temperature range covered was 120–340°K, with occasional measurements outside this range. The extremely complex results obtained will not be described in complete detail, rather the main features will be illustrated and the most important conclusions derived from the results will be discussed.

Although 35 per cent of the protons in the sample were surface hydroxyl groups, having a T_1 value of several seconds, the longitudinal relaxation curves were in general single component, except below 140°K, where they were multicomponent; the components not being experimentally separable however. The most important features of the T_1 data were a broad minimum around room temperature and another minimum at much lower temperatures. This low temperature minimum was assigned to benzene molecules adsorbed on the low energy sites. A T_1 value calculated assuming intramolecular interactions only, for weakly adsorbed benzene molecules freely rotating about their hexad axes, agreed well with the observed minimum value.

The high temperature T_1 minimum was interpreted as showing the existence of strongly adsorbed benzene molecules rotating about their hexad

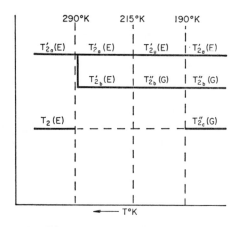

FIG. 11. A schematic representation of the transverse relaxation behaviour of protons in benzene adsorbed on silica gel E ≡ exponential; G ≡ Gaussian. (Woessner.[27])

axes with a broad distribution of correlation times for this motion. The observation of only single component longitudinal relaxation above 143°K was ascribed to either the near equality of T_1 for weakly adsorbed benzene molecules, with T_1 for strongly adsorbed benzene molecules undergoing cross relaxation with the hydroxyl protons, or to molecular exchange between the two types of benzene environment. Woessner points out that life times of as long as 0·1 sec would be short enough to cause this apparent single component behaviour.

The transverse relaxation data as a function of temperature were extremely complex in their behaviour. The essential features are outlined diagramatically in Fig. 11. Woessner's notation for the various relaxation times has been retained. At room temperature the relaxation was two component, one due to the surface hydroxyl protons, the other, with relaxation time T'_{2a}, due to benzene protons. The latter relaxation was exponential and remained so until the very lowest temperature. At temperatures below room temperature a second benzene proton relaxation state appeared with relaxation time T'_{2b}. At higher temperatures this component exhibited exponential behaviour whilst at lower temperatures it became Gaussian, characterized by the time constant T''_{2b}. A third benzene proton relaxation state, not distinguished from the hydroxyl proton relaxation, appeared at the lowest temperatures. This followed a Gaussian decay law and was represented by the relaxation time T''_{2c}.

Woessner has interpreted these three distinct relaxation states, their change from exponential to Gaussian behaviour and their apparent populations in terms of anisotropic motion of benzene molecules at the surface and the known surface characteristics. The three relaxation states were rationalized in terms of the two adsorption states which were indicated by the site energy distribution for benzene adsorption.

Models were considered in which benzene molecules rotated rapidly about their hexad axes whilst these axes wobbled at much lower frequencies. The conclusions reached were that state $a(T'_{2a})$ is characterized by the greatest motion, which as far as the relaxation data are concerned appears to be isotropic. Woessner points out that this does not necessarily mean that the motion is isotropic. Except for the lowest temperature studied, the relaxation of state a was exponential. State $b(T'_{2b}, T''_{2b})$ was explained as being dominated by slower motions than state a, these motions being pseudo-isotropic at higher temperatures and anisotropic with considerable hexad axis wobble at lower temperatures. The third state, c, was dominated by anisotropic motion throughout its temperature range, the hexad axis wobble not being particularly important as shown by the small temperature dependence of T''_{2c}. Below 200°K, T''_{2c} was essentially temperature independent and had a value consistent with that calculated for benzene molecules reorienting about fixed hexad axes.

The changes of transverse relaxation with temperature were discussed in terms of the known adsorption states by considering the benzene molecules in three categories. The first two, characterized by transverse relaxation times T_{2W} and T_{2S}, were respectively weakly adsorbed benzene molecules and the fraction of strongly adsorbed benzene molecules with rotational axis correlation times, such that $\sigma_0^2 \tau_c^2 < 1$. The third category were the strongly adsorbed molecules with correlation times such that $\sigma_0^2 \tau_c^2 > 1$ (T''_{2c}). The division of the strongly bound molecules into these two groups arose from the existence of the distribution of correlation times for these molecules and the point of division is essentially the same as the critical correlation time τ_k in Resing's apparent phase change theory, which was discussed earlier.

The changes in the apparent state populations and relaxation times with temperature were explained in terms of these three categories by considering that at higher temperatures part of the T_S molecules had correlation times long enough to be indistinguishable from T_W molecules, the expected changes occurring as the temperature was lowered, e.g. as the temperature is lowered some T_S molecules enter the T_{2c}'' state and the T_S molecules originally associated with T_W molecules are no longer associated in this way.

In summary it may be stated that although no quantitative data on surface interactions were obtained in this paper, considerable success was achieved in explaining the observed results which might augur well for the future of the technique in this field.

5.6. *The Silica Gel–Alcohol System*

Geschke and Pfeifer[60] have discussed the interpretation of high resolution NMR observations of line broadening effects (T_2) with specific reference to microporous adsorbents. They defined two situations; that in which the observed average relaxation time of the molecules is directly proportional to the relaxation time of the surface adsorbed molecules in the micropores and that in which the observed relaxation depends only on the diffusion coefficient for the micropore liquid and the liquid content of the micropores.[61] These conditions, together with others defining whether individual lines in a spectrum are resolvable in the presence of adsorbent, were used to suggest the optimum conditions for study of a system by high resolution techniques. The results were applied briefly and qualitatively to ethanol adsorbed, at two surface coverages, on a silica gel. It was shown, as expected, that the ethanol molecule bonds to the surface via its hydroxyl group.

5.7. *The Magnesium Oxide–Water System*

The ability of second moment studies to give a substantial amount of valuable information on surface phenomena is amply illustrated by the study by Webster et al.[62] of water adsorbed on magnesium oxide. The major part of this work was concerned with chemisorbed water, present as surface hydroxyl groups. The spectra were measured at room temperature as no substantial difference was observed between spectra obtained at this temperature and at −150°C, indicating the absence of any motional narrowing effects.

Magnesium hydroxide, dehydrated at 300°C, eventually gives an oxide with 0·08 molecule of water per square Ångstrom of surface. The second moment of this system was found to be 4·5 gauss². A model was considered, based on the crystal structure of magnesium oxide, in which hydroxyl groups were added to surface magnesium atoms, and hydrogen atoms to surface oxygen atoms. This system, which consists of a double layer of hydroxyl groups, has a theoretical second moment of 4·7 gauss² and a monolayer coverage corresponding to 0·11 molecule of water per square Ångstrom.

A study of the second moment of the magnesium oxide as a function of surface coverage yielded an extrapolated value (at 0.11 molecule $Å^2$) within ± 0.2 gauss^2 of 4.7 gauss^2. Thus the simple model reproduced the observed second moment extremely well. The variation of the second moment with surface coverage enabled some statements to be made about the mode of desorption. The main conclusion reached was that desorption resulted in the formation of small clusters of hydroxyl groups rather than pairs or a random distribution.

A sample of the oxide, dehydrated at 1000°C and then rehydrated, showed a second moment of 7.4 gauss^2 at a surface coverage of 0.04 molecule/$Å^2$. Comparison of the shapes of the resonance curves for this sample and the previous one showed that this was a real broadening of the resonance due to smaller proton–proton distances, rather than the presence of a separate much broader component in the resonance line. It was suggested that the treatment at 1000°C modifies the surface such as to bring the two planes of hydroxyl groups closer together.

Studies of the surface produced initially when the hydroxide was dehydrated at 300°C confirmed the existence of a metastable state which has a resonance line having an extra, very broad component, thus raising the second moment considerably. Infrared spectra showed that this was not due to residual magnesium hydroxide particles (second moment ~ 20 gauss^2) and it was concluded that again some surface was produced having smaller proton–proton distances than the original surface.

The irreversible change of physisorbed water into chemisorbed water observed on rehydrated magnesium oxide (dehydrated at 1000°C) over a period of two years was shown to be due to localized bulk rehydration, because of a considerable increase in second moment during this time. This increase paralleled the decrease in the intensity ratio of the narrow to the broad resonance lines observed in these systems in the presence of both physisorbed and chemisorbed water.

5.8. Miscellaneous Studies

There have been many other studies which have covered one or more aspects of the subject matter of this article. Aston et al.[83] have studied the proton magnetic resonance line widths in methane in the presence of finely divided rutile. This formed part of an investigation into the nature of the λ-point of methane. A similar study has been reported for carbon tetrafluoride.[84]

Odajima et al.[85–7] have studied the proton resonance of water in cellulose. The results led to the suggestion that the adsorbed molecules were undergoing anisotropic motion.

E

An area in which a great deal of work has been done but which will not be discussed in detail here is that of the magnetic resonance properties of nuclei in molecules contained in zeolites. Much of this work has been concerned with the partial orientation of the molecules in the channels of the zeolites and the consequent splitting of the resonance lines but some workers have investigated the relaxation properties of the nuclei in such systems and a few representative references are included.[28, 87-91]

Reuben et al.[40, 25] have published brief reports of nuclear spin relaxation studies of molecules adsorbed on a porous Vycor glass.

5.9. Colloidal Systems and Systems of Biological Interest

In biological systems and colloidal suspensions, interactions occur between solvent molecules and the much larger "solute" molecules (biopolymers, micelles, dispersed polymers, etc.). The relaxation times of nuclei contained in the small solvent molecules will obviously be influenced by these intera-actions, which in many cases will be qualitatively similar to those considered earlier for solid inorganic-type adsorbents. Usually, the major difference is that the solvent or adsorbate is present in larger proportions in these systems, i.e. larger surface coverages are involved if the phenomena are interpreted from a surface interaction point of view.

The review by Jardetzky,[1] mentioned earlier, covers a considerable amount of the work done using nuclear relaxation techniques to study interactions in biological systems.

Several groups of workers have studied the magnetic resonance character-istics of water absorbed in keratin fibres.[26, 63-7] It is found in general that the water molecules have considerably lower mobility than in bulk water. Clifford and Sheard[26] have measured T_1 and T_2 for water protons in natural albino human hair. They interpreted their results quantitatively in terms of the existence of a broad distribution of correlation times. A log-normal distribution function was chosen and it was found necessary to use a large width parameter to fit the experimental results. A plot of the median correla-tion time for the distribution against $(T°K)^{-1}$ gave a good straight line. It was pointed out, however, that to interpret the slope of this line in terms of an activation enthalpy necessitated placing all the distribution in the correlation times onto the activation entropy term, as discussed originally by Resing.[5]

The results of the existence of the distribution were discussed briefly in terms of the model, due to Feughelman and Haly,[67] of water–protein binding that postulates the existence of five types of bound water. Whilst the relaxation measurements reflect activation parameters only, it seems likely that the existence of a distribution of correlation times implies some form of spectrum of energy levels for the absorbed water molecules. This is consistent with Feughelman and Haly's recognition that their five-state model was an approximation and that a continuous distribution of states was more likely to be correct.

Various other spin relaxation studies have been made of colloidal systems and a few will be mentioned here. Fawcett et al.[68] have studied proton relaxation times in aqueous silver iodide sols. They suggested that at room temperature the correlation time for the motion of a molecule near a colloidal particle was longer than in bulk water and that it increased with decreasing surface charge. Concurrent studies of flocculation rates and relaxation times in polyvinylacetate sols have been interpreted by Johnson et al.[69] in terms of the clustering of water about the colloidal particles. It was suggested that the ordering of the water molecules arose at least in part from the effect of dispersion forces acting between two colloidal particles. In this work, it was noted that $T_1 > T_2$ and it was suggested that this was due to specific surface effects.

Clifford,[71] and Clifford and Pethica[70] have studied the proton spin–lattice relaxation times in long chain alkyl sulphate solutions and have interpreted the data in terms of a tightening of the water structure about the dissolved alkyl chains. It was shown quite conclusively that the restricted motion was not in the alkyl chain itself but was certainly associated with the water molecules.

Daszkiewicz et al.[72] have studied protein hydration in ovalbumin solutions by means of relaxation times. They interpreted their results in terms of an irrotationally bound layer of water adjacent to the protein molecule, these molecules undergoing rapid exchange, from the point of view of the relaxation time measurement, with what was essentially free bulk water.

A technique which is finding increasing use in the study of the binding of paramagnetic metal ions to macromolecules, especially those of biological significance, is the so-called relaxation enhancement technique. In this, the relaxation rates of nuclei in solvent molecules is enhanced by the increase in the rotational correlation time of the solvated metal ion when it becomes attached to the large macromolecule. The technique has been investigated in detail and has been developed to quantitative levels in terms of determining the number of binding sites, the number of solvent molecules remaining in the ion's solvation shell on binding to the macromolecule, etc. It has been used to investigate the binding of ions to nucleic acids,[73, 74] adenosine triphosphate,[75, 76] bovine serum albumin,[77] the citrate lyase of streptococcus diacetilactis[78] and other systems. It has also been used to demonstrate ternary complex formation between manganous ion, creatine kinase and ATP and ADP.[79] The results of a study of the proton relaxation times in living muscle[21] have been interpreted in terms of the release of some strongly bound water during contraction and at death. In the case of contraction the release was reversible, in death it was irreversible. A discussion of the relaxation times was given in terms of a two-phase model with rapid nuclear exchange between phases. Although these studies have not been aimed at evaluating surface phenomena in the accepted sense of the phrase the information obtained is of direct relevance to the subject in its widest meaning.

6. CONCLUSIONS

The fact which stands out from the preceding sections is the complexity of the relaxation behaviour exhibited by nuclei contained in adsorbed molecules. This is not surprising, however, in view of the extreme environment-sensitive nature of nuclear spin relaxation times and the inherent complexity of surface phenomena. One lesson which emerges from a consideration of the literature on this subject is the need to interpret the results of spin relaxation investigations in the light of information obtained by other spectroscopic techniques and by classical adsorption studies of the same system. The investigation of a high surface-area adsorbent solely by nuclear spin relaxation is unlikely to yield much unambiguous information but in conjunction with other methods it can give detailed information about the adsorbate–adsorbent interactions.

In carrying out a study of nuclear magnetic relaxation times in such systems it is necessary to vary surface coverage, temperature and the frequency at which the measurements are made (ω_0) over the widest possible ranges. Only then will the maximum amount of information be obtained and this is required if unambiguous interpretation of the results is to be accomplished. It is also apparent that paramagnetic impurities often play an important, if in most cases an unwanted, role in promoting relaxation. Indeed, as noted earlier, it is not clear in some instances whether these impurities are dominant or not. Second moment studies obviously have the capabilities of revealing quite considerable details of the distribution of nuclei on surfaces and this type of study should form a natural partner for direct measurements of T_1 and T_2 in any system.

Finally, the author would like to suggest that future work on well characterized, diamagnetic surfaces will lead to these techniques becoming useful and powerful additions to the various methods already available for the study of surface phenomena.

ACKNOWLEDGEMENTS

The author wishes to thank Drs. B. A. Pethica, S. M. A. Lecchini, B. Sheard, J. Clifford and other members of the Chemical Physics section of the Unilever Research Laboratories, Port Sunlight, for extremely useful discussions and for permission to discuss the results of unpublished work. Professor N. Sheppard is also thanked for his comments on the draft manuscript.

Permission to reproduce diagrams from the *Journal of Chemical Physics* and the *Journal of Physical Chemistry* is gratefully acknowledged.

REFERENCES

1. O. JARDETZKY, *Advances in Chemical Physics*, vol. VII, ch. 13, Interscience, 1964.
2. J. R. ZIMMERMAN and W. E. BRITTIN, *J. Phys. Chem.* **61**, 1328 (1957).
3. D. E. WOESSNER, *J. Chem. Phys.* **35**, 41 (1961).
4. D. E. WOESSNER, *J. Chem. Phys.* **36**, 1 (1962).
5. H. A. RESING, *J. Chem. Phys.* **43**, 669 (1965).
6. H. WINKLER, *Proc. 10th Colloq. Ampère, Leipzig, Sept.* 1961, p. 219.
7. F. BLOCH, *Phys. Rev.* **70**, 460 (1946).
8. P. S. HUBBARD, *Rev. Mod. Phys.* **33**, 249 (1961).
9. C. P. SLICHTER, *Principles of Magnetic Resonance*, ch. 5, Harper & Row, 1963.
10. A. ABRAGAM, *The Principles of Nuclear Magnetism*, ch. 7, O.U.P., 1961.
11. R. KUBO and K. TOMITA, *J. Phys. Soc. Japan* **9**, 888 (1954).
12. J. M. DEUTCH and J. S. WAUGH, *J. Chem. Phys.* **43**, 1914 (1965).
13. N. BLOEMBERGEN, E. M. PURCELL and R. V. POUND, *Phys. Rev.* **73**, 679 (1948).
14. D. K. C. MCDONALD, *Noise and Fluctuations*, Wiley, 1962.
15. J. A. POPLE, W. G. SCHNEIDER and H. J. BERNSTEIN, *High-resolution Nuclear Magnetic Resonance*, McGraw-Hill (1959).
16. J. W. EMSLEY, J. FEENEY and L. H. SUTCLIFFE, *High Resolution Nuclear Magnetic Resonance Spectroscopy*, vol. 1, Pergamon (1965).
17. Reference 10, chs. 3 and 4.
18. J. H. VAN VLECK, *Phys. Rev.* **74**, 1168 (1948).
19. Reference 9, p. 154.
20. E. R. ANDREW and R. G. EADES, *Proc. Roy. Soc.* (*London*) A **218**, 537 (1953).
21. C. S. BRATTON, A. L. HOPKINS and J. W. WEINBERG, *Science* **147**, 738 (1965).
22. J. R. ZIMMERMAN, B. G. HOLMES and J. A. LASATER, *J. Phys. Chem.* **60**, 1157 (1956).
23. J. R. ZIMMERMAN, and J. A. LASATER, *J. Phys. Chem.* **62**, 1157 (1957).
24. H. M. MCCONNELL, *J. Chem. Phys.* **28**, 430 (1958).
25. D. FIAT, *Proc. XXIX meeting Israel Chem. Soc.*, p. 41 (1961).
26. J. CLIFFORD and B. SHEARD, private communication.
27. D. E. WOESSNER, *J. Phys. Chem.* **70**, 1217 (1966).
28. J. P. COHEN, ADDAD, *Proc. Colloq. Ampère*, 1962, p. 164.
29. A. ODAJIMA, *Progr. Theoret. Phys.*, (*Kyoto*) *Suppl.* **10**, 142 (1959).
30. P. S. HUBBARD, *Phys. Rev.* **109**, 1153 (1958); **111**, 1746 (1958).
31. D. E. WOESSNER and J. R. ZIMMERMAN, *J. Phys. Chem.* **67**, 1590 (1963).
32. D. E. WOESSNER, *J. Chem. Phys.* **39**, 2783 (1963).
33. V. I. KVLIVIDZE, *Dokl. Akad. Nauk. SSSR* **157**, 673 (1964).
34. D. E. O'REILLY, H. P. LEFTIN and W. K. HALL, *J. Chem. Phys.* **29**, 970 (1958).
35. W. K. HALL, H. P. LEFTIN, F. J. CHESELSKE and D. E. O'REILLY, *J. Catalysis* **2**, 506 (1963).
36. J. CLIFFORD and S. M. A. LECCHINI, private communication.
37. V. I. KVLIVIDZE, N. M. IEVSKAYA, T. S. EGOROVA, V. F. KISELEV and N. D. SOKOLOV, *Kin. and Catalysis* (*USSR*) **3**, 73 (1962).
38. D. MICHEL, *Z. Naturforsch* **21a**, 366 (1966).
39. C. KITTEL and E. ABRAHAMS, *Phys. Rev.* **90**, 238 (1953).
40. J. REUBEN, D. FIAT and M. FOLMAN, *Israel J. Chem.* **1**, 276 (1963).
41. D. E. O'REILLY and C. P. POOLE, *J. Phys. Chem.* **67**, 1762 (1963).
42. H. WINKLER, *Z. Naturforsch* **16a**, 780 (1961).

43. S. REBALL and H. WINKLER, *Z. Naturforsch* **19a**, 861 (1964).
44. W. S. BREY, Jnr. and K. D. LAWSON, *J. Phys. Chem.* **68**, 1474 (1964).
45. H. KIMMEL, *Z. Naturforsch* **16a**, 1058 (1961).
46. H. A. RESING, J. K. THOMPSON and J. J. KREBS, *J. Phys. Chem.* **68**, 1621 (1964).
47. J. K. THOMPSON, J. J. KREBS and H. A. RESING, *J. Chem. Phys.* **43**, 3853 (1965).
48. M. H. COHEN and D. TURNBULL, *J. Chem. Phys.* **31**, 1164 (1959).
49. W. E. BLUMBERG, *Phys. Rev.* **119**, 79 (1960).
50. H. C. TORREY, J. KORRINGA, D. O. SEEVERS and J. UEBERSFELD, *Phys. Rev. Letters*, **3**, 418 (1959).
51. E. ERB, J. L. MOTCHANE and J. UEBERSFELD, *Compt. Rend.* **246**, 2121 (1958).
52. E. ERB, J. L. MOTCHANE and J. UEBERSFELD, *Compt. Rend.* **246**, 3050 (1958).
53. J. UEBERSFELD, J. L. MOTCHANE and E. ERB, *J. Phys. Rad.* **19**, 843 (1958).
54. M. JACUBOWICZ and J. UEBERSFELD, *Compt. Rend.* **249**, 2743 (1959).
55. J. UEBERSFELD, *J. Chim. Phys.* **56**, 805, (1959).
56. J. L. MOTCHANE and J. UEBERSFELD, *J. Phys. Rad.* **21**, 801 (1960).
57. J. L. MOTCHANE and J. UEBERSFELD, *Arch. Sci.* **13** (fasc. spéc.), 682 (1960).
58. A. P. LEGRAND, J. AUVRAY and J. UEBERSFELD, *J. Chim. Phys.* **61**, 210 (1964).
59. D. FREUDE, H. PFEIFER and H. WINKLER, *Z. Phys. Chem.* (Leipzig), in press.
60. D. GESCHKE and H. PFEIFER, *Z. Phys. Chem.* (Leipzig), in press.
61. D. MICHEL and H. PFEIFER, *Z. Naturforsch* **20a**, 220 (1965).
62. R. K. WEBSTER, T. L. JONES and P. J. ANDERSON, *Proc. Brit. Ceram. Soc.* **5**, 153 (1965).
63. T. M. SHAW, R. H. ELSKEN and R. E. LUNDIN, *J. Textile Inst.* **51**, T562 (1960).
64. G. W. WEST, A. R. HALY and M. FEUGHALMAN, *Text. Res. J.* **31**, 899 (1961).
65. J. H. BRADBURY, W. F. FORBES, J. D. LEIDER and G. W. WEST, *J. Polymer Sci.* **A2**, 3191 (1964).
66. J. LYNCH and A. MARSDEN, *J. Text. Inst.* **57**, T1 (1966).
67. M. FEUGHELMAN and A. R. HALY, *Text. Res. J.* **62**, 966 (1962).
68. A. S. FAWCETT, G. D. PARFITT and A. L. SMITH, *Nature* **204**, 775 (1964).
69. G. A. JOHNSON, S. M. A. LECCHINI, E. G. SMITH, J. CLIFFORD and B. A. PETHICA, private communication.
70. J. CLIFFORD and B. A. PETHICA, *Trans. Faraday Soc.* **61**, 1 (1965).
71. J. CLIFFORD, *Trans. Faraday Soc.* **61**, 1276 (1965).
72. O. K. DASZKIEWICZ, J. W. HENNEL and B. LUBAS, *Nature* **200**, 1006 (1963).
73. J. EISINGER, F. F. ESTRUP and R. G. SHULMAN, *J. Chem. Phys.* **42**, 43 (1965).
74. R. G. SHULMAN, H. STERNLICHT and B. J. WYLUDA, *J. Chem. Phys.* **43**, 3116 (1965).
75. H. STERNLICHT, R. G. SHULMAN and E. W. ANDERSON, *J. Chem. Phys.* **43**, 3123 (1965).
76. H. STERNLICHT, R. G. SHULMAN and E. W. ANDERSON, *J. Chem. Phys.* **43**, 3133 (1965).
77. A. S. MILDVAN and M. COHN, *Biochem.* **2**, 910 (1963).
78. R. L. WARD and P. A. SRERE, *Biochem. and Biophys. Acta* **99**, 270 (1965).
79. M. COHN and J. S. LEIGH, *Nature* **193**, 1037 (1962).
80. H. WINKLER, *Wiss. Zeit. Karl-Marx-Univ. Leipzig, Math-Naturwiss* **4**, 913 (1965).
81. D. BECKERT and H. PFEIFER, *Ann. der Phys.* **16**, 262, (1965).
82. A. A. KOKIN and A. A. IZMEST'EV, *Russ. J. Phys. Chem.* **39**, 309 (1965).
83. J. G. ASTON and H. W. BERNARD, *J. Amer. Chem. Soc.* **85**, 1573 (1963).
84. Q. R. STOTTLEMEYER, G. R. MURRAY and J. G. ASTON, *J. Amer. Chem. Soc.* **82**, 1284 (1960).
85. A. ODAJIMA, J. SOHMA and S. WATANABE, *J. Chem. Phys.* **31**, 276 (1959).
86. A. ODAJIMA, J. SOHMA and S. WATANABE, *J. Phys. Soc. Japan* **14**, 308 (1959).
87. M. SASAKI, T. KAWAI, A. HIRAI, T. HASHI and A. ODAJIMA, *J. Phys. Soc. Japan* **15**, 1652 (1960).
88. Y. AYANT, E. BELORIZKY, X. PARE and J. ROSSET, *J. de Phys.* **25**, 696 (1964).
89. I. V. MATYASH, M. A. PIONTKOVSKAYA and L. M. TARASENKO., *Zhur. Strukt. Khim* **4**, 92 (1963).
90. S. P. GABUDA and G. M. MIKHAILOV, *Zhur. Strukt. Khim.* **4**, 404 (1963).
91. Y. AYANT, *Compt. Rend.* **255**, 3400 (1962).

CHAPTER 4

RELAXATION PROCESSES IN SYSTEMS OF TWO NON-IDENTICAL SPINS

E. L. MACKOR and C. MACLEAN†

Koninklijke/Shell Laboratorium, Amsterdam,
The Netherlands

(Shell Research N.V.)

CONTENTS

LIST OF SYMBOLS

S, I	Nuclear spins in two-spin systems
J_{SI} or J	Scalar coupling constant between spins S and I
$\mid \pm)\mid \pm >$	Eigenfunctions of system of two non-identical spins in strong magnetic field
$N_{\pm\pm}$	Occupation number of energy states $\mid \pm)\mid \pm >$
S	Total magnetization of S spins
S^+	Magnetization of S spins with neighbouring I spin $\mid + >$

† Present address: Scheikundig Laboratorium van de Vrije Universiteit, Amsterdam.

S^-	Magnetization of S spins with neighbouring I spin $\vert - >$
$S^+ - S^-$	Difference magnetization of S spins
$I^+ - I^-$	Difference magnetization of I spins
$\omega_0, \omega_2, \omega_S, \omega_I$	Relaxation transitions in the two-spin system
ω_S^+, ω_S^-	Relaxation transitions of S spins with neighbouring I spins $\vert + >$ and $\vert - >$, respectively
Δ_S	Deviation of ω_S^+ and ω_S^- from mean value ω_S
	Relaxation transitions caused by intermolecular interactions:
$\omega_k'(k = 0, 2, S, I)$	intermolecular SI
$\omega_{kS}''(k = 0, 2, S)$	intermolecular SS
$\omega_{kI}''(k = 0, 2, I)$	intermolecular II
$J(\omega_\alpha)$	Spectral density at frequency ω_α
	Rate constants describing relaxation of magnetization:
Ω	intramolecular SI
Ω'	intermolecular SI
Ω_S''	intermolecular SS
Ω_I''	intermolecular II
O_S	Nuclear Overhauser effect on S spins (irradiating I spins)
T_{2S}	Transverse relaxation time of S spins
T_Δ	Relaxation time of difference magnetization
τ_c	Rotational correlation time
$\sigma_\parallel, \sigma_\perp$	Parallel and perpendicular components of chemical shift tensor
$\Delta\sigma$	Anisotropy in the chemical shift $(\sigma_\parallel - \sigma_\perp)$
θ_0	Angle between CF bond and intermolecular FH vector in $CHFCl_2$
H_0	External magnetic field
H_1	Radiofrequency field
γ_S	Gyromagnetic ratio of S spins
η	Viscosity
N	Number of spins per cc
a	Molecular radius
b	Distance between spins in molecule

SUMMARY

Nuclear relaxation in molecules with two non-identical spins (S and I) may originate from intra- and intermolecular interactions between like and/or unlike spins. The spins ($S = \frac{1}{2}$, $I = \frac{1}{2}$) are supposed to be coupled by a small scalar interaction; four magnetizations, S^+, S^-, I^+ and I^- can be measured. Experiments are described from which the various contributions to the relaxation mentioned above may be determined.

It is shown that the relaxation rate of the longitudinal difference magnetizations, $(S^+ - S^-)$ and $(I^+ - I^-)$, is a useful parameter. The latter is related to the question whether the interactions that are responsible for the relaxation are intra- or intermolecular. Transverse relaxation rates are likewise influenced by the relative magnitude of these two types of interaction. Theoretical formulae are derived for transverse relaxation rates in systems of scalar coupled nuclear spins.

Attention is paid to anomalous relaxation in nuclear magnetic resonance. This phenomenon may occur if a nuclear spin is relaxed by two relaxation mechanisms with the same correlation time. It has been utilized in the determination of the sign of spin–spin coupling constants.

The theory has been applied to liquid $CHFCl_2$ for which experimental data are reported.

1. INTRODUCTION

Relaxation in nuclear magnetic resonance is induced by interaction of the spins with their surroundings. An important mechanism in liquids, on which we will concentrate our attention, is provided by the dipole–dipole interaction between the spins. The Brownian motion introduces a time dependence, which causes relaxation to take place.

In this chapter an exhaustive discussion of nuclear relaxation in liquids will not be attempted. Rather attention will be confined to the special topic of nuclear relaxation by dipolar interaction between the spins. The cross-term effects between other relaxation mechanisms and the dipole–dipole interaction will also be discussed.

This chapter deals with nuclear magnetic relaxation in liquids consisting of molecules with two non-identical spins ($S = I = \frac{1}{2}$). The interactions that induce the relaxation may be intra- or intermolecular. Four such dipolar interactions may be distinguished: one intramolecular interaction between spins S and I and three intermolecular interactions: between spins S and I, between two S spins and between two I spins. In principle, all four interactions have different time dependences and hence make different time contributions to the total relaxation. Specific examples of molecules containing two non-identical spins with spin quantum numbers of one-half are HF, CHFO and $CHFCl_2$.

As was shown by Solomon[1, 2] and Bloembergen,[2] one can deduce from the nuclear Overhauser effect to what extent the relaxation is caused by interactions between like or unlike nuclei. Non-identical spins which show a time dependent interaction influence each other's relaxation. Saturation of the S spins modifies the intensity of the I resonance and vice versa. This phenomenon, called the Overhauser[3] effect, can be used to decide how far the relaxation of the S spins is affected by I spins. Obviously, the Overhauser effect cannot distinguish between intra- and intermolecular interactions between non-identical spins; nor can it provide information concerning the relative importance of the two types of interactions between identical spins.

In this contribution we wish to investigate the relaxation, both longitudinal and transverse, of two-spin systems in some detail, in order to demonstrate that the above contributions to the relaxation can be completely determined. This determination may be based on a combination of two types of data,[4] i.e. the steady-state Overhauser effect and the relaxation of the longitudinal magnetization. A new aspect is the use of the "difference magnetization" in our problem.

We also investigate how far transverse relaxation can be used to advantage. Transverse relaxation rates are affected by the presence of a time independent scalar coupling between the two spins in the molecule.[5] Appropriate equations have been derived on the basis of current theories[6] for the T_2's of scalar coupled two-spin systems.

Finally we discuss anomalous relaxation in two-spin systems. This phenomenon, well known in electron resonance, takes into account the effect of cross terms between relaxation mechanisms that operate simultaneously on the spins in the molecule. In $CHFCl_2$ these relaxation mechanisms are provided by the dipole–dipole interaction and the anisotropy of the fluorine chemical shift.[7] An interesting application is the determination of the sign of the coupling constant between the spins in the molecule.[8] It will appear that this sign can only be expressed in terms of the sign of the chemical shift anisotropy ($\Delta\sigma$). A brief discussion is presented concerning the measurement of such anisotropies.[9]

The theoretical considerations have been applied to the molecule $CHFCl_2$. Even simpler two-spin molecules exist (e.g. HF, HT), but the relaxation may be complicated by other factors. For instance, HF is liable to rapid chemical exchange which causes fusion of the NMR lines.[2, 10]

2. THEORY

2.1. Longitudinal Relaxation[4]

Consider a system of two non-identical spins ($S = \frac{1}{2}$, $I = \frac{1}{2}$) coupled by a scalar interaction $J\vec{S}.\vec{I}$; J is assumed to be much smaller than the chemical shift between S and I. Four NMR lines are observed in the spectrum, corresponding to four longitudinal magnetizations (see Fig. 1(a)). These magnetizations are proportional to differences in occupation numbers $N_{\pm\pm}$ of the energy states $|\pm\rangle|\pm\rangle$:

$$S_+ = \text{const } [N_{++} - N_{-+}]$$
$$S_- = \text{const } [N_{+-} - N_{--}]$$
$$I_+ = \text{const } [N_{++} - N_{+-}]$$
$$I_- = \text{const } [N_{-+} - N_{--}]$$

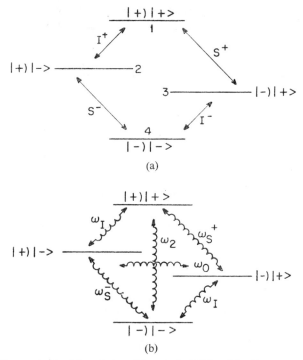

FIG. 1. Energy levels (a) and transition probabilities (b) of a system of two non-identical spins $S = \frac{1}{2}$ and $I = \frac{1}{2}$.

The transition probabilities ω_0, ω_2, ω_S and ω_I between the energy levels are indicated in Fig. 1(b). The transition probabilities $\omega_S^+ (++ \rightarrow -+)$ and $\omega_S^- (+- \rightarrow --)$ have been taken as being equal. In Section 2.3 we drop this limitation and treat the general case.[7]

In dealing with intermolecular relaxation one has to introduce, in an analogous fashion, the occupation numbers and transition probabilities for a system of two two-spin molecules. The two molecules may be considered to form a collision complex. If the complex tumbles as a whole the movement of the four spins is correlated. Hubbard[11] has shown that deviations from exponential behaviour occur when four identical spins at the corners of a square relax by dipolar interaction; these deviations are small, however, and can be neglected. In an analogous four-spin system comprising two pairs of non-identical spins the effect of the correlated motions is larger.[12] In our collision complex it is assumed that the force of interaction is weak, which will be an excellent approximation for small, non-planar molecules (e.g. $CHFCl_2$). Correlation effects can then be neglected; this assumption simplifies the calculations considerably.

In this combined system there are sixteen available energy levels between which relaxation transitions are possible. As stated in Section 1, we must also consider intermolecular SI, SS and II interactions. The contribution to the total relaxation rate may vary from one interaction to another since the time dependences are different in principle. The transition probabilities are denoted by

$$w'_k(k = 0, 2, S, I) \quad \text{intermolecular} \quad SI$$

$$w''_{kS}(k = 0, 2, S) \qquad \text{,,} \qquad SS$$

$$w''_{kI}(k = 0, 2, I) \qquad \text{,,} \qquad II.$$

The equations of motion of the total magnetization $S = (S^+ + S^-)$ and $I = (I^+ + I^-)$ were derived by Solomon.[1] For intramolecular SI interactions:

$$\frac{d}{dt} S = -(\omega_2 + 2\omega_S + \omega_0)(S - S^0) - (\omega_2 - \omega_0)(I - I^0) \qquad (1a)$$

$$\frac{d}{dt} I = -(\omega_2 - \omega_0)(S - S^0) - (\omega_2 + 2\omega_I + \omega_0)(I - I^0) \qquad (1b)$$

S^0 and I^0 are the equilibrium magnetizations of the S and I spins. These equations can be obtained by treating the population and depopulation of the energy levels as a rate process and taking suitable combinations of the occupation numbers. As far as longitudinal relaxation is concerned, this procedure is completely equivalent to more sophisticated treatments (see Section 2.2).

If intermolecular SI interactions provide the relaxation mechanism, then the approach to the steady state is described by the same equations, the only difference being that primed transition probabilities are to be used instead of the unprimed ones. This means that the relaxation of the longitudinal magnetization cannot be used to distinguish between intra- and intermolecular SI interactions.

Apart from SI interactions, the relaxation of the S magnetization is also affected by intermolecular interactions between S spins (where II interactions do not contribute):

$$\frac{d}{dt} S = -2(w''_S + 2w''_{2S})(S - S^0) \qquad (2a)$$

and also

$$\frac{d}{dt} I = -2(w''_I + w''_{2I})(I - I^0) \qquad (2b)$$

It will now be shown that the relaxation of the "difference magnetizations", $(S^+ - S^-) = (I^+ - I^-)$, is very sensitive to whether the interactions are intra- or intermolecular. For a system of two non-identical spins interacting intramolecularly, Shimizu and Fujiwara[13] found the following time dependence of the difference magnetization:

$$\frac{d}{dt}(S^+ - S^-) = -2(\omega_S + \omega_I)(S^+ - S^-) \tag{3a}$$

$$\frac{d}{dt}(I^+ - I^-) = -2(\omega_S + \omega_I)(I^+ - I^-) \tag{3b}$$

The double processes, ω_2 and ω_0, do not enter the time constant. That this should be so can be seen if we write the difference magnetization $(S^+ - S^-)$ in terms of the populations of the energy levels; we then have

$$(S^+ - S^-) = \text{const } [(N_{++} + N_{--}) - (N_{+-} + N_{-+})] \tag{4}$$

This shows that the processes ω_2 and ω_0 do not affect $(S^+ - S^-)$. However, the intermolecular double transitions, ω_2' and ω_0', do not conserve the difference magnetization. Consider, for instance, two molecules both in the state $|++>$; an ω_2' process will cause the final states to be $|+->$ and $|-+>$, showing that $(S^+ - S^-)$ is changed. In fact, the relaxation is described by[4]

$$\frac{d}{dt}(S^+ - S^-) = -2(\omega_I' + \omega_S' + \omega_2' + \omega_0')(S^+ - S^-) \tag{5a}$$

$$\frac{d}{dt}(I^+ - I^-) = -2(\omega_I' + \omega_S' + \omega_2' + \omega_0')(I^+ - I^-) \tag{5b}$$

Intermolecular interactions between identical spins contribute to the relaxation of the difference magnetizations in the following manner:[4]

$$\frac{d}{dt}(S^+ - S^-) = -2(\omega_S'' + \omega_{2S}'' + \omega_I'' + \omega_{2I}'')(S^+ - S^-) \tag{6a}$$

$$\frac{d}{dt}(I^+ - I^-) = -2(\omega_S'' + \omega_{2S}'' + \omega_I'' + \omega_{2I}'')(I^+ - I^-) \tag{6b}$$

2.2. Dipole–dipole Interaction

The previous equations have a general validity since the transition probabilities have not been specified. From now on we consider relaxation by dipole–dipole interaction where the following proportionalities[14, 15] hold:

$$\omega_2 : \omega_0 : \omega_S : \omega_I = \omega_2' : \omega_0' : \omega_S' : \omega_I' = 12 : 2 : 3 : 3$$
$$\omega_{2S}'' : \omega_S'' = \omega_{2I}'' : \omega_I'' = 12 : 3 \tag{7}$$

In order to have a consistent notation throughout, we mention, without proof, the connection between the transition probabilities and the spectral densities[15] $J(\omega_\alpha)$, $J'(\omega_\alpha)$, $J_S''(\omega_\alpha)$, $J_I''(\omega_\alpha)$. The notation is evident: $J(\omega_\alpha)$ denotes a spectral density resulting from intramolecular dipolar interaction at the frequency ω_α. We have

$$\omega_2 = \tfrac{9}{16}\gamma_S^2\gamma_I^2\hbar^2 J(\omega_S + \omega_I)$$
$$\omega_0 = \tfrac{1}{16}\gamma_S^2\gamma_I^2\hbar^2 J(\omega_S - \omega_I)$$
$$\omega_S = \tfrac{9}{16}\gamma_S^2\gamma_I^2\hbar^2 J(\omega_S)$$
$$\omega_I = \tfrac{9}{16}\gamma_S^2\gamma_I^2\hbar^2 J(\omega_I) \tag{8}$$

and so on.

The spectral densities are interrelated:[15]

$$J(\omega_S + \omega_I) : J(\omega_S - \omega_I) : J(\omega_S) : J(\omega_I) = 4 : 6 : 1 : 1 \tag{9}$$

Expressed in terms of spectral densities, equation (1) is, for instance,

$$\frac{d}{dt} S = -\tfrac{3}{4}\gamma_S^2\gamma_I^2\hbar^2 [\{\tfrac{3}{4}J(\omega_S + \omega_I) + \tfrac{3}{2}J(\omega_S) + \tfrac{1}{12}J(\omega_S - \omega_I)\}(S - S^0)$$
$$+ \{\tfrac{3}{4}J(\omega_S + \omega_I) - \tfrac{1}{12}J(\omega_S - \omega_I)\}(I - I^0)] \tag{10a}$$

$$\frac{d}{dt} I = -\tfrac{3}{4}\gamma_S^2\gamma_I^2\hbar^2 [\{\tfrac{3}{4}J(\omega_S + \omega_I) - \tfrac{1}{12}J(\omega_S - \omega_I)\}(S - S^0)$$
$$+ \{\tfrac{3}{4}J(\omega_S + \omega_I) + \tfrac{3}{2}J(\omega_I) + \tfrac{1}{12}J(\omega_S - \omega_I)\}(I - I^0)] \tag{10b}$$

The proportionalities (equation (7)) allow the introduction of four rate constants, which have the significance of weight factors; they are

$$\Omega \quad \text{(intramolecular} \quad SI)$$
$$\Omega' \quad \text{(intermolecular} \quad SI)$$
$$\Omega_S'' \quad \text{(intermolecular} \quad SS)$$
$$\Omega_I'' \quad \text{(intermolecular} \quad II)$$

The total relaxation rates of the sum and difference magnetizations, expressed in terms of the rates Ω, are[4]

$$\frac{d}{dt} S = -(20\Omega + 20\Omega' + 30\Omega_S'')(S - S^0) - (10\Omega + 10\Omega')(I - I^0) \tag{11a}$$

$$\frac{d}{dt} I = -(10\Omega + 10\Omega')(S - S^0) - (20\Omega + 20\Omega' + 30\Omega_I'')(I - I^0) \tag{11b}$$

$$\frac{d}{dt} (S^+ - S^-) = -(12\Omega + 40\Omega' + 30\Omega_S'' + 30\Omega_I'')(S^+ - S^-) \tag{11c}$$

$$\frac{d}{dt} (I^+ - I^-) = -(12\Omega + 40\Omega' + 30\Omega_S'' + 30\Omega_I'')(I^+ - I^-) \tag{11d}$$

To illustrate the use of these equations we have plotted in Fig. 2 the relaxation of the S spins by dipole–dipole interaction with the I spins (i.e. $\Omega''_S = \Omega''_I = 0$). The time dependence of the total magnetization has been normalized. The dashed lines represent the time dependence of the difference magnetization for intramolecular ($\Omega' = 0$) and intermolecular interactions ($\Omega = 0$).

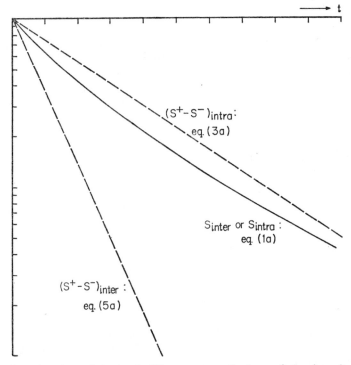

Fig. 2. Relaxation of sum and difference magnetizations of S spins after an adiabatic passage through S^+. The relaxation has been calculated for intramolecular and intermolecular dipolar coupling between S and I spins.

The equations (11) do not suffice to obtain the four unknowns (Ω; Ω'; Ω''_S; Ω''_I) from experiment. In the first place the time constants of the decays of the difference magnetizations, $(S^+ - S^-)$ and $(I^+ - I^-)$, T_Δ, are identical. Furthermore, the magnetic moments of the S and I spins may be the same; in this case the decays of the S and I spins are identical. Later on we shall identify S with the spin of the fluorine nucleus and I with the proton; the magnetic moments differ by only 5 per cent. We need two additional pieces of experimental data. The nature of the problem immediately suggests the nuclear Overhauser effect, which has been discussed briefly in the introduction.

Expressed in terms of the weight factors, the steady state Overhauser effect is

$$O_I = \frac{\Omega + \Omega'}{2\Omega + 2\Omega' + 3\Omega_I''} \tag{12a}$$

$$O_S = \frac{\Omega + \Omega'}{2\Omega + 2\Omega' + 3\Omega_S''} \tag{12b}$$

O_I is the Overhauser effect on the spins I, saturating the spins S. These expressions are obtained in a straightforward manner from equations (11a) and (11b), respectively. To obtain equation (12a) from (11b) one has to utilize the fact that in a steady state $dI/dt = 0$ and that $S = 0$ if the S spins are strongly irradiated.

2.3. Transverse Relaxation

We now want to determine the extent to which transverse relaxation can supply information in the elucidation of the problem under investigation,[5] i.e. the analysis of the dipolar interactions that induce the relaxation.

In order to interpret experimental data we need a set of equations for transverse relaxation rates analogous to equations (10 a, b). One has to realize that the spin systems we have in mind consist of scalar coupled spins to which the normally used expressions for transverse relaxation[15] do not apply. The presence of a scalar coupling between the two spins in the molecule affects the rate of transverse relaxation and a new set of equations has to be derived. This has been done[5] on the basis of current theories[6] using the notational simplicity of Redfield's treatment. We shall, from the outset, confine ourselves to relaxation by dipole–dipole interaction. The calculations have been described in detail in reference 5. Briefly, the introduction of the spin–spin coupling constant modifies the time dependence of the elements of the spin density matrix.

A few remarks concerning notation are in order. To calculate the relaxation of the longitudinal magnetization one may treat the population and depopulation of the energy levels as a rate process described by the master equation. By suitable combination of the various populations one then arrives in a straightforward manner at equations (1–6). It is natural to express the relaxation in terms of the transition probabilities (ω, ω', ω_S'' and ω_I''), since these denote transitions between eigenstates of the energy. In transverse relaxation, however, one studies the spin system while precessing in a plane perpendicular to the magnetic field and there is no advantage in denoting the transverse relaxation rate in terms of the transition probabilities. It is preferable to express T_2^{-1} in terms of spectral densities.

For intramolecular relaxation between the two non-identical spins the rates of transverse relaxation are[5]

$$T_{2S}^{-1}(J_{SI} = 0) = \tfrac{3}{4}\gamma_S^2\gamma_I^2\hbar^2[\tfrac{1}{6}J(0)$$
$$+ \tfrac{1}{24}J(\omega_I - \omega_S) + \tfrac{3}{4}J(\omega_S) + \tfrac{3}{2}J(\omega_I) + \tfrac{3}{8}J(\omega_S + \omega_I)] \quad (13a)$$

$$T_{2S}^{-1}(J_{SI} \neq 0) = \tfrac{3}{4}\gamma_S^2\gamma_I^2\hbar^2[\tfrac{1}{6}J(0)$$
$$+ \tfrac{1}{24}J(\omega_I - \omega_S) + \tfrac{3}{4}J(\omega_S) + \tfrac{3}{4}J(\omega_I) + \tfrac{3}{8}J(\omega_S + \omega_I)] \quad (13b)$$

The introduction of the scalar coupling constant (J_{SI}) causes the transverse relaxation to decrease by a factor of $\tfrac{17}{20}$.

The corresponding relations are for intermolecular dipolar interactions, again between non-identical spins,

$$T_{2S}^{-1}(J_{SI} = 0) = \tfrac{3}{4}\gamma_S^2\gamma_I^2\hbar^2[\tfrac{1}{6}J'(0)$$
$$+ \tfrac{1}{24}J'(\omega_S - \omega_I) + \tfrac{3}{4}J'(\omega_S) + \tfrac{3}{2}J'(\omega_I) + \tfrac{3}{8}J'(\omega_S + \omega_I)] \quad (14a)$$

$$T_{2S}^{-1}(J_{SI} \neq 0) = \tfrac{3}{4}\gamma_S^2\gamma_I^2\hbar^2[\tfrac{1}{6}J'(0)$$
$$+ \tfrac{1}{12}J'(\omega_S - \omega_I) + \tfrac{3}{4}J'(\omega_S) + \tfrac{9}{4}J'(\omega_I) + \tfrac{3}{4}J'(\omega_S + \omega_I)] \quad (14b)$$

Introduction of the coupling constant is tantamount to an inverted three-halves effect:

$$T_{2S}^{-1}(J_{SI} \neq 0)/T_{2S}^{-1}(J_{SI} = 0) = \tfrac{3}{2}.$$

The rate of transverse relaxation of S spins, as far as caused by SS interactions, is independent of the spin coupling constant:

$$T_{2S}^{-1}(J_{SI} = 0) = T_{2S}^{-1}(J_{SI} \neq 0) = \tfrac{3}{4}\gamma_S^4\hbar^2[\tfrac{3}{8}J_S''(0) + \tfrac{15}{4}J_S''(\omega_S) + \tfrac{3}{8}J_S''(2\omega_S)] \quad (15)$$

The II interactions do not affect T_{2S}^{-1} if $J_{SI} = 0$. But if $J_{SI} \neq 0$, an inversion of an I spin broadens an S line:

$$T_{2S}^{-1}(J_{SI} \neq 0) = \tfrac{3}{4}\gamma_I^4\hbar^2[\tfrac{1}{24}J_I''(0) + \tfrac{3}{4}J_I''(\omega_I) + \tfrac{3}{8}J_I''(2\omega_I)] \quad (16)$$

We can now introduce in a straightforward manner the weight factors Ω to obtain:

$$T_{2S}^{-1}(J_{SI} = 0) = 20\Omega + 20\Omega' + 30\Omega_S'' \quad (17)$$

$$T_{2S}^{-1}(J_{SI} \neq 0) = 17\Omega + 30\Omega' + 30\Omega_S'' + 10\Omega_I'' \quad (18)$$

Equations (17) and (18) complement those describing longitudinal relaxation (equations (11 a–d)).

In Fig. 3 we have plotted transverse relaxation rates of two scalar coupled non-identical spins relaxing by dipole–dipole interaction. For comparison the rates of longitudinal relaxation have been added. Figure 3 shows that the rates of transverse relaxation are not so sensitive to the origin of the interactions as the relaxation rates of the difference magnetization.

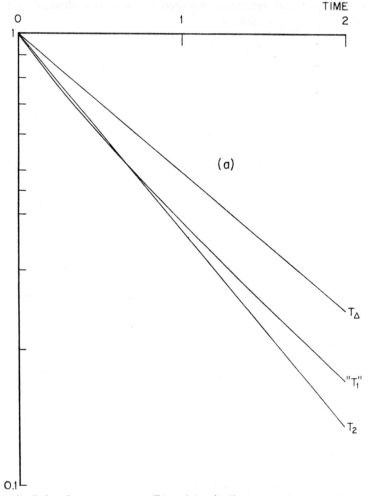

FIG. 3a. Relaxation os transverse (T_2) and longitudinal magnetizations ("T_1" and T_Δ) calculated for intramolecular dipolar interactions ($J_{SI} \neq 0$). T_2 has been scaled arbitrarily to 1 second.

2.4. *Anomalous Relaxation*

We shall now drop the restriction of Section 2.1 that the transition probabilities $\omega_S^+(++ \to -+)$ and $\omega_S^-(+- \to --)$ are equal. A difference between ω_S^+ and ω_S^- is to be expected if the nuclear spin S is relaxed by two mechanisms with the same correlation time.[7, 8] Then a cross term between these two interactions may introduce an asymmetry into the relaxation. As a specific example consider relaxation of the spins S by intramolecular dipole–

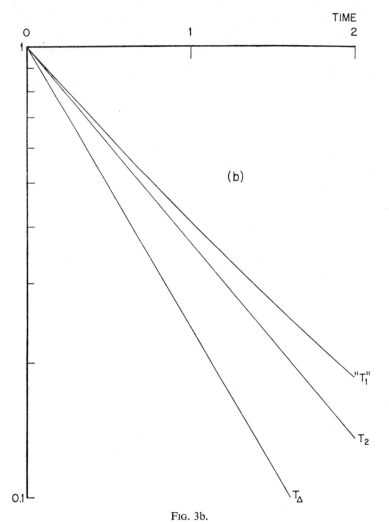

FIG. 3b.

FIG. 3. Relaxation of transverse (T_2) and longitudinal magnetizations ("T_1" and T_Δ) calculated for intermolecular dipolar interactions ($J_{SI} \neq 0$, $\Omega' = \Omega''_S = \Omega''_I$). T_2 has been scaled arbitrarily to 1 second.

dipole interaction and by an anisotropy in the chemical shift. The spins I are supposed to relax only by dipole–dipole interaction. We then have

$$\omega_S^+ = \omega_S(1 + \varDelta_S)$$
$$\omega_S^- = \omega_S(1 - \varDelta_S) \qquad (19)$$
$$\omega_I^+ = \omega_I = \omega_I.$$

The cross term $\omega_S \Delta_S$ affects the relaxation of the total and difference magnetizations as far as it is determined by intramolecular interactions. The motions of spins on different molecules are not correlated and there is no effect of cross terms on the intermolecular part of the relaxation. Hence, out of equations (1–6) only equations (1 a, b) have to be modified. For intramolecular interactions the total and difference magnetizations obey the following equations:

$$\frac{d}{dt} S = -(\omega_2 + 2\omega_S + \omega_0)(S - S^0)$$

$$- (\omega_2 - \omega_0)(I - I^0) - 2\Delta_S \omega_S(S^+ - S^-) \quad (20a)$$

$$\frac{d}{dt} I = -(\omega_2 - \omega_0)(S - S^0) - (\omega_2 + 2\omega_I + \omega_0)(I - I^0) \quad (20b)$$

$$\frac{d}{dt}(S^+ - S^-) = -2\Delta_S \omega_S(S - S^0) - 2(\omega_S + \omega_I)(S^+ - S^-) \quad (20c)$$

If the attention is confined to dipolar interactions one may, as before, use the proportionality relations (7). The total relaxation rates, expressed in terms of the rates Ω, taking into account a small contribution by the cross-relaxation term are:

$$\frac{d}{dt} S = -(20\Omega + 20\Omega' + 30\Omega''_S)(S - S^0)$$

$$- (10\Omega + 10\Omega')(I - I^0) - 6\Delta_S \Omega(S^+ - S^-) \quad (21a)$$

$$\frac{d}{dt} I = -(10\Omega + 10\Omega')(S - S^0)$$

$$- (20\Omega + 20\Omega' + 30\Omega''_I)(I - I^0) \quad (21b)$$

$$\frac{d}{dt}(S^+ - S^-) = -(12\Omega + 40\Omega' + 30\Omega''_S + 30\Omega''_I)(S^+ - S^-)$$

$$- 6\Delta_S \Omega(S - S^0) \quad (21c)$$

From these equations the relaxation of the lines in the NMR spectrum can be calculated (for instance, $S^+ = \frac{1}{2}[S + (S^+ - S^-)]$). As an example S^+ and S^- have been plotted in Fig. 4 for initial conditions $S - S^0 = -2S^0$, $I - I^0 = S^+ - S^- = I^+ - I^- = 0$. These values hold after a 180° pulse on the S doublet. The curves have been calculated for $\Omega'/\Omega = \Omega''_{S,I}/\Omega = 0.5$ and $\Delta_S = \frac{2}{3}$. It is seen that the lines S^+ and S^- relax differently to the steady state.

2.5. Application of Anomalous Relaxation to the Determination of Signs of Spin–spin Coupling Constants

In this section we assume that the cross term, through which an asymmetry in the relaxation of the S lines arises, is caused by the combined effect of

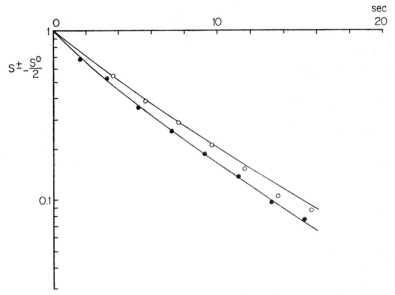

FIG. 4. Relaxation of S lines: S^- (top curve) and S^+ (bottom curve) for initial conditions $S - S^0 = -2S^0$; $I - I^0 = S^+ - S^- = I^+ - I^- = 0$. These values hold after a 180° pulse on the S doublet. The curves have been calculated for $\Omega'/\Omega = \Omega''/\Omega = 0.5$ and $\Delta = \frac{2}{3}$. Experimental points are for fluorine in CHFCl₂, cf. Fig. 11.

intramolecular dipolar interaction and an anisotropy of the chemical shift of the S spins ($\sigma_\parallel - \sigma_\perp = \Delta\sigma$). One can then verify[7] that

$$\omega_S \Delta_S = \tfrac{3}{2}\gamma_S\gamma_I\hbar r_S^{-3} \times \Delta\sigma_S H_0 \tau_c (3\cos^2\theta_0 - 1) \tag{22}$$

where r_s is the distance between the spins; θ_0 is the angle between the CF bond and the internuclear FH vector (for CHFCl₂: $\theta_0 \sim 30°$). It is seen that the sign of Δ_S is related to the sign of the anisotropy of the chemical shift ($\Delta\sigma$).

If $H_0(i \to j)$ is the resonance field at which the transition $i \to j$ is observed keeping the radiofrequency constant, one has for the fluorine transitions

$$H_0(1 \to 3) > H_0(2 \to 4) \quad \text{if } J > 0 \tag{23a}$$

$$H_0(2 \to 4) > H_0(1 \to 3) \quad \text{if } J < 0. \tag{23b}$$

The numbers refer to the energy levels of Fig. 1. It follows from equation (22) that the relaxation rate of the NMR line $H_0(1 \to 3)$ is higher than $H_0(2 \to 4)$, provided $\Delta\sigma$ is positive and vice versa. Hence, in order to obtain the sign of J_{HF} unambiguously, the sign of the anisotropy of the chemical shift is required. The manner in which $\Delta\sigma$ has been measured experimentally[9] will be discussed in the experimental part.

3. EXPERIMENTAL

The experiments were performed on $CHFCl_2$ in which the fluorine (S) and the hydrogen (I) spins form a system of two non-identical spins. The $CHFCl_2$ sample in the liquid state was kept under its own vapour pressure. The instrument used was a Varian NMR spectrometer, equipped with a low temperature insert, operating at 14·3, 40 and 56·4 Mc/s.

We shall first describe the measurements on longitudinal relaxation, then discuss the experimental aspects of transverse relaxation, and finally we shall point out the conditions required for observing anomalous relaxation.

3.1. *Longitudinal Relaxation*

It is intended to measure S and I, and $(S^+ - S^-)$ and $(I^+ - I^-)$. The initial conditions obtained by a reversal of the populations of two (out of four) energy levels are particularly attractive. Experimentally this reversal is achieved by means of an adiabatic fast passage through the transition that connects these levels. The rate of change of the magnetic field H_0, expressed in units of the amplitude of the radiofrequency field (H_1), should be small in relation to γH_1 but larger than the rate of transverse relaxation:

$$T_2^{-1} \ll \frac{1}{H_1} \frac{dH_0}{dt} \ll \gamma H_1$$

The time needed is about 0·1 s, which is short in comparison with the relaxation times $(T_2 = 0·92$ s at $-142°C)$. The coupling constant J_{HF} in $CHFCl_2$ is larger (53·6 c/s) and the population reversal can be made without affecting the populations of the other levels.

In order to get a complete set of data from which the decays of S, I and $(S^+ - S^-) = (I^+ - I^-)$ can be obtained one measures the time dependence of $S^\pm$ and $I^\pm$. The observation of the time dependence of, say the $S^\pm$ magnetizations after an adiabatic fast passage through an S line poses no problem. The power of a Varian oscillator (V 4311) is sufficient for a population reversal. After passage through resonance of one line the radiofrequency power is reduced to a small value and the decays are observed by scanning periodically through the line (Fig. 5).

The effect of an adiabatic fast passage through the line S^+ or S^- on the I lines and vice versa is more difficult to observe. The correct resonance frequencies of the S and I spins in the magnetic field are found by adjusting the frequency of a spin decoupler, operating, for instance, near the S resonance, for collapse of the doublet splitting in the I resonance; in the runs in question the correct frequency was adjusted with an accurate electronic counter (HP 5254 L). Following this, the power of the spin decoupler is

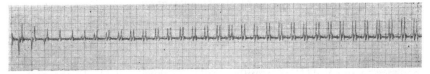

FIG. 5. Change of the magnetization of hydrogen I^+ and I^- after an adiabatic flip through I^+; hydrogen in $CHFCl_2$ at 20°C. I^+ is the low field component of the doublet. The magnetic field is swept with the linear sweep periodically (1·8 sec). The tail at the left of the figure corresponds with the reduction of power immediately after the passage of the line I^-; it practically coincides with the start of the experiment. Spin rotation is the dominant relaxation mechanism of fluorine (S).

reduced to a level suitable for a rapid passage through one of the S lines. The I resonance is observed at the moment of the adiabatic passage through the S line. The power of the spin decoupler is then switched off and the decays are observed by periodically scanning the I spectrum (Fig. 6). The dynamic Overhauser effect can also be determined from these experiments. Although the values obtained were in agreement with expectation the experimental accuracy is not high.

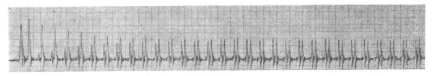

FIG. 6. Change of fluorine magnetization S^+ and S^- after adiabatic flip of I^+ (hydrogen); $CHFCl_2$ at 20°C. The moment of the adiabatic flip corresponds with the first peak at the left of the figure. Period of linear sweep: 1 second. Compare this diagram with Fig. 5.

The decay of $(I^+ - I^-)$ after reversing S^+ or S^- should be identical with the decay of $(I^+ - I^-)$ measured after reversing I^+ or I^- as described in the previous section. This was verified experimentally. At the low temperatures the relaxation rate is high and the experiment must be performed in a short time. Several repetitions of the same experiment were needed.

The steady state Overhauser effect was measured in the following way. One nuclear species, say S, was strongly irradiated with radiofrequency power while the I spins were observed at low power. After the strong irradiation had been removed the doublet of the I spins was recorded as a function of time. The change in intensity of the doublet is a direct measure of the magnitude of the Overhauser effect. The determination from the intensity of the collapsed doublet is liable to give erratic results because double quantum transitions may add significantly to the intensity.

3.2. Transverse Relaxation

For the measurement of T_2 the Varian NMR spectrometer was modified in the way described by Grünwald, Jumper and Meiboom.[16] Their set up employs three Tektronix A 161 pulse units, two of which supply a pulse sequence to the V 4311 transmitter while the third unit blocks the receiver during a radiofrequency pulse. In general the maximum output of the transmitter is used; the minimum length of a 90° pulse is 0·3 msec which is sufficiently short for our purpose.

The free precession signal of a two-spin system is a super-position of two precessions at Larmor frequencies J c/s apart. It is represented by $(1 + \cos Jt)^{1/2} F(t)$. The function $F(t)$ would be the shape of the signal in the absence of spin coupling. Its length depends—as is well known—on the homogeneity of the external field. Using a modification of the Carr–Purcell method,[5] we perform the T_2 measurement in a homogeneous field so as to have a long free precession signal. Following the 90° pulse the sequence of 180° pulses is applied *during* the free precession. A simple resultant signal, from which the "real" T_2 can directly be determined, is obtained if the 180° pulses are applied in the minima (or the maxima) of the free precession. The minima occur at times $t = (2n + 1)/J$ seconds after the 90° pulses ($n = 0, 1, 2, \ldots$). One may apply the 180° pulses at times $t = 3/J$, $5/J$, $7/J$, ... or, if more convenient, at time $t = 3/J$, $7/J$, ..., etc. An important advantage of this procedure is that the influence of molecular diffusion does

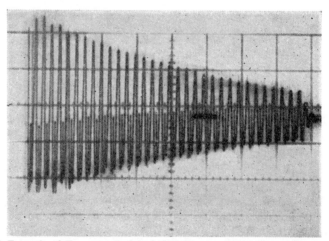

Fig. 7. Example of T_2 measurement of CHFCl$_2$ at -142°C (fluorine resonance at 56·4 Mc/s). The induced signal is detected with a phase-sensitive detector. The 180° pulses are applied in the minima of the free precession. T_2 can be obtained directly from the time dependence of the maxima.

not easily come into play because of the homogeneity of the external field. Figure 7 may serve as an example (CHFCl$_2$; $-142°$C; 56·4 Mc/s). The coupling constant is 53·6 c/s; the minima in the free precession signal are 38 msec apart, which, in this particular experiment, is also the interval between 180° pulses.

3.3. *Anomalous Relaxation*

The experimental problem is to observe a difference in the relaxation of the two S lines. Intermolecular and spin rotational interactions may contribute to the relaxation of CHFCl$_2$ in the neat liquid.[17] Since these interactions tend to mask the effect looked for, they should be suppressed. A 30 per cent volume solution of CHFCl$_2$ in COS was therefore studied[7] in the temperature range $-150°$C to $-120°$C. A crucial test is the field dependence of the anomalous relaxation effect and experiments were accordingly performed at two radiofrequencies, viz. 56·4 Mc/s and 14·3 Mc/s.

4. RESULTS

Brown, Gutowsky and Shimomura[17] have measured the rates of longitudinal and transverse relaxation of CHFCl$_2$ (m.p. $-155°$C) over a considerable temperature range. Their data show that at temperatures close to the melting point the relaxation of both fluorine and hydrogen is dipolar in nature. This conclusion is confirmed by our data on the longitudinal relaxation of the difference magnetization and of the steady state Overhauser effects, as will appear presently. The data of Brown, Gutowsky and Shimomura[17] show also that at low temperatures the chlorine nuclei reorient at such a high rate that they can be neglected in a discussion of the fluorine and hydrogen relaxation. In the high temperature range this no longer holds and the rates of transverse relaxation of both fluorine and hydrogen are affected by the quadrupolar relaxation of the chlorine nuclei.

4.1. *Longitudinal Relaxation*

The relaxation rates of both the total and the difference longitudinal magnetizations have been plotted in Figs. 8 (a) and (b) ($-142°$C) for fluorine and hydrogen, respectively. It is seen that the fluorine magnetization ($S - S^0$), after inverting one component of the doublet, behaves similarly to the hydrogen magnetization ($I - I^0$) after inverting a proton line. This symmetry between fluorine and hydrogen is also apparent from the Overhauser effects (see Table 1). Evidently, as mentioned above, the relaxation mechanism is provided by dipole–dipole interaction between the spins. We now first

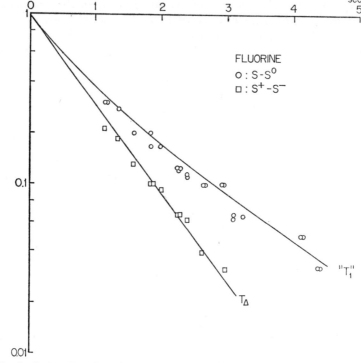

FIG. 8a. Relaxation of longitudinal total and difference magnetizations in CHFCL$_2$ for fluorine after inverting a fluorine line. Magnetizations were scaled to $S^0 = I^0 = 1$.

TABLE 1. MAGNITUDES OF RATES OF RELAXATION AND OVERHAUSER
EFFECTS OF CHFCl$_2$ AT $-142°$C (56·4 Mc/s)

$$T_\Delta^{-1} = 1·19 \pm 0·15 s^{-1}$$
$$T_2^{-1}(H) = 1·09 \pm 0·15 s^{-1}$$
$$T_2^{-1}(F) = 1·09 \pm 0·15 s^{-1}$$
$$O_H = 0·40 \pm 0·05$$
$$O_F = 0·40 \pm 0·05$$

(T_Δ denotes the time constant of the relaxation of the difference magnetization.)

discuss the determination of the rates Ω, Ω' and Ω'' as obtained from longitudinal relaxation for which the analysis of Section 2.1, which has been described in reference 4, may be applied. The relative relaxation rates of the total and difference magnetizations can be expressed in terms of the ratios Ω'/Ω and Ω''/Ω, which can be determined by a curve-fitting procedure. One finds:

$$\Omega'/\Omega \simeq 0·4; \quad \Omega''/\Omega = 0·24 \tag{24}$$

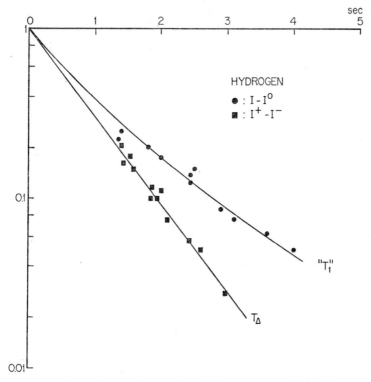

F$_{IG}$. 8b. Relaxation of longitudinal total and difference magnetizations in CHFCl$_2$ for hydrogen after inverting a hydrogen line. Magnetizations were scaled to
$$S^0 = I^0 = 1.$$

This shows that intramolecular relaxation dominates at $-142°$C. The absolute values of the rates Ω can likewise be obtained from the experiment; the results have been collected in Table 2.

4.2. Transverse Relaxation

We shall now investigate the extent to which transverse relaxation rates can be used in this problem. Application of the data of Table 1 to equations (11), (12) and (18) lead to the relations

$$\Omega'/\Omega + 2{\cdot}3\Omega''/\Omega = 0{\cdot}9$$
$$\Omega'/\Omega - 6{\cdot}0\Omega''/\Omega = -1{\cdot}0 \tag{25}$$

These relations represent straight lines in a plot of Ω'/Ω vs. Ω''/Ω (Fig. 9). The dashed lines in Fig. 9 correspond to the estimated margins of error.

TABLE 2. COMPARISON OF MEASURED RATES OF RELAXATION WITH
THE RATES CALCULATED FROM BPP THEORY

	$-142°C$ s^{-1}		$20°C$ s^{-1}	
	Calc.†	Obs.	Calc.†	Obs.
20Ω	0·6	0·4	0·06	<0·006
$20\Omega'$	0·03	0·1⁵	0·003	~0·006
$20\Omega''$	0·03	0·2⁵	0·003	~0·006
Ω'/Ω	0·048	0·4	0·048	>1
Ω'/Ω''	1·0	1·7	1·0	—

† Intramolecular distance $b = 2·04$ Å. The radius estimated
(i) from density gives $a = 3·1$ Å, (ii) from molecular geometry using
van der Waals radii $a = 2·4$ Å. The value $a = 2·8$ Å was taken. An
extrapolated value $\eta = 1·5$ cP at $-142°C$ was used.

According to the figure the possible values of Ω''/Ω that are compatible with
the experimental limits of error are given by

$$0·1 < \Omega''/\Omega < 0·5, \tag{26}$$

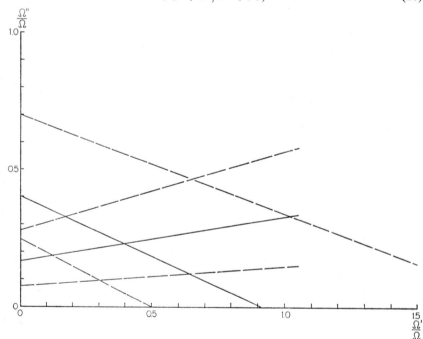

FIG. 9. Relative weights of interactions between identical spins versus the one
between non-identical spins. Dashed lines: estimated margins of error for CHFCl₂
at $-142°C$, 56·4 Mc/s.

while the uncertainty in Ω'/Ω is even greater. The ratios (24) are compatible with those of Fig. 9.

In order to combine the available information, we normalize the transverse relaxation rate to unity and in a semilogarithmic plot construct the relaxation of total and difference magnetizations for a set of the parameters Ω'/Ω and Ω''/Ω (Fig. 10). The set corresponds with a steady-state Overhauser effect of

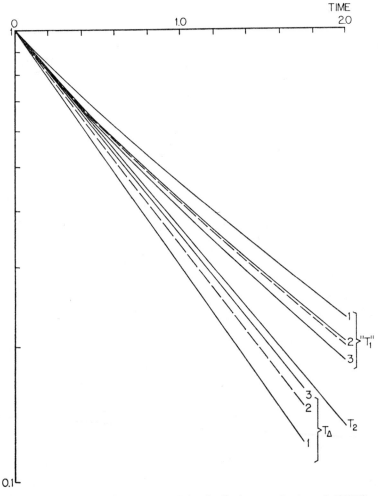

FIG. 10. Relaxation of transverse and longitudinal magnetization of $CHFCl_2$ at $-142°C$. T_2 has been scaled to 1 second. Relaxation of total longitudinal and difference magnetizations (1–3) have been calculated for three different values of the ratios Ω'/Ω and Ω''/Ω; (1) $\Omega'/\Omega = 1·00$, $\Omega''/\Omega = 0·33$; (2) $\Omega'/\Omega = 0·40$, $\Omega''/\Omega = 0·23$, (3) $\Omega'/\Omega = \Omega''/\Omega = 0·20$ keeping the Overhauser effects $O_H = O_F = 0·40$. The experimental points fall on dashed curves (compare Figs. 8(a), (b)).

0·40. The dashed curves have been measured experimentally. The agreement is closest for $\Omega'/\Omega = 0\cdot4$ and $\Omega''/\Omega = 0\cdot23$, i.e. within the limits of error of the previous result.

4.3. *Anomalous Relaxation*

The results of the anomalous relaxation experiment[7] are shown in Fig. 11 (CHFCl$_2$ in COS at $-142°$C; 56·4 Mc/s). A 180° pulse is applied to the fluorine doublet at the beginning of the experiment, and its subsequent relaxation to the steady state is recorded. Immediately after the pulse the doublet component first scanned is farthest removed from its steady state value, in both top and bottom spectra. After a few seconds (see markers in Fig. 11) the situation is reversed in the bottom spectra. The arrows under the

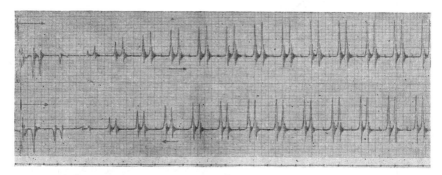

FIG. 11. Experimental behaviour of the fluorine lines S^+ and S^- after applying a 180° pulse to the fluorine doublet. After the pulse the spectrum is scanned periodically at low radiofrequency power. Top: direction of scanning from low to high field; bottom: from high to low field. The high field line relaxes fastest (CHFCl$_2$ in COS, 30 per cent by volume, $-142°$C, 56·4 Mc/s). Compare with Fig. 4.

spectra indicate the direction of the scan of the magnetic field. Thus at 56·4 Mc/s the high field line proves to relax faster than the low field line. When the experiment is repeated at 14·3 Mc/s under identical conditions both fluorine lines behave identically, which one would expect for an effect caused by an anisotropy in the chemical shift.

We can now express the sign of the coupling constant in terms of the sign of the anisotropy of the chemical shift; if the anisotropy of the chemical shift $(\sigma_{\parallel} - \sigma_{\perp} = \Delta\sigma)$ is positive, then the sign of the coupling constant J_{HF} is negative and vice versa. On the basis of theoretical arguments the sign of $\Delta\sigma$ is expected to be positive.[18] This has been confirmed for a number of cases, but recently at least three compounds have become known for which a negative value of $\Delta\sigma$ is reported.[9, 20] Hence, in order to obtain the sign of J_{HF} in $CHFCl_2$ unambiguously, it is necessary to measure the sign of $\Delta\sigma$; we shall briefly indicate how the anisotropy of the chemical shift of $CHFCl_2$ has been measured.

The method followed was to study the wide line NMR spectrum of the molecule.[20] At sufficiently high values of the external magnetic field this spectrum shows a pronounced asymmetry caused by the anisotropy of the chemical shift and the sense of the asymmetry is related to the sign of $\Delta\sigma$. Since dipolar interactions tend to mask this asymmetry we have investigated the wide line NMR spectrum of the deuterated compound[9] (Fig. 12). In

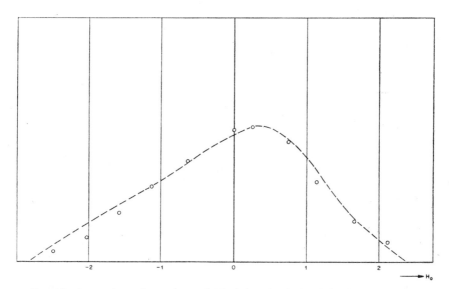

FIG. 12. Comparison of experimental (circles) and calculated (broken curve) line shapes: $CDFCl_2$ in hexadeuteroacetone (50 per cent vol.) at $-196°C$ and 56·4 Mc/s. In the calculation the anisotropy of the chemical shift was assumed to be -205 ppm; the superposed Gaussian broadening corresponded to a linewidth of 1·60 gauss.

order to interpret this spectrum we have plotted in Fig. 13 the NMR line shape of a polycrystalline sample where the spins have an axially symmetric chemical shift; broadening of the component lines is superposed in the bottom spectrum. Comparison of the experimental spectrum with Fig. 13 then shows that the anisotropy of the chemical shift of the C—F bond in $CHFCl_2$ has the negative sign. We now can give the sign of the coupling constant J_{HF} unambiguously:

$$J_{HF} = +53·6 \text{ c/s}. \tag{27}$$

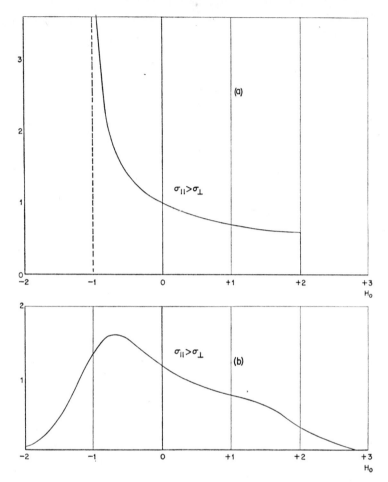

Fɪɢ. 13. (a) Line shape in a polycrystalline specimen due to an axially symmetric chemical shift. (b) The same as Fig. 13(a), after superposition of Gaussian broadening of the component lines.

5. DISCUSSION

It is of interest to compare the results of the previous section with what is expected from theory. Several approaches to nuclear relaxation are known.[15] In these theories the problem is in the last stage reduced to the calculation of a correlation time. For dipolar relaxation this may be a rotational or a translational correlation time; for relaxation by spin-rotation interaction in liquids the correlation time is associated with the decay of angular velocity.

As has been discussed in the previous section, the relaxation of $CHFCl_2$ at $-146°C$ is dipolar in nature and a comparison of the experimental results with the theory of Bloembergen, Purcell and Pound[14] suggests itself. In this approach the correlation times for rotation and translation are obtained from the Debye theory; the molecule is treated as a sphere rotating in a viscous liquid and the validity of the Stokes–Einstein relation is assumed.

The intra- and intermolecular contributions to the relaxation, according to the BPP theory, are

$$20\Omega = (\pi^{-1}\gamma_I^2\gamma_S^2h^2/3)(a^3\eta/b^6kT) \tag{28a}$$

$$20\Omega' = (\gamma_I^2\gamma_S^2h^2/5)(N\eta/kT) \tag{28b}$$

$$20\Omega''_S = (\gamma_S^4h^2/5)(N\eta/kT) \tag{28c}$$

$$20\Omega''_I = (\gamma_I^4h^2/5)(N\eta/kT) \tag{28d}$$

$$\Omega/\Omega' = \Omega/\Omega'' = (\tfrac{5}{3\pi})(a^3/b^6N), \quad \text{if } \gamma_I = \gamma_S \tag{28e}$$

In Table 2 the experimental values at $-142°C$ are compared with the ones predicted from equations (28 a–e). The observed value of the rotational contribution to the relaxation is too small by a factor of $\tfrac{2}{3}$; the experimental translational contribution is at least three times the calculated one. The agreement is as good as can be expected in view of the crudeness of the model, the use of ill-defined parameters (e.g. the molecular radius) and of an extrapolated viscosity. Furthermore, in accordance with the BPP theory it is found that the rotational contribution to the relaxation is larger than the translational one. The ratio Ω'/Ω'' is expected to be ~ 1, in reasonable agreement with experiment. On the whole the agreement is not unsatisfactory.

It is of interest to investigate whether the agreement is as satisfactory in the high-temperature range. The data recorded at $+20°C$ (see Table 2) have been obtained from the longitudinal relaxation of the protons as the fluorine spins are relaxed by spin–rotation interaction.[17] The relaxation mechanism for the longitudinal magnetization of hydrogen at $+20°C$ is the dipole–dipole interaction.

At $+20°C$ the rotational contribution to the relaxation proves much smaller than one would expect from the BPP theory; the translational part is about correctly predicted. Thus at high temperatures the rotational relaxation, as predicted by the BPP theory, is too efficient.

This high temperature behaviour of $CHFCl_2$ ties in with measurements on other molecules, published in the literature.[21, 22] For low viscosity liquids the rotational correlation times predicted with the Stokes–Einstein relation generally prove too long, in other words, the role of rotational friction is overestimated. According to Steele[22] the rotational relaxation is then determined not by the viscosity but by the moment of inertia I_m. For a

F

spherical top molecule this contribution to the longitudinal relaxation is

$$T_{\text{rot}}^{-1} = (3\gamma^4 h^2/16\pi^2 b^6)(\pi I_m/kT)^{1/2} \tag{29}$$

These considerations rationalize our experimental results: at room temperature Steele's approach[21, 22] is probably valid whereas at $-142°C$ the viscosity is so high that the frictional model, which underlies the BPP[14] theory, gives an adequate description of the results.

Finally we turn to the results of our experiments connected with anomalous relaxation. Tiers[23] has shown by double resonance that the ^{13}CH and the ^{13}CF coupling constants in $CHFCl_2$ are of opposite sign, whereas the ^{13}CH and HF coupling constants are of like sign. Since we have found the coupling constant between hydrogen and fluorine in $CHFCl_2$ to be positive it must be concluded that

$$J_{\text{HF}} = +53 \cdot 6 \text{ c/s}$$
$$J_{^{13}\text{CH}} = +220 \cdot 0 \text{ c/s}$$
$$J_{^{13}\text{CF}} = -293 \cdot 8 \text{ c/s} \tag{30}$$

Thus the directly bonded ^{13}CH coupling constant in $CHFCl_2$ is positive, which can be compared with similar results published.

The positive sign of *ortho* and *meta* J_{HH} in benzene has been established by Saupe and Englert[24] from the NMR spectrum of benzene when dissolved in an anisotropic environment.

Buckingham and McLauchlan[25] have measured *ortho*. J_{HH} in *para*-nitrotoluene using the elegant electric field method; the authors conclude that the sign is positive.

The above results and those of our own relaxation measurements are in agreement with theoretical predictions (cf. reference 26).

ACKNOWLEDGEMENT

The measurements have been performed by Mr. C. W. Hilbers; his expert experimental assistance is gratefully acknowledged.

REFERENCES

1. I. SOLOMON, *Phys. Rev.* **99**, 559 (1955).
2. I. SOLOMON and N. BLOEMBERGEN, *J. Chem. Phys.* **25**, 261 (1956).
3. A. OVERHAUSER, *Phys. Rev.* **91**, 476 (1953).
4. E. L. MACKOR and C. MACLEAN, *J. Chem. Phys.* **42**, 4254 (1965).
5. C. MACLEAN and E. L. MACKOR, *ibid.* **44**, 2708 (1966).
6. A. G. REDFIELD, *IBM J. Research Develop.* **1**, 19 (1957).
7. E. L. MACKOR and C. MACLEAN, *J. Chem. Phys.* **44**, 64 (1966).
8. H. SHIMIZU, *J. Chem. Phys.* **40**, 3357 (1964).

9. C. MacLean and E. L. Mackor, *Proc. Phys. Soc.* **88**, 341 (1966).
10. C. MacLean and E. L. Mackor, *Proc. Coll. Ampère, Eindhoven*, 1962, North Holland Publ. Co., Amsterdam.
11. P. S. Hubbard, *Phys. Rev.* **109**, 1153 (1958).
12. G. W. Flynn and J. D. Baldeschwieler, *J. Chem. Phys.* **38**, 226 (1963).
13. H. Shimizu and S. Fujiwara, *J. Chem. Phys.* **34**, 1501 (1961).
14. N. Bloembergen, E. M. Purcell and R. V. Pound, *Phys. Rev.* **73**, 679 (1948).
15. A. Abragam, *The Principles of Nuclear Magnetism*, Oxford University Press (1961).
16. E. Grünwald, C. F. Jumper and S. Meiboom, *J. Amer. Chem. Soc.* **84**, 4664 (1962).
17. R. J. C. Brown, H. S. Gutowsky and K. Shimomura, *J. Chem. Phys.* **38**, 76 (1962).
18. A. Saika and C. P. Slichter, *J. Chem. Phys.* **22**, 26 (1954).
19. E. Hunt and H. Meyer, *J. Chem. Phys.* **42**, 353 (1964); Z. Veksli, *8th European Congress on Molecular Spectroscopy, August* 1965, Copenhagen.
20. E. R. Andrew and D. P. Tunstall, *Proc. Phys. Soc.* **81**, 986 (1963).
21. W. B. Moniz, W. A. Steele and J. A. Dixon, *J. Chem. Phys.* **38**, 2418 (1963); D. E. O'Reilly and G. E. Schacher, *ibid* **39**, 1768 (1963); R. W. Mitchell and M. Eisner, *ibid* **33**, 86 (1960).
22. W. A. Steele, *J. Chem. Phys.* **38**, 2411 (1963).
23. G. V. D. Tiers, *J. Amer. Chem. Soc.* **84**, 3972 (1962); *J. Phys. Chem.* **85**, 1373 (1963).
24. A. Saupe and G. Englert, *Phys. Rev. Letters*, **11**, 462 (1962); A. Saupe, *Z. Naturforsch.* **20a**, 572 (1965).
25. A. D. Buckingham and K. A. McLauchlan, *Proc. Chem. Soc.* **1963**, 144.
26. M. Karplus, *J. Amer. Chem. Soc.* **84**, 2458 (1962).

CHAPTER 5

MICRODYNAMIC BEHAVIOUR OF LIQUIDS AS STUDIED BY NMR RELAXATION TIMES

H. G. HERTZ

Institut für Physikalische Chemie und Elektrochemie
der Technischen Hochschule, Karlsruhe Germany

CONTENTS

GLOSSARY OF SYMBOLS

A	coupling constant of scalar interaction
$A(\rho)$	Fourier transform of $\mathscr{P}_1(\mathbf{r})$
a	molecular diameter
a^*	molecular radius
$\mathring{a}$	distance of closest approach between equal spins
$\mathring{a}_{IS}$	distance of closest approach between spins I and S

b	constant spin–spin distance within a molecule
b_{ij}	constant intramolecular distance between spins i and j
$C_i\ i = 1, 2, 3 \ldots$	rotation-step constant
$C_{1a}, C_{1b} \ldots$	rotation-step constant in sub-liquids $a, b \ldots$
C_{1ab}	rotation-step constant for the transition from sub-liquid a to sub-liquid b and vice versa
C_I, C_S	concentrations of spin I and S respectively (spins cm^{-3})
D	translational self-diffusion coefficient
D_I, D_S	translational self-diffusion coefficient of spins I and S, respectively (the indices I and S are added in all cases where different spins are present in the system)
$\tilde{D}_I, \tilde{D}_S$	concentration dependent translational self-diffusion coefficient of spins I and S respectively
$\tilde{D}_{IS}$	$= \frac{1}{2}(D_I + D_S)$
D_r, D_{r_1}, D_{r_2}	rotational diffusion coefficients
E	activation energy
E_r	activation energy for the rotational motion of a molecule
E_b	activation energy for the exchange process of a solvent molecule in the solvation sphere
e	elementary charge
$f_0(t), f_1(t)$	time correlation functions of spherical harmonics of order 1
$g_1(t), g_2(t)$	time correlation functions of spherical harmonics of order 2
$\tilde{g}_1(t), \tilde{g}_2(t)$	time correlation functions of the functions $Y_2^m(\theta,\phi)/r^3, m = 1, 2$
H_0	static magnetic field
^2H	isotope of hydrogen with mass 2, sometimes denoted as D as well
h	Planck's constant; $\hbar = h/2\pi$
I	spin of a nucleus
k	Boltzmann constant
$\mathbf{M}$	vector of the nuclear magnetization
$\mathbf{M}_I, \mathbf{M}_S$	vector of the nuclear magnetization of spins I and S respectively
$\mathscr{M}_x, \mathscr{M}_y, \mathscr{M}_z$	components of the nuclear magnetization vector
$\mathbf{M}^0$	vector of the nuclear magnetization in thermal equilibrium
$\mathscr{M}_\perp$	transversal component of the magnetization vector

M_0	molar mass
m	moles of solute per kg of solvent (molality)
N, N_I	number of spins (spins I) in the liquid
N_L	Avogadro's number
n	number of like spins in a molecule
n_h, n_h^b	general hydration (or solvation) number of solute particles (e.g. geometrically defined)
n_h^+, n_h^-	general hydration (or solvation) number of cation or anion respectively
n_c	number of molecules rigidly bound in the first coordination sphere of an ion
$P, \mathscr{P}, \mathit{p},$	probabilities as indicated by the argument in the parentheses added
$P_A, \mathscr{P}_A, P_B, \mathscr{P}_B$	probabilities concerning a particle of species A, B respectively
$\mathscr{P}_1$	probability concerning one single translational step
Q	nuclear quadrupole moment
R	gas constant
R'	shortest intermolecular spin–spin distance within the rigid first solvation sphere
R'_{IS}	closest distance of approach between solute and solvent spin within the first solvation sphere
R_{IS}	$= R'_{IS}$ if $n_c = 0$, otherwise $=$ closest distance of approach between solute and solvent spin in the second solvation sphere
$\mathbf{r}$	vector characterizing position or displacement of a particle
r	magnitude of vector $\mathbf{r}$
$\langle r^2 \rangle$	mean square displacement of a particle under one translational jump
S	spin of a nucleus or electron spin on a paramagnetic particle
T	absolute temperature
T_1	longitudinal relaxation time
T_2	transversal relaxation time
$(1/T_{1,\,2})_{\text{intra}}$	intramolecular relaxation rate
$(1/T_{1,\,2})_{\text{inter}}$	intermolecular relaxation rate
$1/T_{ij}$	intramolecular relaxation rate of a spin pair ij within a molecule
$1/T_{ia}, 1/T_{ib}, 1/T_{ic} \ldots$	$i = 1, 2$ relaxation rates in sub-liquids $a, b, c \ldots$
$1/T'_{ia}, 1/T'_{ib}, 1/T'_{ic} \ldots$	$i = 1, 2$ intramolecular relaxation rates in sub-liquid $a, b, c \ldots$

$(1/T_{1,\,2})_{\text{capt}}$, $(1/T_{1,\,2})_{\text{capt}}^{(IS)}$ contribution to the relaxation rate due to solvent molecules captured by the rigid solvation sphere during a time of order of the correlation time

$(1/T_{1,\,2})_{\text{sc}}$ relaxation rate by scalar interaction

$(1/T_{1,\,2})_{\text{transl}}^{(IS)}$ translational contribution to the intermolecular relaxation caused by dipolar interaction between spins I and S

$1/T_{1,\,2}^{\pm}$ relaxation rate in the cationic or anionic solvation spheres apart from scalar contributions

$(1/T_{1,\,2}^{+})_{\text{ion}}$ total relaxation rate in the cationic solvation sphere (dipolar + scalar interaction with $\tau_s \ll \tau_b$)

$[1/T_{1,\,2}^{+}]_{\text{ion}}$ total relaxation rate in the cationic solvation sphere (scalar interaction with $\tau_s \ll \tau_b$)

V volume of the system under consideration

V_m molar volume

$\partial^2 V/\partial z'^2$ electrical field gradient at the nuclear site within a molecule

$\partial^2 V^0/\partial_{z'^2}$, $\partial^2 V^{\pm}/\partial_{z'^2}$ electrical field gradient at the nuclear site within a solvent molecule in the free solvent, the cationic and anionic solvation sphere respectively

x_A, x_B mole fractions of particles A and B respectively

x_a, x_b, x_c ...   mole fraction of solvent molecules in sub-liquids a, b, c ...

x^+, x^- mole fractions of solvent molecules in the cationic and anionic solvation sphere respectively

Y_l^m, Y_l^{m*} spherical harmonics

α proportionality factor between $1/D$ and $1/T_1$

β' angle between field gradient and intramolecular axis of rotation

γ gyromagnetic ratio

γ_I, γ_S gyromagnetic ratio of spins I and S respectively

Δ angle between spin–spin vector and intramolecular axis of rotation

$\Delta H_{1/2}$ line width (half width)

ΔH line width between points of maximum slope

$\Delta \bar{\omega}_{1/2}$ line width (half width) measured in (angular) frequency scale

$\Delta \bar{\omega}$ line width between points of maximum slope measured in (angular) frequency scale

$\Delta \omega$ measured chemical shift of the solvent spin in (angular) frequency scale

$\Delta\omega_b$	chemical shift in (angular) frequency scale of the solvent molecule on a site in the first coordination sphere
η	viscosity
$\bar{\eta}$	asymmetry factor of the electrical field gradient
$\bar{\eta}^0, \eta^\pm$	asymmetry factor of the electrical field gradient in a solvent molecule in the free solvent, the cationic and anionic solvation sphere respectively
$\boldsymbol{\rho}$	vector occurring in the three dimensional Fourier-transform
ρ	magnitude of vector $\boldsymbol{\rho}$
Ω	rotation of a molecule
ω	NMR resonance frequency
ω_I, ω_S	resonance frequency of spins I and S respectively
ω_k	$k = 1, 2, 3$ components of the angular velocity vector
θ^*	temperature (°C)
θ, θ_0	angular coordinates of the spin–spin vector
ϕ, ϕ_0	angular coordinates of the spin–spin vector
τ	time constant of finite rotational step motion
$\tau_a, \tau_b, \tau_c \ldots$	mean life-time of a solvent molecule in sub-liquids $a, b, c \ldots$
$\tau_{b'}$	mean life-time of a solvent molecule in the rigid first solvation sphere
τ_c	correlation time
$\tau_c^0 \equiv \tau_{ca}, \tau_{cb}, \tau_{cc}$	correlation times for a solvent molecule in sub-liquids $a, b, c \ldots$
τ_c^+, τ_c^-	correlation time of the solvent molecule in the cationic and anionic solvation sphere respectively
τ_{c_1}	correlation time of a spin pair undergoing intramolecular reorientation in the limit of infinitely slow intramolecular reorientation
τ_r	molecular reorientation time
$\tau_{ra}, \tau_{rb}, \tau_{rc} \ldots$	molecular reorientation time of a solvent molecule in sub-liquids $a, b, c \ldots$
$\tau_r^o, \tau_r^+, \tau_r^-$	molecular reorientation time of a solvent molecule in the free solvent, the cationic and anionic solvation sphere respectively
τ_d	time constant of translational jump motion
$\tau_d^{(I)}, \tau_d^{(S)}$	time constant of translational jump motion for spins I and S respectively
$\bar{\tau}_d^{(I)}, \bar{\tau}_d^{(S)}$	concentration dependent translational jumping time of spins I and S respectively

$\bar{\tau}_d^{(IS)}$	$\frac{1}{2}(\bar{\tau}_d^{(I)} + \bar{\tau}_d^{(S)})$
τ_{rs}	correlation time for spin-rotational interaction
τ_s	electron spin relaxation time
τ_e	$\dfrac{1}{\tau_e} = \dfrac{1}{\tau_s} + \dfrac{1}{\tau_b'}$
$\tau_1, \tau_2,$	time constants of infinitesimal step rotational diffusion
τ_2'	time constant of infinitesimal step rotational diffusion of the rigid first solvation sphere
τ_0	time constant for intramolecular rotation about an axis fixed in the molecule
τ_c'	$\dfrac{1}{\tau_c'} = \dfrac{1}{\tau_2'} + \dfrac{1}{\tau_b'}$
ζ	number of neighbour spins at distance R' in the rigid first solvation sphere

1. INTRODUCTION

If we consider a snapshot of the molecular arrangement in a liquid in internal and chemical equilibrium at time $t = 0$, we will find the liquid to be a "heap"[1] made up of a very great number of molecules. We assume that the molecular system which forms the liquid is wholly subject to the laws of classical mechanics and thus we are able to ascribe a well-defined position to each molecule and, if the particles are not of complete spherical symmetry, to determine for each molecule the orientation of a given set of vectors fixed within the molecule relative to a coordinate system fixed in space (the laboratory system). After a time t we perform a second snapshot and generally we will find that the positions and the orientations of the molecules have changed. For instance, we count the number of particles which are displaced by a vector $\mathbf{r}$ after the time t has elapsed and divide this number by the total number (very large) of molecules in the system. If the system is sufficiently large the result is the probability $\mathscr{P}(\mathbf{r}, t)$ that a given particle is displaced by $\mathbf{r}$ after a time t. In the same way we may determine the probability $P(\Omega, t)$ that a given molecule has performed a rotation Ω after the time t—where it is immaterial for the present discussion in which way we specify the rotation Ω. Conversely, if we know the probabilities $\mathscr{P}(\mathbf{r}, t)$ and $P(\Omega, t)$, we would be able to predict the average results of the microscopic observations just mentioned, or, in other words, we would know some important details of the microdynamic behaviour of the liquid.

It is the purpose of the present article to show the extent to which nuclear magnetic resonance relaxation time measurements have so far contributed to our knowledge of the microdynamic behaviour of liquids in the sense just defined. It may be stated right at the beginning that no precise presentation of any of the explicit functions $\mathscr{P}(\mathbf{r}, t)$ or $P(\Omega, t)$ are available from NMR relaxation times. We can, however, obtain certain more limited quantities which are related to $\mathscr{P}(\mathbf{r}, t)$ and $P(\Omega, t)$ as will be seen below.

If the liquid under consideration is a mixture of different molecules, say A, B, C, ..., new aspects appear. In the first place we have now probabilities $\mathscr{P}_A(\mathbf{r}, t)$, $\mathscr{P}_B(\mathbf{r}, t)$ ... for each kind of molecule and probabilities $P_A(\Omega, t)$, $P_B(\Omega, t)$ ... for all those particles which are not of complete spherical symmetry. Moreover, we may be able to define different classes of molecules of the same species, say A, the different classes being characterized by different probability functions. To be more specific, we consider a rather dilute solution of particles B in A. If on determining $\mathscr{P}_A(\mathbf{r}, t)$ and $P_A(\Omega, t)$ we sort out only those molecules which are in immediate contact with particles B, i.e. which are members of the first solvation sphere, we obtain functions $\mathscr{P}'_A(\mathbf{r}, t)$ and $P'_A(\Omega, t)$ which are different from those which we would observe for molecules far away from any particle B. A typical example is the determination of the microdynamic behaviour in the hydration sphere of ions, where NMR relaxation time measurements have been applied to the study of molecular motions in electrolyte solutions. The present article will present some of the results so far available.

There is another question of interest: if we have a pair of individual molecules which are closest neighbours at zero time, what is the probability that after a time t has elapsed they are still closest neighbours? In pure liquids, it has not yet been possible to determine this quantity reliably by relaxation time measurements although such information would be of great interest, while for ionic solutions, relaxation time measurements have been successful. Thus, average life-times of molecules in the first solvation shell of ions can be measured. The results of these investigations will be reported here.

However, it is not possible to describe all processes of this type and the discussion will be limited to those cases where the particle in the first coordination shell of an ion is a molecule of the solvent A. This means that all microdynamical processes which are conventionally called chemical reactions are excluded here. They form the subject of chemical kinetics of reactions (generally fast) in solution. By discussing only processes of solvent molecules we introduce rather arbitrarily a borderline to chemical kinetics which, of course, in itself is not clearly defined. Moreover, we omit processes where the molecule known as a stable entity in the gas phase exchanges one or more of its atoms or nuclei with other molecules of the same kind in the liquid (for example, protolytic exchange reactions). Finally, microdynamic events where the molecule changes its geometrical configuration from one possible state to another or where only the geometrical configuration of a specified atom relative to other specified atoms in the same molecule changes will not be treated here. Both these subjects as well as the foregoing one are topics of reaction kinetics. A great number of NMR investigations have been performed to study these processes but relaxation time measurements in a more restricted sense are of minor importance.

So far the changes of positions and orientations of molecules with time in the liquid have been mentioned. Of course, in the same way the changes of velocity and angular momentum with time are of interest and the probability $f(\mathbf{v}_0, \mathbf{v}, t)$ that a particle known to have the velocity $\mathbf{v}_0$ at $t = 0$ has the velocity $\mathbf{v}$ at time t and the probability $\phi(\mathbf{J}_0, \mathbf{J}, t)$ that a particle which had an angular momentum $\mathbf{J}_0$ at $t = 0$ has an angular momentum $\mathbf{J}$ at time t form important details of the complete description of the microdynamic behaviour of the liquid. The connection of NMR relaxation time measurements with these quantities will be briefly touched, the information available being rather limited.

We conclude this introduction with a brief consideration of the following question. Suppose we have the desired complete knowledge of the microdynamic behaviour of a liquid, is this knowledge of any value beyond the satisfaction which it gives in itself? Probably the most important point is that if we have information concerning the microdynamic behaviour we may hope to correlate this information with the structure of the liquid. The structure of a liquid is given by a set of molecular distribution functions which generally include the relative positions and orientations of molecules within a group of n molecules selected at random from the liquid ($n = 2$ pair distribution function). Thus slower or faster molecular motions would indicate the presence of more or less structure in the liquid.[2]

In a shorthand or "chemical" notation the knowledge of the microdynamical behaviour would be a tool to study association equilibria or weak complex equilibria in liquids. However, quantitative expressions are not yet available to relate microdynamical properties to true structural properties. Another important aspect which is related to the microdynamic behaviour as well as to the molecular structure of the liquid is the question concerning the interaction energy of a molecule with another one in the isolated state or with a group of other molecules in the liquid as a function of their relative positions and orientations. For example, if we find that a molecule performs a very rapid rotational motion we will conclude that the interaction energy with its surroundings is almost independent of its orientation.

2. BASIC THEORY

The magnetization $\mathbf{M}$ of a sample of N nuclear magnetic moments with spin I in the state of thermal equilibrium is a vector parallel to the static magnetic field of magnitude H_0. Identifying the direction of the static magnetic field as the z-direction we have

$$\mathscr{M}_z^0 = \frac{N\gamma^2\hbar^2 I(I+1)}{3kT} H_0,$$ (1)

$$\mathscr{M}_x^0 = \mathscr{M}_y^0 = 0.$$

If by any physical method we produce a nuclear magnetization different from that given by equation (1), then on removing the external stimulus the magnetization moves back to its equilibrium value. This motion consists of a contribution $\gamma\mathbf{M} \times \mathbf{H}_0$ caused by the torque of the static magnetic field on the magnetization and of a contribution due to the interaction of the nuclear moments with other particles in their molecular environment. The second of the two contributions implies the relaxation process of the magnetization. Especially in liquids the interaction energy of the nuclei with their neighbouring particles fluctuates statistically with time, since the relative position of the particles changes continuously due to their thermal motion. Now, qualitatively the rate of approach of the magnetization towards its equilibrium value increases with increase of the spectral intensity of the power spectrum of the fluctuating interaction energy at the resonance frequency $\omega = \gamma H_0$. The power spectrum itself is related to the time constants of the molecular motion or, in a somewhat modified version, to the microdynamic behaviour of the liquid and from this stems the possibility of obtaining information concerning the microdynamic behaviour of liquids from NMR relaxation studies.

We will not give here any further details of the theory but will simply present the formulae necessary for the discussion to be given in the next sections. Introductions to the theory may be found in a number of textbooks.[3-6]

2.1. *Intramolecular Relaxation Rate of a Two-spin System*

If we have a spin system where every molecule carries two like spins and if there exists only a magnetic dipole–dipole interaction between the nuclei—as, for instance, in water—the change with time of the z-component of the magnetization is given by the formula:

$$\frac{d\mathscr{M}_z}{dt} = -\frac{1}{T_1}(\mathscr{M}_z - \mathscr{M}_z^0).$$ (2)

The time constant T_1 of the exponential approach of $\mathcal{M}_z$ towards its equilibrium value is called the longitudinal relaxation time. $1/T_1$ is usually called the longitudinal relaxation rate and is composed of two parts, an intramolecular and an intermolecular part:

$$\frac{1}{T_1} = \left(\frac{1}{T_1}\right)_{\text{intra}} + \left(\frac{1}{T_1}\right)_{\text{inter}}. \tag{3}$$

In the next section we will show by which experimental procedure it is possible to separate the intramolecular relaxation rate from the intermolecular one. We discuss first the intramolecular contribution which originates from the interaction between the two spins within the same molecule. It may be shown that the following relation holds for $(1/T_1)_{\text{intra}}$[6]†

$$\left(\frac{1}{T_1}\right)_{\text{intra}} = \frac{4\pi}{5} \frac{\gamma^4 \hbar^2}{b^6} I(I+1) \left[\int_{-\infty}^{+\infty} e^{-i\omega t} g_1(t) dt + 4 \int_{-\infty}^{+\infty} e^{-i2\omega t} g_2(t) dt \right]. \tag{4}$$

Here b is the distance between the two nuclei with spin I, γ is their gyromagnetic ratio, ω is the resonance frequency of the nucleus, $g_{1,\,2}(t)$ is the time correlation function of the normalized spherical harmonics $Y_2^{1,\,2}(\theta, \phi)$, i.e.

$$g_m(t) = \overline{Y_2^m(\theta(0), \phi(0)) Y_2^{m*}(\theta(t), \phi(t))}, \quad m = 1, 2. \tag{5}$$

The bar indicates the average over a representative statistical ensemble of systems. Equation (5) may be written

$$g_m(t) = \frac{1}{4\pi} \int \int Y_2^{m*}(\theta, \phi) Y_2^m(\theta_0, \phi_0) P(\theta_0, \phi_0, \theta, \phi, t)$$
$$\times \sin \theta_0 \sin \theta \, d\theta_0 \, d\theta \, d\phi_0 \, d\phi, \quad m = 1, 2 \tag{6}$$

where $P(\theta_0, \phi_0, \theta, \phi, t)$ is the probability that the vector connecting the two spins in the molecule has an orientation relative to the laboratory system in the range $\sin \theta \, d\theta \, d\phi$ at θ and ϕ at time t when it had an orientation θ_0, ϕ_0 at the time $t = 0$.

$$\theta(0) \equiv \theta_0, \quad \phi(0) \equiv \phi_0, \quad \theta(t) \equiv \theta, \quad \phi(t) \equiv \phi.$$

The appearance of the quantity $P(\theta_0, \phi_0, \theta, \phi, t)$ in equation (6) is of crucial importance for our purpose. As mentioned in the introduction, the function $P(\Omega, t)$—the probability that a molecule has performed a rotation Ω after a time t—is one of the important elements for the description of the microscopic behaviour of a liquid. In equation (6) we find a similar function. It is not the

† This expression holds exactly for $I = \frac{1}{2}$, in other cases equation (4) is a good approximation.

probability function for the rotational motion of a rigid body but the less general probability function for the rotational motion of a vector fixed in the molecule which is the spin–spin vector in our case.

We summarize the connection between the diffusive rotational motion of a molecule and the intramolecular relaxation rate in the following way: all rotations of the molecule which leave the spin–spin vector unaffected do not contribute to the relaxation rate. The rotational motion of the spin–spin vector is described by a probability function $P(\theta_0, \phi_0, \theta, \phi, t)$ which serves to construct the time correlation function of the second order spherical harmonics (of the arguments θ and ϕ, equations (5) and (6)). In the next step the Fourier transform of this correlation function is calculated and the result is connected with the intramolecular relaxation rate by well-known properties of the nucleus and by the constant spin–spin distance.

It is clear from the foregoing that the measurement of one single parameter $(1/T_1)_{\text{intra}}$ cannot yield the complete function $P(\theta_0, \phi_0, \theta, \phi, t)$ and the available information about $P(\theta_0, \phi_0, \theta, \phi, t)$ must of necessity be very rudimentary. Thus, what one does is to propose a model for the rotational motion of the vector, the model being characterized by a set of parameters, one or more of which will appear in equation (4) and if one knows all other quantities the desired parameter can be calculated.

We will begin by considering the behaviour of a molecule with spherical symmetry, in which case the starting point for the indicated procedure is the diffusion equation for the *isotropic* rotational diffusion of a vector.

$$\partial P(\theta, \phi, t)/\partial t = D_r \Delta_S P(\theta, \phi, t), \tag{7}$$

Δ_S is the Laplacian operator on the surface of a sphere and D_r is the rotational diffusion coefficient. The solution of equation (7) which converges to a δ-function at $t = 0$ is

$$P(\theta_0, \phi_0, \theta, \phi, t) = \sum_{l, m} Y_l^{m*}(\theta_0, \phi_0) Y_l^m(\theta, \phi) e^{-tD_r l(l + 1)}. \tag{8}$$

The condition for equations (7) and (8) to hold is that the net rotation of the particle carrying the vector is the result of a very great number of erratic infinitesimal rotational steps. This condition will be satisfied for a big particle with almost macroscopic dimensions. However, it is doubtful to what an extent this description can be applied to the motion of a molecule in the liquid. To the author's knowledge, no one has attempted a rotational random walk calculation which has a true finite step mechanism.

A certain generalization of equation (8) is given by the following method: suppose that the molecule resides in an equilibrium position in the liquid for a certain time $0 < t' < t_1$. Within this time interval $P(\theta_0, \phi_0, \theta, \phi, t)$ is determined by equation (8). At $t = t_1$ the molecule performs a finite rotational jump, and immediately after this jump $P(\theta_0, \phi_0, \theta, \phi, t)$ is supposed to be

$$P(\theta_0, \phi_0, \theta, \phi, t_1) = \sum_{l, m} Y_l^{m*}(\theta_0, \phi_0) Y_l^m(\theta, \phi) e^{-D_r l(l + 1)(t_1 + \xi)} \tag{9}$$

$$= \sum_{l, m} Y_l^{m*}(\theta_0, \phi_0) Y_l^m(\theta, \phi) C_l e^{-D_r l(l + 1)t_1} \tag{10}$$

i.e.

$$C_l = e^{-D_r l(l + 1)\xi}; \quad 0 \leqslant \xi. \tag{11}$$

Thus the rotational jump at time t_1 is assumed to cause a probability distribution which is identical with the distribution that would have been developed at the time $t = t_1 + \xi$ if only true infinitesimal step diffusion were effective all the time. In the special case $C_l = 0$ for $l \geqslant 1$ we have at $t > t_1$

$$P(\theta_0, \phi_0, \theta, \phi, t) = 1/4\pi. \tag{12}$$

For times $t > t_1$ then the vector is assumed to continue a diffusive motion according to equation (8) until after a certain time it performs another finite rotational jump and so on. It may be shown[7] that the whole effect of these finite rotational jumps is a probability function:

$$P(\theta_0, \phi_0, \theta, \phi, t) = \sum_{l, m} Y_l^{m*}(\theta_0, \phi_0) Y_l^m(\theta, \phi) e^{-t\{l(l + 1)D_r + (1 - C_l)/r\}} \tag{13}$$

or

$$P(\theta_0, \phi_0, \theta, \phi, t) = \sum_{l, m} Y_l^{m*}(\theta_0, \phi_0) Y_l^m(\theta, \phi) e^{-t\{1/\tau_2 + 1 - C_l)/\tau\}} \tag{14}$$

where we have put

$$1/\tau_l = D_r l(l + 1). \tag{15}$$

Here τ is the average time between two finite rotational jumps. By equation (11) $0 \leqslant C_l \leqslant 1$; for $C_l = 1$ we have infinitesimal step diffusion, for $C_l = 0$ the probability distribution is determined by the relative magnitudes of τ_l and τ. If we introduce equations (14) and (15) into equation (6) we find the time correlation function of the spherical harmonics of second order:

$$g_{1, 2}(t) = \frac{1}{4\pi} e^{-t\{1/\tau_2 + C_1 - C_l/\tau\}} = \frac{1}{4\pi} e^{-t/\tau_c} \tag{16}$$

τ_c may be called here simply the correlation time, it is the time constant which characterizes the time dependence of the correlation function to be used here. For the derivation of equation (16) use has been made of the orthogonality relations of the spherical harmonics and it is seen that from the whole probability distribution only the term with $l = 2$ is picked out.

The Fourier transform of equation (4) is easily calculated, the result is

$$\left(\frac{1}{T_1}\right)_{intra} = \frac{2}{5} \frac{\gamma^4 \hbar^2}{b^6} I(I + 1) \left\{\frac{\tau_c}{1 + \omega^2 \tau_c^2} + \frac{4\tau_c}{1 + 4\omega^2 \tau_c^2}\right\}. \tag{17}$$

For systems with equal spins we are interested in the great majority of applications only in the case of extreme narrowing, i.e. $\omega^2 \tau_c^2 \ll 1$ and we obtain from equation (17)

$$\left(\frac{1}{T_1}\right)_{\text{intra}} = \frac{2\gamma^4 \hbar^2}{b^6} I(I+1)\tau_c. \tag{18}$$

We recall that the components (in spherical coordinates) of the spin–spin vector are proportional to the spherical harmonics of first order $Y_1^{0,1}(\theta, \phi)$. Thus, if we calculate the time correlation functions $f_{0,1}(t)$ of $Y_1^{1,0}(\theta,\phi)$ we find:

$$f_m(t) = \frac{1}{4\pi} \int \int Y_1^{m*}(\theta, \phi) Y_1^m(\theta_0, \phi_0) P(\theta, \phi, \theta_0, \phi_0, t)$$

$$\times \sin \theta \sin \theta_0 \, d\theta \, d\theta_0 \, d\phi \, d\phi_0, \quad m = 0, 1 \tag{19}$$

$$= \frac{1}{4\pi} e^{-t\{1/\tau_1 + (1-C_1)/\tau\}} = \frac{1}{4\pi} e^{-t/\tau_r}.$$

Since τ_r is the time after which the correlation between the values of a vector component is lost, we call τ_r the reorientation time or tumbling time of the molecule. By equation (15) we have $\tau_2 = \frac{1}{3}\tau_1$ and by equation (11) $C_2 = C_1^3$, furthermore, by using equations (16) and (19), one deduces the relation

$$\frac{1}{\tau_r} = \frac{1}{3}\frac{1}{\tau_c} + \frac{1}{\tau}\left\{\frac{2}{3} - C_1\left(1 - \frac{1}{3} C_1^2\right)\right\} \tag{20}$$

between the reorientation time τ_r and the correlation time τ_c. We see that $\tau_r = 3\tau_c$ if $C_1 = 1$ and

$$\tau_c = \frac{\tau_r \tau}{3\tau - 2\tau_r} = \frac{\tau_1 \tau}{3\tau + \tau_1} \tag{21}$$

if $C_1 = 0$ and finally, if $C_1 = 0$ and $\tau_1 \gg \tau$ we have $\tau_c = \tau$. Summarizing, we find that the knowledge of τ_c from the measured relaxation time and the spin–spin distance gives us the desired reorientation time of the molecule if reasonable assumptions about C_1 and τ can be made. Equation (20) yields a final result which gives us a piece of information concerning the microdynamic behaviour of the liquid. However, compared with the full probability function the information is rather rudimentary.

For τ_c the Debye approximation has been used often

$$\tau_c = \frac{4\pi a^{*3}\eta}{3kT} \tag{22}$$

where a^* is the radius of the molecule and η is the macroscopic viscosity. This approximation can easily lead to results which are wrong by one order of magnitude. The experimental value of τ_c is usually smaller than that given by equation (22). There are other theories which try to predict τ_c or similar quantities. Particularly for pure liquids almost all work in the literature is devoted to the comparison of the experimental τ_c with the results of these theories. However, the discussion of these theories is out of the scope of the present article which is dedicated to the experimental determination of τ_c.

In the literature the case has been tieated where the rotational diffusion coefficient becomes very great or, what means the same thing, when the rotational friction constant approaches zero.[8–10] If the friction constant is exactly zero, we would have a dynamically coherent reorientation process. For friction coefficients close to zero the correlation functions are given essentially by Gaussian functions[10]

$$f_0(t) \approx e^{-\tau^{*2}}$$

$$g_m(t) \approx e^{-3\tau^{*2}} \quad m = 0, 1, 2$$

$$\tau^* = t(kT/\bar{I})^{1/2}.$$

$\bar{I}$ is the average moment of inertia of the molecule. At extreme narrowing the relaxation rate is[10] ($I = \frac{1}{2}$):

$$\left(\frac{1}{T_1}\right)_{\text{intra}} = \frac{3}{2}\frac{\gamma^4\hbar^2}{b^6} \cdot \frac{1}{2}\left(\frac{\pi\bar{I}}{3kT}\right)^{1/2}.$$

Thus, in principle, the form of the relation for the relaxation rate remains unchanged, only a special expression for τ_c is introduced. If we calculate τ_c by equation (18) we may compare the result with the time $\frac{1}{2}(\pi\bar{I}/3kT)^{1/2}$. Usually, the latter will be smaller than τ_c by a factor 3 to 10. On the other hand, if the friction constant decreases more and more, then the spin-rotational interaction becomes the dominant relaxation mechanism (see below) and makes the pure dipolar part unobservable. Thus we consider the situation just described as a certain limiting case but in the following we do not mention it any further.

So far the correlation time τ_c occurring in equation (18) has a well-defined microdynamic meaning in terms of a molecular reorientation process only for isotropic rotational diffusive motion. This means that the particle has to be of essentially spherical shape and that the torques exerted by the molecular surroundings in the average must also have essentially spherical symmetry relative to the centre of the molecule. If the shape of the molecule deviates strongly from a sphere, there are generally three rotational diffusion coefficients about the three molecular axes or, in other words, three time constants describing the reorientation of the molecules about the three axes. Woessner[11] and Shimizu[12a] calculated independently the effect of the anisotropic rotational motion of a rigid body on the relaxation rate (see also references 8 and 9). Both authors applied the diffusion equation as given by Perrin,[13] Shimizu used an explicit solution of the diffusion equation whereas Woessner did not. (Another calculation of the rotational diffusive motion has been published by Favro,[14] while the special case of isotropic motion of a rigid body has been treated by Furry[15]).

It will be sufficient here to discuss an axially symmetric ellipsoid, i.e. a molecule to which two rotational time constants have to be ascribed. The measured correlation time now represents a mixture of contributions from both these times. Besides the relative magnitude of the two time constants the direction of the spin–spin vector relative to the long axis of the ellipsoid is of great importance. Clearly, if the spin–spin vector is parallel to the latter axis, τ_c measures directly the reorientation time of this axis if equation (20) is applied. If the spin–spin vector is perpendicular to the long axis—which will apply to the majority of examples—both authors[11, 12] give explicit expressions of $(1/T_1)_{intra}$ demonstrating the contributions of the two rotational diffusion coefficients D_{r_1} and D_{r_2} (their formulae apply for any direction of the spin–spin vector as well). Thus, making reasonable assumptions about D_{r_1}/D_{r_2} one is able to determine both D_{r_1} and D_{r_2} from the intramolecular relaxation rate. The following situation should be kept in mind: to describe the microdynamic behaviour of the liquid we are not primarily interested in the correlation times (of the second order spherical harmonics) but rather in the true molecular reorientation times. But from equation (20) we see that the calculation of the tumbling time from the correlation time introduces an uncertainty which may be estimated to be about a factor of 2. Now, only for "very long" molecules will the quotient D_{r_1}/D_{r_2} differ appreciably from unity. But it is these molecules which are very likely to undergo more or less complicated internal motions which invalidate the theory mentioned here. For smaller molecules the available quantitative results predict smaller effects and, of course, they make use of the infinitesimal step rotational diffusion equation. But for a molecule this is only a rough approximation and if the reorientation times about the two axes are different—i.e. the diffusion coefficients differ—it could still be that the correlation times are equal since the jumping mechanism is different. Hence for a molecule of moderate size and slightly ellipsoidal shape it seems wisest to consider as the *reorientation time of the molecule* a reorientation time as estimated by equation (20) from τ_c, keeping in mind that probably the reorientation about the longer axis is slightly faster and that about the shorter axis is somewhat more delayed.

There is another situation which is of interest for the interpretation of the correlation time of a spin–spin pair: suppose the spin–spin vector performs a rotational motion about an axis which is fixed in the molecular frame and the molecular frame itself undergoes an isotropic (or anisotropic) rotational diffusive motion. In what way is the observed correlation time τ_c connected to the time constants which determine these two separate and supposedly independent motions? Obviously, in those cases it is extremely important to know the answer to this question when we have to take into account the possibility that the reorientation of a certain group of atoms about an intramolecular axis is as fast as or even faster than the tumbling of the whole molecule. Finding a certain τ_c we thus would not know whether it concerns only the group which carries the spin pair or if we had any information about the tumbling of the whole molecule or at least of a vector within the molecule which is defined to represent its axis. Woessner[16] has given an answer to this question. We present here his result for $\omega^2\tau_c^2 \ll 1$:

$$(\tfrac{1}{T_1})_{intra} = \tfrac{3}{2}\gamma^4\hbar^2 b^{-6} I(I+1)[\tfrac{1}{3}(3\cos^2\varDelta - 1)^2\tau_{c_1} + (\sin^2 2\varDelta)\tau_{c_2} + (\sin^4\varDelta)\tau_{c_3}] \tag{23}$$

with

$$\frac{1}{\tau_{c_2}} = \frac{1}{\tau_{c_1}} + \frac{1}{\tau_0}$$

$$\frac{1}{\tau_{c_3}} = \frac{1}{\tau_{c_1}} + \frac{4}{\tau_0}$$

for infinitesimal step rotation about the internal axis,

$$\frac{1}{\tau_{c_3}} = \frac{1}{\tau_{c_2}}$$

for finite rotational jumps about the internal axis between three equilibrium positions. Here τ_0 is the time constant which describes the intramolecular rotation of the spin–spin vector, τ_{c_1} is the correlation time one would observe if the spin–spin vector lies in the fixed molecular axis, $\varDelta$ is the angle between the spin–spin vector and the fixed axis. Equation (23) concerns isotropic motion of the whole molecule. Woessner has generalized his results for anisotropic motion of the molecule, the result is analogous to equation (23) and will not be given here.[11] From equation (23) we observe that for $\varDelta = 0°$ we get the same result as equation (18) with $\tau_c = \tau_{c_1}$, thus we observe the isotropic motion of the whole molecule. As another limiting case we take $\varDelta = 90°$, we have from equation (23) for finite jump internal rotation of the spin pair:

$$\left(\frac{1}{T_1}\right)_{intra} = \tfrac{3}{2}\gamma^4\hbar^2 b^{-6} I(I+1)\tau_{c_1}\left(\frac{1}{3} + \frac{\tau_0}{\tau_{c_1} + \tau_0}\right) \tag{24}$$

giving

$$\left(\frac{1}{T_1}\right)_{intra} = 2\gamma^4\hbar^2 b^{-6} I(I+1)\tfrac{1}{4}\tau_{c_1} \quad \text{for } \tau_0 \ll \tau_{c_1} \tag{24a}$$

and

$$\left(\frac{1}{T_1}\right)_{\text{intra}} = 2\gamma^4\hbar^2 b^{-6}I(I+1)\tfrac{5}{8}\tau_{c_1} \quad \text{for } \tau_0 = \tau_{c_1} \tag{24b}$$

and

$$\left(\frac{1}{T_1}\right)_{\text{intra}} = 2\gamma^4\hbar^2 b^{-6}I(I+1)\tau_{c_1} \qquad \text{for } \tau_0 \gg \tau_{c_1}. \tag{24c}$$

Thus we get the correct limiting case (equation (24c)) for the completely rigid molecule. For a molecule where the motion about the fixed internal axis is very fast compared with τ_{c_1} the true reorientation time of the whole molecule is four times as long as the one we would obtain if we calculated it from the correlation time not knowing that intramolecular reorientation is present. We see that according to Woessner's theory the uncertainty in a molecular reorientation time which stems from a possible intramolecular rotation of a (single) spin pair with $\Delta = 90°$ is maximal at a factor of 4. For angles $\Delta \sim 55°$, the uncertainty is greater.

It must be mentioned, however, that Woessner's theory is only an approximation. Instead of calculating the correct modified probability distribution function $P(\theta_0, \phi_0, \theta, \phi, t)$ the orientation dependent functions occurring in the theory of dipolar interaction are split up in one part referring to the intramolecular motion and another part referring to the motion of the whole molecule. Then, for the averaging process the two independent probability distributions of the two different types of motion are used. It can be shown that Woessner's method does not yield the correct limiting value as indicated by equation (24c) if a consequent calculation of the correlation functions is performed.[17] Thus, no accurate theory is as yet available to be substituted for equations (23) and (24)—however, we have to use them in the absence of more accurate equations.

2.2. The Intermolecular Relaxation Rate

We ask next, what information concerning the microdynamic behaviour is given by the intermolecular relaxation rate of a system of like spins interacting through their magnetic dipoles (see equation (3)). We write the formula for the intermolecular relaxation rate[6]

$$\left(\frac{1}{T_1}\right)_{\text{inter}} = \frac{4\pi}{5}\gamma^4\hbar^2 N_I\left(\int_{-\infty}^{+\infty} \tilde{g}_1(t)e^{-i\omega t}\,dt + 4\int_{-\infty}^{+\infty} \tilde{g}_2(t)e^{-2i\omega t}\,dt\right). \tag{25}$$

Equation (25) refers to a system of N_I spins, i.e. we have the total relaxation of a spin I caused by the uncorrelated motion of $N_I - 1$ other spins with which $N_I - 1$ spin pairs are formed (all partner spins with spin I). The term $\tilde{g}_{1,\,2}(t)$ is the time correlation function of the functions:

$$Y_2^m(\theta, \phi)/r^3, \quad m = 1, 2,$$

i.e.

$$\tilde{g}_m(t) = \int \int \frac{Y_2^{m*}(\theta, \phi) Y_2^m(\theta_0, \phi_0)}{r^3 r_0^3} p(\mathbf{r}_0) \mathscr{P}(\mathbf{r}_0, \mathbf{r}, t) \, d\mathbf{r}_0 \, d\mathbf{r}. \qquad (26)$$

Here $\mathscr{P}(\mathbf{r}_0, \mathbf{r}, t)$ is the probability that the vector connecting a pair of two spins has the value $\mathbf{r}$ (with polar coordinates r, θ, ϕ) at time t if it had the value $\mathbf{r}_0$ (with polar coordinates r_0, θ_0, ϕ_0) at time $t = 0$. The term $p(\mathbf{r}_0)$ is the probability that the spin–spin vector takes on the value $\mathbf{r}_0$ at $t = 0$, and apart from short distance ordering effects it is equal to $1/V$, the reciprocal volume of the system. The difference of equation (26) as compared with equation (4) and equation (6) is that in the latter case b is a constant distance whereas here r is also a fluctuating quantity.

Clearly the probability $\mathscr{P}(\mathbf{r}_0, \mathbf{r}, t)$ is one of the functions which characterize the microdynamic behaviour of the liquid. However, we must be prepared to obtain not very accurate information from equations (25) and (26), since $\mathscr{P}(\mathbf{r}_0, \mathbf{r}, t)$ concerns all possible vectors $\mathbf{r}_0$ and $\mathbf{r}$, that is, also vectors of macroscopic length. For vectors of this length we expect that the motion of the vector is determined simply by the macroscopically measurable self-diffusion coefficient. On the other hand, for small values of $\mathbf{r}$, the term $\mathscr{P}(\mathbf{r}_0, \mathbf{r}, t)$ would yield interesting information. But here the question arises as to what extent the nearby vectors contribute to $\tilde{g}_m(t)$ since all vectors $\mathbf{r}_0$ and $\mathbf{r}$ have to be considered and, even if there is an appreciable contribution, how good is the approximation of uncorrelated translational motion of all spins implied in equation (25)? Other difficulties will be discussed below.

Torrey,[18] employing the results of the random walk theory given by Chandrasekhar,[19] has calculated $\mathscr{P}(\mathbf{r}_0, \mathbf{r}, t)$ in terms of two other, more elementary microscopic quantities: (i) the average time between two translational jumps τ_d, i.e. the probability that the particle performs a jump in dt at t, is

$$e^{-t/\tau_d} \frac{dt}{\tau_d};$$

(ii) $\mathscr{P}_1(\mathbf{r})$, the probability that a given jump displaces the particle by a vector $\mathbf{r}$. One could identify τ_d with the time τ already used for the discussion of the rotational jumps since it is reasonable to assume that a translational jump causes a rotational jump at the same time. For identical spins we are again only interested in the case of extreme narrowing $(\omega^2 \tau_d^2 \ll 1)$. Torrey[18] obtained the following results:

$$\mathscr{P}(\mathbf{r}, t) = \frac{1}{8\pi^3} \int \exp\{-i\mathbf{r} \cdot \boldsymbol{\rho} - (t/\tau_d)[1 - A(\boldsymbol{\rho})]\} \, d\boldsymbol{\rho}$$

$$\mathscr{P}(\mathbf{r}_0, \mathbf{r}, t) = \mathscr{P}(\mathbf{r} - \mathbf{r}_0, 2t)$$

$$\left(\frac{1}{T_1}\right)_{\text{inter}} = \gamma^4 \hbar^2 I(I + 1) \frac{4\pi C_I \tau_d}{\mathring{a}^3} \int_0^\infty \frac{(J_{3/2}(a\rho))^2}{1 - A(\rho)} \cdot \frac{d\rho}{\rho}. \qquad (27)$$

C_{I_3} is the concentration of spins (i.e. the number of spins per cm³), $J_{3/2}$ is the Bessel function of order $\frac{3}{2}$ and $\mathring{a}$ is the distance of closest approach between two spins. $A(\rho)$ is the three-dimensional Fourier transform of $\mathscr{P}_1(\mathbf{r})$:

$$A(\rho) = \int \mathscr{P}_1(\mathbf{r}) \exp (i\rho \cdot \mathbf{r}) \, d\mathbf{r}.$$

We give some examples for $A(\rho)$:

$$\mathscr{P}_1(r) = \frac{1}{(2\pi\langle r^2\rangle/3)^{3/2}} \exp (-3r^2/2\langle r^2\rangle) \tag{28}$$

$$A(\rho) = \exp (-\rho^2\langle r^2\rangle/6) \tag{29}$$

$\langle r^2\rangle$ is the mean square displacement of the particle caused by one jump.

$$\mathscr{P}_1(r) = \delta(r - l)/4\pi l^2 \tag{30}$$

$$A(\rho) = (\sin l\rho)/l\rho, \tag{31}$$

l is the constant length of flight associated with one jump.

$$\mathscr{P}_1(r) = \frac{1}{(4\pi D\tau_d r)} \exp (-r/(D\tau_d)^{1/2})$$

$$= \frac{1}{2\pi\langle r^2\rangle r/3} \exp (-r(6/\langle r^2\rangle)^{1/2}) \tag{32}$$

$$A(\rho) = \frac{1}{1 + D\tau_d\rho^2} = \frac{1}{1 + \langle r^2\rangle\rho^2/6}. \tag{33}$$

D is the self-diffusion coefficient of the particle. The last formula for $\mathscr{P}_1(r)$, equation (32), has the advantage that it yields a closed expression for $(1/T_1)_{\text{inter}}$. After substitution of equation (33) into equation (27) one has the result:

$$\left(\frac{1}{T_1}\right)_{\text{inter}} = \frac{4\pi}{3} \gamma^4\hbar^2 I(I + 1) \frac{C_I\tau_d}{\mathring{a}^3} \left(1 + \frac{2\mathring{a}^2}{5D\tau_d}\right). \tag{34}$$

Thus with known distance of closest approach and self-diffusion coefficient the microdynamical quantity τ_d may be calculated from $(1/T_1)_{\text{inter}}$. Since $D\tau_d = \langle r^2\rangle/6$ one sees from equation (34) that the information concerning τ_d improves the smaller $\mathring{a}^2/\langle r^2\rangle$ is. If one inserts one of the other expressions for $A(\rho)$ into equation (27) the numerical calculation shows that the value of the resulting relaxation rate is essentially unchanged. Thus in the limit of extreme narrowing, $(1/T_1)_{\text{inter}}$ is not sensitive to the details of the translational jump mechanism. The same is true if the molecular pair distribution function $g(r)$ is introduced into equation (26) instead of $p(r_0) = V^{-1}$, which has been done by Oppenheim and Bloom.[20] These authors start from a generalized Langevin equation to describe the motion of a pair of particles. They find a remarkably small dependence of $(1/T_1)_{inter}$ on variations of the pair distribution function. This general insensitivity to microdynamical and structural details was anticipated by the comment following equation (26). It may be remarked that for slow molecular motions far away from the condition $\omega^2 \tau_d^2 \ll 1$, this situation no longer holds.

Equation (34) is based on the model of the uncorrelated motions of pairs of like spins where one spin, the reference spin, is common to all pairs. With regard to this reference spin the system has complete spherical symmetry. The distance of closest approach $\overset{\circ}{a}$ defines a sphere into which none of the pair-partners of the reference spin can penetrate.† Now, the intramolecular relaxation rate as given by equation (18) also concerns a spin pair where both spins are fixed in the same molecule. Thus if one of these two spins is considered to be the reference spin of the translational problem, the model of translational motion cannot have spherical symmetry and the distance of closest approach between the spins is a function of the coordinates of the molecular encounter specified by a coordinate system fixed in the molecule. It follows that the effective distance of closest approach will lie somewhere between a distance which is really the closest distance to which the two spins can approach and the molecular diameter. Furthermore, the translational motion of the two spins sitting on the same molecule is always correlated, i.e. for all distances relative to the reference spin. This has the consequence that the time constant of the translational motion as determined from the macroscopic self-diffusion coefficient is longer than the one the reference spin "sees" in its close neighbourhood, because here the translational motion is superposed on the rotational diffusion. As already stated, the motion of two spins is correlated, but for the close neighbourhood of the reference spin this is also the case for spherical particles with one spin in their centre. The special nature of correlation is, of course, different but, since we saw that for the relaxation rate under conditions of extreme narrowing the special nature of translational jumps is of little importance, it is very likely that there is only a slight decrease of the relaxation rate as compared with that resulting from equation (34) with a equal to the effective distance of closest approach. And indeed this qualitative result is in agreement with a calculation concerning the same subject performed by Hubbard.[21] Hubbard's treatment, however, is not strictly correct, since the author does not introduce a new probability

† According to Torrey, $\mathscr{P}(\mathbf{r}_0, \mathbf{r}, t)$ is derived by forming the product $\mathscr{P}(\mathbf{r}_i, t)\mathscr{P}(\mathbf{r}_j - \mathbf{r}_0, t)$. Indices i and j denote the two pair partners, $\mathbf{r} = \mathbf{r}_j - \mathbf{r}_i$. But $\mathscr{P}(\mathbf{r}_i, t)$ and $\mathscr{P}(\mathbf{r}_j - \mathbf{r}_0, t)$ are calculated under the condition that each point of space can be taken on by the respective particle. Thus, in spite of the fact that for calculating $(1/T_1)_{inter}$ we confine $\mathbf{r} - \mathbf{r}_0$ to values with magnitude $\geqslant \overset{\circ}{a}$, the corresponding probability actually originates from a process where both particles can be simultaneously at the same position. This shows once more the delicate character of Torrey's theory neglecting correlation effects in the motion of the particles. Other theories of the intermolecular relaxation rate are not free from this difficulty.

function $\mathscr{P}(\mathbf{r}_0, \mathbf{r}, t)$ which takes account of the rotational motion superposed on the translational one. Essentially, he splits the position functions occurring in the expression for the interaction energy into two parts, and then introduces independent ordinary free rotational and translational diffusive motion for these two parts. The procedure is similar to that used by Woessner.[16]

2.3. *The Transversal Relaxation Time of a System of Pairs of Like Spins*

On the previous pages we described the parameter T_1 which determines the motion of the z-component of the magnetization to its equilibrium value. The description of the change in time of the transversal components is very much facilitated if we observe it in a coordinate system rotating with the resonance frequency about the z-axis, z being the direction of the static magnetic field. Then it may be shown that the following relations hold:[6]

$$\frac{d\mathscr{M}_x}{dt} = -\frac{1}{T_2}\mathscr{M}_x \qquad (35)$$

$$\frac{d\mathscr{M}_y}{dt} = -\frac{1}{T_2}\mathscr{M}_y$$

if we have again magnetic dipole–dipole interaction between a pair of two like spins. T_2 is called the transversal relaxation time. For conditions of extreme narrowing[6] $T_1 = T_2$ and thus measurements of T_2 yield no additional information.

2.4. *The Intramolecular Relaxation Rate for Molecules with more than Two Like Spins*

In a molecule with n equal spins (all spin I) each nucleus interacts with $n-1$ other nuclei in the same molecule. Again the formulae for the relaxation of a spin pair would be applicable were there not the complication that all the $n-1$ spin pairs perform correlated motions during the tumbling process of the whole molecule. A great number of calculations based on the semi-classical density matrix theory have been undertaken with the result that the decay of any non-equilibrium component of the nuclear magnetization is no longer exponential as given by equations (2) and (35) for the two-spin system (like spins). Instead of this, the change in time of $\mathscr{M}_z - \mathscr{M}_z^0$ and $\mathscr{M}_\perp$ is represented by a sum of several exponentials—and in certain cases these are even modulated by sine and cosine functions[22]—where the number of terms depends on (i) the symmetry of the molecular arrangement, (ii) the correlation time as compared to the resonance frequency, (iii) the degree of anisotropy of the molecular motion.

Fortunately, the calculations showed that in all cases of interest for us (extreme motional narrowing) the exponentials except one are only very small corrections to just this leading one. And indeed deviations from the exponential decay have never been observed in the pertinent experiments. Thus to a very good approximation the decay of the magnetization is determined by one exponential with the time constant

$$\frac{1}{T_{1,\,2}} = \frac{1}{n} \sum_{i<j} \frac{2}{T_{ij}}. \tag{36}$$

Here T_{ij} is the relaxation time one would calculate for the spin pair formed from spin i and j according to equation (18) and the sum includes all pairs of like spins in the molecule. For each of the rates $1/T_{ij}$ one has the same correlation time τ_c. For a molecule which undergoes isotropic rotational diffusion and where no intramolecular reorientations occur, τ_c is connected to the reorientation time of the molecule by equation (20). Thus if the geometry of the molecule is known and the rigidity of the molecule warranted, the reorientation time can be calculated (or estimated) from the measured intramolecular relaxation rate by equations (36), (18), and (20).

The detailed form of the decay of the magnetization is not of interest here. Three and four spin systems with different degrees of symmetry have been treated by Hubbard,[23, 24] Kattawar and Eisner,[25] Schneider,[22, 26, 27] Runnels,[28] Hilt and Hubbard.[29] Formulae for the exact time dependence of the components of the magnetization may be found in their papers. Generally the deviation from the pure exponential decay is greater for less symmetrical spin arrangements and for slower and more anisotropic motions. Fenzke[30] was able to show that for any arrangement of n like spins in a molecule the decay of the longitudinal magnetization for small times (up to the term linear in t) is determined by a time constant according to equation (36). Moreover, for all proton configurations in benzene (partly substituted by deuterium) he gives numerical values for the deviation of the actual term in t^2 from the one to be expected if an exact exponential decay would prevail. A maximum value of 0·44 for the relative difference in the quadratic terms occurs in his table for $(\omega\tau_c)^2 \ll 1$. Longer correlation times give no greater deviations.†

Finally, we should remark that for the intramolecular rotation of a group of spins the same situation holds as described for the spin pair[16, 27] and that Woessner[31] treated the case that within a molecule two interacting spins perform motions relative to one another, i.e. one spin is fixed in the "rigid" molecule and the other moves with a rotating group. As is to be expected the author finds that the correlation time is shortened as compared with the one resulting from the reorientation of the whole molecule. The corresponding formulae are complicated.

For the intermolecular relaxation rate of molecules with several spins no

† *Note added in proof.* See also a recent paper by T. G. Powles and R. Figgins, *Ann. Physik,* 7. Flg. **19**, 84 (1967).

new aspects appear. Again the approximation of uncorrelated motion of all spins has to be used and for the distance of closest approach a reasonable mean value has to be assumed.

2.5. *The Relaxation in Molecules with Non-identical Spins*

We consider a molecule which carries two different spins I and S interacting by magnetic dipole–dipole interaction. The sum of all spins I forms the magnetization $\mathbf{M}^{(I)}$ and the sum of all spins S forms the magnetization $\mathbf{M}^{(S)}$. Then the theory shows that the change with time of the z-component of the two magnetizations is given by[6, 32]

$$\frac{d\mathcal{M}_z^{(I)}}{dt} = -\frac{1}{T_1^{II}}(\mathcal{M}_z^{(I)} - \mathcal{M}_z^{(I)°}) - \frac{1}{T_1^{IS}}(\mathcal{M}_z^{(S)} - \mathcal{M}_z^{(S)°}) \tag{37a}$$

$$\frac{d\mathcal{M}_z^{(S)}}{dt} = -\frac{1}{T_1^{SS}}(\mathcal{M}_z^{(S)} - \mathcal{M}_z^{(S)°}) - \frac{1}{T_1^{SI}}(\mathcal{M}_z^{(I)} - \mathcal{M}_z^{(I)°}). \tag{37b}$$

The subscript zero indicates the equilibrium value.

For the time constants occurring in equation (37) the theory yields the following results if only the intramolecular part is considered:

$$\left(\frac{1}{T_1^{II}}\right)_{\text{intra}} = \frac{2\gamma_I^2\gamma_S^2\hbar^2 S(S+1)}{5b^6}$$

$$\times \left\{\frac{1}{3}\frac{\tau_c}{1+(\omega_I-\omega_S)^2\tau_c^2} + \frac{\tau_c}{1+\omega_I^2\tau_c^2} + \frac{2\tau_c}{1+(\omega_I+\omega_S)^2\tau_c^2}\right\} \tag{38a}$$

$$\left(\frac{1}{T_1^{IS}}\right)_{\text{intra}} = \frac{2\gamma_I^2\gamma_S^2\hbar^2 S(S+1)}{5b^6}$$

$$\times \left\{\frac{2\tau_c}{1+(\omega_I+\omega_S)^2\tau_c^2} - \frac{1}{3}\frac{\tau_c}{1+(\omega_I-\omega_S)^2\tau_c^2}\right\} \tag{38b}$$

and similar expressions for $(T_1^{SS})_{\text{intra}}$ and $(T_1^{SI})_{\text{intra}}$ where the indices I and S are interchanged. ω_I and ω_S are the respective observed resonance frequencies of spin I and S:

$$\omega_I = \gamma_I H_0$$

$$\omega_S = \gamma_S H_0.$$

For two spins to be unlike it is necessary that the effective resonance frequencies at a given H_0 are sufficiently different, i.e.

$$|\omega_I - \omega_S| \gg \frac{1}{T_2^{(I)}}, \quad \frac{1}{T_2^{(S)}} \quad \text{(see below)}$$

τ_c is again the correlation time describing the decay of the correlation between the second order spherical harmonics as the time goes on. It is the same quantity which appears in equation (17). I and S are the spins of the two nuclei, b is the inter-spin distance.

On the one hand we see from equation (37) that the motions of the two magnetizations are inter-related which, of course, complicates the evaluation of the parameters containing the correlation times. On the other hand, the coupling between the two magnetizations causes interesting effects which may be studied by double irradiation methods. However, these experiments will not be described here.

Fortunately there are many important cases where equation (37a) simplifies. This holds true if the spins not directly under observation—the spins S, say—interact strongly with their molecular surroundings (the "lattice" in the usual language of magnetic resonance spectroscopy) via interactions different from the magnetic dipole–dipole interactions with spin I. This has the consequence that $\mathscr{M}_z^{(S)} - \mathscr{M}_z^{(S)\circ}$ will always be practically zero. The strong interactions mentioned may be dipolar interactions with other magnetic species, quadrupole interactions, or, if S represents an electronic spin, the interaction of the latter with the "crystal field" within the molecular species bearing it. Apart from this, if spin S has an electric quadrupole moment and if $I = \frac{1}{2}$, S is greater than I and under conditions of extreme narrowing the following relation holds:

$$\left(\frac{1}{T_1^{II}}\right)_{\text{intra}} \bigg/ \left(\frac{1}{T_1^{IS}}\right)_{\text{intra}} = \frac{2S(S+1)}{I(I+1)} \gg 1.$$

Thus, in many cases of practical interest the second term may be dropped from the right-hand side of equation (37a) to a very good approximation. However, for equation (38a) to be correct the interaction of the spin S with its surroundings must be very much weaker than its Zeeman interaction with the static field.[32a] In particular, for certain paramagnetic ions, e.g. N_i^{2+} C_0^{2+}, F_e^{2+}, the extremely short electron spin relaxation time gives an experimental indication that the interaction of S with the crystal field is too strong. In these cases the interaction of S with the crystal field has to be included in the perturbation Hamiltonian of the two spin systems and equations (37) and (38) no longer hold (see below).

A comment is needed for the situation in which the difference in effective resonance frequencies is caused by a difference in "chemical" shielding of the same nuclear species. Then very often the measuring device does not distinguish between the magnetization $\mathbf{M}^{(I)}$ and $\mathbf{M}^{(S)}$ and measures the sum of both (e.g. T_1 measurements by $90°$–$90°$ pulse sequence). In this event regarding the longitudinal relaxation the spin system shows the behaviour of a system of like spins. However, the observed T_1 is $\neq T_2$ even if $\omega^2 \tau_c^2 \ll 1$.

From equation (17) we proceed immediately to the case of extreme narrowing since only this case is of interest for the present article. As already mentioned for unlike spins S can represent an electron spin and if this is so ω_S is about three orders of magnitude greater than ω_I and thus, even in highly fluid liquids, we easily attain the dispersion region of $1/T_1$ with ordinary magnetic fields, where the relaxation rate is frequency dependent according to equation (38a). This fact is of great importance, because by studying the frequency dependence of $1/T_1$ we are now able to measure directly the correlation time τ_c whereas under conditions of extreme narrowing, which holds in highly fluid liquids if both spins are nuclear spins, according to equations (18) or (38c) we always measure the product of an interaction term and the correlation time. For $(\omega_I + \omega_S)^2\tau_c^2 \ll 1$, equation (38a) reduces to

$$\left(\frac{1}{T_1^{(I)}}\right)_{\text{intra}} = \frac{2}{3} \cdot \frac{2\gamma_I^2\gamma_S^2\hbar^2 S(S+1)}{b^6} \tau_c. \tag{38c}$$

The decay of the transversal magnetization (in the rotating coordinate system) is exponential even for unlike spins.[6] Thus, for example,

$$\frac{d\mathscr{M}_x^{(I)}}{dt} = -\frac{1}{T_2^{(I)}} \mathscr{M}_x^{(I)}, \tag{39}$$

with an intramolecular contribution to $1/T_2^{(I)}$

$$\left(\frac{1}{T_2^{(I)}}\right)_{\text{intra}} = \frac{4\gamma_I^2\gamma_S^2\hbar^2 S(S+1)}{15\, b^6} \left\{\tau_c + \frac{1}{4}\frac{\tau_c}{1+(\omega_I-\omega_S)^2\tau_c^2}\right.$$
$$\left. + \frac{3}{4}\frac{\tau_c}{1+\omega_I^2\tau_c^2} + \frac{3}{2}\frac{\tau_c}{1+\omega_S^2\tau_c^2} + \frac{3}{2}\frac{\tau_c}{1+(\omega_I+\omega_S)^2\tau_c^2}\right\}. \tag{40}$$

By interchanging the indices I and S in equation (40) one obtains $(1/T_2^{(S)})_{\text{intra}}$. It may be seen from equations (38c) and (40) that for extreme narrowing one has again $T_1 = T_2$.

A quantitative calculation is not yet available for the intramolecular relaxation rate of a system of several identical and non-identical spins in one molecule. Approximations may be used taking account of the number of like and unlike spin pairs and of the reduction factor $\frac{2}{3}$ for the contribution of unlike spins.

The intermolecular relaxation rate of spin system I in a liquid with unlike spins under conditions of extreme narrowing is obtained if one adds to the term for like spins in equation (34) a term

$$\left(\frac{1}{T_1^{(I)}}\right)_{\text{inter}} = \frac{2}{3}\pi\gamma_I^2\gamma_S^2\hbar^2 S(S+1)\, \frac{C_S\tau_d^{(IS)}}{(\mathring{a}_{IS})^3}\left(1 + \frac{2}{5}\frac{(\mathring{a}_{IS})^2}{D_{IS}\tau_d^{(IS)}}\right). \tag{41}$$

It is reasonable to put $\tau_d^{(IS)} = (\tau_d^{(I)} + \tau_d^{(S)})/2$ and $D_{IS} = (D_I + D_S)/2$ where the indices I and S indicate the particles bearing spin I and spin S respectively. $\mathring{a}_{IS}$ is the (effective) closest distance of approach between spin I and S. For a pure liquid we have, of course, $\tau_d^{(I)} = \tau_d^{(S)}$ and $D_I = D_S$, in mixtures $\tau_d^{(I)}, \tau_d^{(S)}, D_I$, and D_S may be concentration dependent (see below).

For equation (41) to be correct the conditions for neglecting the coupling to the S-magnetization mentioned above have to be fulfilled. Again, if S is an electron spin, generally the case of extreme narrowing is not verified. Pfeifer[33] has derived the full frequency dependent expression for the inter-molecular relaxation rates $1/T_1$ and $1/T_2$ for unlike spins including the effect of electron spin relaxation. The resulting formulae are very complicated, for details the literature reference should be consulted (see also below).

2.6. Relaxation by Quadrupole Interaction

If $I > \frac{1}{2}$ the charge distribution of the nucleus is no longer spherically symmetric. Thus the nucleus possesses an electric quadrupole moment which interacts with the electric field gradient at the position of the nucleus produced by the surrounding electric charges. If the nucleus has an electrical environment which has cubic symmetry, the quadrupole interaction vanishes. In all other cases the quadrupole interaction causes an additional relaxation mechanism. However, as a consequence of molecular collisions in liquids the instantaneous electric field gradient will never have exact cubical symmetry. Consequently all nuclei with spin $I > \frac{1}{2}$ show a certain contribution to the total relaxation rate which originates from quadrupole interaction. In fact, there are a great number of systems where the relaxation effect due to magnetic dipole–dipole interaction is completely negligible when compared with the quadrupolar part. If the respective atom is bound covalently, nuclei with large quadrupole moments such as iodine and bromine are so strongly relaxed by quadrupole interaction that it is impossible to detect their magnetic resonance at all, unless their environment has approximate cubic symmetry.

The reason for the spin relaxation by quadrupole interaction is always the statistical fluctuation of the electric field gradient at the nuclear site caused by molecular motions. Again the spectral intensity of the interaction energy at the magnetic resonance frequency determines the rate of relaxation. In the present article we consider only the case of an intramolecular field gradient, i.e. the components of the field gradient tensor are fixed quantities in the molecular frame. Of course, in the laboratory system which defines the axis of quantization of the spin system, the tensor components fluctuate if the molecule undergoes reorientation processes. Under conditions of extreme narrowing—all other cases are not of interest for us—the following formula has been deduced:[6]

$$\frac{1}{T_1} = \frac{1}{T_2} = \frac{3}{40} \frac{2I + 3}{I^2(2I - 1)} \left(1 + \frac{\tilde{\eta}^2}{3}\right)\left(\frac{eQ}{\hbar} \cdot \frac{\partial^2 V}{\partial z'^2}\right)^2 \tau_c. \tag{42}$$

Here Q is the electric quadrupole moment of the nucleus, $\partial^2 V/\partial z'^2$ is the z-component of the field gradient in a coordinate system fixed in the molecule which is chosen to form the principal axes of the field gradient tensor. The asymmetry parameter, $\tilde{\eta}$, is given by

$$\tilde{\eta} = \frac{\partial^2 V/\partial x'^2 - \partial^2 V/\partial y'^2}{\partial^2 V/\partial z'^2},$$

τ_c is the time constant characterizing the correlation function of the eigenfunctions with quantum number $L = 2$ for the symmetrical top. Apart from a normalization factor, the latter functions are identical with the matrix elements of the rotation matrix $D_{m,m'}^L(\alpha, \beta, \gamma)$ of dimension $L = 2$[34] (α, β, γ = Euler angles describing the rotation of the rigid body constituting the molecule). If we assume that the molecule performs isotropic rotational motion, τ_c may be shown† to be the same correlation time which occurs in equation (18), that is, the correlation time of the spherical harmonics of order 2 for the isotropic motion of a vector. Thus, again equation (20) (or a better analogous one if available) connects the correlation time as determined by equation (42) with the reorientation time of any fixed vector inside the rigid molecule itself.

Equation (42) has one disadvantage and one advantage in comparison with equation (18). The disadvantage comes from the fact that $\partial^2 V/\partial z'^2$ and $\tilde{\eta}$ are unknown quantities, whereas equation (18) and the related formula for a molecule containing more than two spins (equation (36)) can be used to find τ_c if the geometry of the molecule is known. Thus, to evaluate equation (42) measured values of the quadrupole coupling constant $eQ\partial^2 V/\partial z'^2/\hbar$ and of $\tilde{\eta}$ have to be available. Very often, the neglect of $\tilde{\eta}$ is a good approximation. The advantage of equation (42) is given by the fact that we have good reasons to assume that an intermolecular contribution caused by quadrupole interaction is small compared with the intramolecular relaxation rate as given by equation (42). Consequently, an isotopic substitution and extrapolation technique to determine $(1/T_1)_{\text{intra}}$ (see below) should not be necessary —and would be quite useless since the field gradient does not depend on isotopic nuclear species. It might be mentioned that for particles where the electric field gradient has essentially cubic symmetry at the nuclear site (spherical ions), it is only the intermolecular relaxation rate which plays a dominant role.

If the nucleus under consideration sits in a group which undergoes intramolecular rotations relative to the rest of the molecule (which we may define as the whole molecule) the correlation time τ_c as calculated from equation

† The solution of the differential equation of isotropic rotational diffusion for a rigid body is an expansion in eigenfunctions of the symmetrical top, each term containing an exponential factor $e^{-L(L+1)D_r t}$, $L = 0, 1, 2, \ldots$. The term D_r is the same rotational diffusion coefficient which appears in equations (8) or (13). On formation of the correlation function due to the orthogonality relations of the indicated wave functions only the term with $L = 2$ is picked out.

(42) no longer refers to the whole molecule but is a mixture of contributions from the rotating group and the whole molecule. Zeidler[35] has treated the problem in question following the approach given in Woessner's paper concerning the same situation with magnetic dipole–dipole interaction.[16] The character of the approximation has been described above (see section 2.1). Zeidler's result for deuterons with spin $I = 1$ is the following:

$$\frac{1}{T_1} = \frac{3}{8} \left(\frac{eQ}{\hbar} \frac{\partial^2 V}{\partial z'^2} \right)^2$$

$$\times \left[\tfrac{1}{4}(3 \cos^2 \beta' - 1)^2 \tau_{c_1} + \{1 - \tfrac{1}{4}(3 \cos^2 \beta' - 1)^2\} \left(\frac{1}{\tau_{c_1}} + \frac{1}{\tau_0} \right)^{-1} \right], \quad (43)$$

$$\tilde{\eta} = 0.$$

β' is the angle between the principal axis z' and the intramolecular axis of rotation, the times τ_{c_1} and τ_0 have the same meaning as in equation (23), i.e. τ_{c_1} is the correlation time in the absence of internal rotation and τ_0 is the time constant of the intramolecular jumping process between three equilibrium positions. For a rotating methyl group one has $\beta' = 180° - 109\cdot5° = 70\cdot5°$ and equation (43) yields:

$$\frac{1}{T_1} = \frac{3}{8} \left(\frac{eQ}{\hbar} \frac{\partial^2 V}{\partial z'^2} \right)^2 \left[0\cdot11\tau_{c_1} + 0\cdot89 \left(\frac{1}{\tau_{c_1}} + \frac{1}{\tau_0} \right)^{-1} \right]. \quad (43a)$$

Thus if $\tau_0 \ll \tau_{c_1}$ the limiting result is

$$\frac{1}{T_1} = \frac{3}{8} \left(\frac{eQ}{\hbar} \frac{\partial^2 V}{\partial z'^2} \right)^2 0\cdot11\tau_{c_1}. \quad (43b)$$

For other values of I, equations (43) and (43a) have to be multiplied by a factor $(2I + 3)/5I^2(2I - 1)$.

Shimizu[12b] has studied the case when the molecule is a rigid ellipsoidal body undergoing an anisotropic reorientation.

2.7. Spin-rotational Interaction

Generally the spin quantum number I, the magnitude of the measured relaxation rate, and the nature of the system under consideration give us a unique indication of the dominant relaxation mechanism for a particular liquid example. If for a liquid diamagnetic sample of normal viscosity the relaxation times are less than about 1 sec and if $I > \tfrac{1}{2}$ we know that we are concerned with quadrupole interaction. If T_1 or T_2 are observed to be greater than about 1 sec, if $I > \tfrac{1}{2}$, and the quadrupole moment is small, then one has to reckon with both quadrupole and dipole interactions. However, even if $I = \tfrac{1}{2}$ one cannot a priori be sure that the magnetic dipole–dipole interaction is the only relaxation mechanism. In many cases known today spin-rotational interaction is present as well and contributes to the total relaxation rate. This effect comes about in the following way: when the molecule rotates, its charge distribution produces a magnetic field at the position of the nucleus and the magnetic moment of the nucleus interacts with this field. Of course, the magnetic field is dependent on the angular velocity of the molecule and since the latter is a quantity fluctuating in time we have a fluctuating interaction between the angular momentum of the molecule and the nuclear spin which causes the relaxation mechanism in question. The only property of this relaxation process which is of importance for us is the correlation time τ_{sr} defined as

$$\overline{\omega_k(0)\omega_k(t)} \sim e^{-t/\tau_{sr}}, \quad k = 1, 2, 3.$$

Here $\omega_k(0)$ is the kth component of the angular velocity vector of the rotating molecule at time zero, $\omega_k(t)$ is the kth component of its angular velocity at time t. The bar denotes an average over a statistical ensemble of identical systems. Thus τ_{sr} is the time constant characterizing the time correlation function of the components of the angular velocity vector. Hubbard[36] has developed the theory of relaxation by spin-rotation interaction in liquids and has shown that under the limiting conditions $\tau_{sr} \ll \tau_c$ (τ_c is the same correlation time used so far for magnetic dipolar relaxation) and $\omega^2 \tau_c^2 \ll 1$, the relaxation rate caused by spin-rotational interaction is proportional to τ_{sr}. Now, by using the Langevin equation for Brownian motion—applied to the angular velocity instead of linear velocity—Hubbard showed that $\tau_{sr} \sim D_r/T$. Thus, since D_r/T increases with increasing temperature, τ_{sr} also increases. This has the consequence that the relaxation rate due to spin-rotational interaction increases as the temperature increases which is opposite to the behaviour for relaxation by dipole–dipole interaction which is proportional to D_r^{-1}. So we get the result that the measurement of the temperature dependence of $1/T_1$ over a sufficiently great temperature range is necessary to decide which of the two interactions is the dominant one for nuclei with $I = \frac{1}{2}$: increasing or decreasing relaxation rate with increasing temperature means spin-rotational or dipole–dipole interaction respectively. The coupling constant for spin-rotational interaction is generally not available, the theoretical results hold only for certain limiting cases. Thus, we shall not report any further results of this topic here.

A relaxation mechanism which has certain common features with that caused by spin-rotational interaction will be mentioned briefly below when the results for liquid xenon are discussed.

The relaxation processes outlined so far are those which are of "practical" interest for us and which appear in pure liquids composed of strictly stable molecules—i.e. of molecules which do not exchange atoms or groups with one another. There are other relaxation mechanisms in pure liquids such as

G

scalar relaxation of the second kind and relaxation through anisotropic chemical shift—the latter has not yet been verified experimentally with certainty—which is not described here. There are, however, other mechanisms such as scalar relaxation of the first kind and other general forms of relaxation which have a great importance in mixtures and we will present the necessary formulae below when treating mixtures.

3. BRIEF OUTLINE OF EXPERIMENTAL METHODS

The experimental technique used to measure NMR relaxation times will only be sketched very briefly. The relaxation times of interest for the present article cover a range from about 10^{-4} sec to about 10^2 sec. The possible experimental methods may be divided into two different groups. In the first group the procedure is based on the possibility of measuring the magnitude of the nuclear magnetization vector—apart from a constant factor—directly. If this can be done, of course its change in time can also be studied. In order to measure the magnitude of the magnetization vector the latter is rotated into a plane perpendicular to the static magnetic field. Here it performs a precessional motion and induces a radiofrequency voltage of the resonance frequency ω in a coil, the axis of which is also perpendicular to the static field and which surrounds the sample containing the spins. The induced r.f. voltage is proportional to the magnitude of the magnetization. The rotation of the nuclear magnetization into the plane perpendicular to the static field is accomplished by the application of a r.f. pulse with frequency ω of appropriate duration (e.g. a "90° pulse"),[37] or by sweeping with the static field through the resonance with proper sweep rate and applying to the sample a steady magnetic r.f. field of frequency ω and of correct amplitude (adiabatic fast passage).[38] Thus if at $t = 0$ a magnetization zero or $-\mathcal{M}_z^0$ is produced (either by one of the methods just mentioned or otherwise) the growth of the magnetization towards equilibrium can be followed at any later time t by rotating it into the plane perpendicular to the static field. The decay of the transverse magnetization is in principle directly given by the approach of the r.f. voltage induced by the precessing magnetization towards the value zero. However, certain difficulties appear as a consequence of the inhomogeneity of the static field and the diffusive motion of the particles carrying the spins. Special methods have been worked out to circumvent these difficulties.[39, 40]

The other large group of measurements concerns all those systems where the field inhomogeneity measured in frequency units is very much smaller than the transversal relaxation rate. In this event T_2 is simply obtained from line width measurements under steady-state conditions. If it is known that $T_1 = T_2$, then the T_1 results are available in the same way. The simple relations hold

$$\frac{1}{T_2} = \tfrac{1}{2}\Delta\bar{\omega}_{1/2}$$ (44a)

where $\Delta\bar{\omega}_{1/2}$ is the line width measured in the (angular) frequency scale, or

$$\frac{1}{T_2} = \tfrac{1}{2}\sqrt{3}\Delta\bar{\omega}$$ (45a)

where $\Delta\bar{\omega}$ is the line width as measured between the two points of maximal slope on the absorption curve. If the line width is given in gauss, one has

$$\frac{1}{T_2} = \tfrac{1}{2}\gamma\Delta H_{1/2}$$ (44b)

$$\frac{1}{T_2} = \tfrac{1}{2}\sqrt{3}\gamma\Delta H.$$ (45b)

The separation of the intramolecular relaxation rate from the inter-molecular relaxation rate in pure liquids is accomplished by mixing the molecules under consideration with the like molecules but—if possible—consisting of isotopic nuclei which have zero or only very small magnetic moments. This method has been applied particularly to proton relaxation time measurements. The deuteron has a very small magnetic moment and consequently the dipolar interaction of the protons with the deuterons is negligible compared with the interaction between the protons. If in the limiting case the concentration of the normal proton-containing molecules gets vanishingly small we are left only with the interaction between the protons of the same molecule. Thus, measuring the relaxation rate of the protons as a function of proton content in a mixture with the fully deuterated but otherwise identical molecule and extrapolating to infinite dilution of the proton-containing molecule yields the intramolecular relaxation rate. The extension of this technique to mixtures is obvious. For measurements of long relaxation times all samples have to be freed carefully from any oxygen (since it is paramagnetic) dissolved in the liquid.

4. RESULTS

4.1. *Pure Liquids*

Xenon possesses one isotope with spin $\tfrac{1}{2}$ of sufficient isotopic abundance to make possible NMR relaxation time measurements. Thus one would suppose that liquid xenon is an ideal system to study the purely intermolecular dipolar relaxation mechanism and to evaluate the results with formula equation (34) or with similar expressions. However, it turns out[41] that the observed relaxation rate of ^{129}Xe is about two to three orders of magnitude larger than one would predict from the simple Bloembergen, Purcell, Pound theory[42] or from Torrey's more quantitative theory (equation (34)). Furthermore, T_1 varies very little with temperature, which is also completely unexpected. It follows that the relaxation mechanism here is quite different from that based on magnetic dipole–dipole interaction. The relaxation mechanism has been shown to be the following:[43, 44] during a molecular collision in the liquid the charge cloud of the xenon atom is distorted. At the same time, of course, this distorted charge cloud rotates as a consequence of the relative motion of the two colliding xenon atoms. The latter rotation produces a fluctuating magnetic field at the nuclear site which causes relaxation. Since the formulae describing this relaxation mechanism do not contain any quantities of microdynamic interest in the sense used here, we omit further discussion.

4.1.1. *Water*

At temperatures below about 200°C in H_2O the protons are relaxed by magnetic dipole–dipole interaction.[45] Above this temperature spin-rotational interaction contributes increasingly.[46-8] In D_2O which contains small amounts of HDO the spin-rotational contribution to the proton relaxation appears at lower temperature because the dipolar interaction between the proton and the deuteron is much weaker.[48]

The longitudinal proton relaxation time of H_2O at 25C° is 3·60 sec (see, for instance, Krynicki,[49] and literature cited therein). The uncertainty of this value is about ± 2 per cent. The water molecule is not a strictly stable molecule, the protons are exchanged in times short compared with the relaxation time. Thus it is not possible to measure the intramolecular relaxation rate separately by the D_2O dilution and extrapolation technique. However, the mean life-time of a given proton in a water molecule in pure (neutral) water at room temperature is about 10^{-3} sec.[50] Since we clearly expect correlation times for the thermal motion of the water molecules much shorter than this, we are allowed to use the formula for the intramolecular relaxation rate of a two-spin system. The longitudinal relaxation does not depend on the magnetic field which means that $\omega^2 \tau_c^2 \ll 1$ (extreme narrowing). In the first place we have to calculate $(1/T_1)_{\text{inter}}$ from equation (34) and then with equation (3), from equation (18) we determine the correlation time τ_c. The self-diffusion coefficient of water at 25°C is $D = 2·5 \times 10^{-5}$ cm^2 sec^{-1}.[51, 52] It is a reasonable estimate to put $\langle r^2 \rangle = a^2$, where $\langle r^2 \rangle$ is the mean square displacement of a water molecule caused by one translational jump, a is the diameter of the water molecule, $a = \mathring{a} = 2·8$ Å. Then, from $\langle r^2 \rangle = 6\tau_d D$, it follows that $\tau_d = 0·5 \times 10^{-11}$ sec,† C_I is $6·667 \times 10^{22}$ cm^{-3}

---

† Cold neutron scattering experiments seem to give slightly smaller values for τ_d, $0·15 \leqslant \tau_d \leqslant 0·4 \times 10^{-11}$, thus $\langle r^2 \rangle < a^2$.[53]

and using equation (34) one finds $(I = \frac{1}{2})$: $(1/T_1)_{inter} = 0.096$ sec^{-1}. The correction factor given by Hubbard[36] is omitted here, it would make $(1/T_1)_{inter}$ about 10 per cent greater. The proton–proton distance b is 1.52 Å, thus equations (3) and (18) yield $\tau_c = 0.27 \times 10^{-11}$ sec. For the calculation of the reorientation time of the water molecule the quantity C_1 in equation (20) has to be known. This is not the case and we have to be content to write the result $0.4 < \tau_r < 0.8 \times 10^{-11}$ sec. To judge the reliability of this result it should be kept in mind that it is based on the formula (34) of Torrey's theory, the reliability of which is not yet well established. However, we know from other liquids that the relative contribution of the inter- and intra-part to the total relaxation rate as calculated here is reasonable and thus we may be quite confident about the results given above. In the literature, slightly different results regarding the detailed description of the intermolecular relaxation rate will be found,[49] for instance Sharma and co-workers[54] used the results of the theory of cold neutron scattering, but essentially the outcome is the same in all cases.

Two brief comments should be made here; firstly, the reorientation time of the water molecule is relatively small, which is of significance for all structural models of water that use more or less extended clusters; secondly, the reorientation time is almost equal to the dielectric relaxation time of water (for details see references 49 and 55).

The plot of log $1/T_1$ against $1/T$ does not yield a straight line. Thus the "activation energy"† of the molecular processes responsible for the spin relaxation is temperature dependent, it varies from about 5 kcal mole^{-1} at 0°C to 3.5 ± 0.2 kcal mole^{-1} in the range between 40° and 100°C. It is interesting to note that the "activation energies" derived from corresponding plots for the viscosity, the dielectric relaxation time and the self-diffusion coefficient are very much alike. Actually, the presentation of activation energies is outside the scope of characterizing the microdynamic behaviour of a liquid as defined in the introduction. The so-called activation energy has no clear physical meaning in our model of a liquid as a heap of molecules—although the empirical existence of such a quantity in a great number of cases is very striking. Thus, we shall occasionally quote experimental values of activation energies to which we will give the somewhat loose meaning of being the change in potential energy of the molecule associated with the dynamical processes which causes the decay of the time correlation of the property under consideration. The change in potential energy occurs *somewhere* on the dynamical "path" of the molecule, it is not necessarily the difference in potential energy between a final and an initial state. The change in potential energy may occur repeatedly during the relevant process. For instance, for

† If not indicated otherwise, the activation energy E of a process is defined in such a way that for the corresponding time constant, the relation holds: $\tau = \tau_0 \exp(E/RT)$.

the time correlation of the components of a vector connecting two particles to vanish, generally very many translational steps of the molecules are necessary.

It is very tempting in certain cases to relate the activation energy to the degree of structure in the liquid. However, at the time being, only qualitative arguments are available to support this idea.[2]

In Table 1 the correlation time τ_c for water is given for different temperatures between 0 and 100°C. The proton relaxation rates as measured by Krynicki[49] have been used. The intermolecular relaxation rate has been calculated using equation (34). The self-diffusion coefficient is taken from Simpson and Carr's paper,[56] their results, however, have been corrected by a constant factor so as to yield $D = 2.5 \times 10^{-5}$ cm^2 sec^{-1} at 25°C. The value of $\langle r^2 \rangle$ was assumed to be constant and equal to a^2 (a = diameter of the water molecule). C_I varies with temperature as the density of water.

TABLE 1. CORRELATION TIME τ_c FOR THE REORIENTATION OF THE H$_2$O MOLECULE IN PURE WATER AS A FUNCTION OF TEMPERATURE θ^* $1.5 \lesssim \tau_r/\tau_c \lesssim 3$

θ^* °C	$D \times 10^5$ cm^2 sec	$\tau_c \times 10^{11}$ sec	θ^* °C	$D \times 10^5$ cm^2 sec	$\tau_c \times 10^{11}$ sec
0	1·14	0·53	50	4·63	0·165
5	1·36	0·44	55	5·13	0·153
10	1·60	0·39	60	5·66	0·140
15	1·86	0·34	65	6·22	0·130
20	2·17	0·30	70	6·79	0·120
25	2·50	0·27	75	7·36	0·113
30	2·89	0·24	80	8·00	0·105
35	3·28	0·22	85	8·52	0·098
40	3·69	0·20	90	9·10	0·092
45	4·13	0·18	95	9·63	0·085
			100	10·16	0·080

The activation energy for τ_c is the same as that given for $1/T_1$

For the longitudinal relaxation time of the deuteron in D$_2$O at 25°C the values 0·45 sec (interpolated[57]) and 0·42 sec $\pm$ 5 per cent[58] have been published. Such a short relaxation time can only be explained by quadrupole interaction. The quadrupole coupling constant $eQ(\partial^2 V/\partial z'^2)/h$ for the gaseous water molecule is known from microwave spectrum measurements to be 315 $\pm$ 0·7 kc/s.[59, 60] The coupling constant in ice is about 30 per cent less than this[61] due to hydrogen bonding.[62] Let us estimate that in liquid water the coupling constant is 25 per cent less than in the gaseous state, that is $eQ(\partial^2 V/\partial z'^2)/h = 237$ kc/s. Thus, from equation (42) with $I = 1$ and $\bar{\eta} = 0$ we find that $\tau_c = 0·29 \times 10^{-11}$ sec. Obviously, the almost exact agreement with the value as derived from the proton relaxation time is fortuitous, but it shows that the results obtained are essentially correct (one has to expect a slight isotopic effect). Woessner[57] reports an "activation energy" characterizing the deuteron relaxation between 40°C and 100°C of 3·9 kcal mole^{-1}, i.e. somewhat higher than found for H_2O. Since the quadrupole coupling constant should increase as the temperature increases, this finding is unexpected, and before attempting an interpretation it may be advisable to check the experimental results.†

The line width $\Delta H_{1/2}$ of the ^{17}O resonance in water at pH $\sim$ 12 at 25°C is 82·5 $\pm$ 2·5 mgauss.[63] At this pH value the proton exchange is fast enough to warrant the validity of $T_1 = T_2$. With $1/T_1 = 1/T_2 = \frac{1}{2}\gamma\Delta H_{1/2}$ we get $T_1 = 6\cdot7 \pm 0\cdot2 \times 10^{-3}$ sec. Luz and Meiboom[64] quote the value $T_1 = 6\cdot3 \times 10^{-3}$ sec. Other results given in the literature are slightly shorter.[65, 66] The somewhat greater line width corresponding to the shorter T_1 is probably due to modulation and field inhomogeneity effects. The nucleus of ^{17}O has spin $\frac{5}{2}$ and its quadrupole coupling constant $eQ(\partial^2 V/\partial z'^2)/h$ in water is not known exactly, however, the quantities $eQ(\partial^2 V/\partial a^2)/h = 8\cdot1 \pm 0\cdot1$ Mc/s and $(\partial^2 V/\partial b^2 - \partial^2 V/\partial c^2)/(\partial^2 V/\partial a^2) = 0\cdot7 \pm 0\cdot1$ for the HD^{17}O molecule have been measured by microwave spectroscopy.[67] The symbols a, b and c indicate the three principal axes of inertia of the HDO molecule. Assuming that the quadrupole coupling constant is not very much different from the value 8·1 Mc/s (the a-axis is in the D____H plane and deviates by 21° 23·5′ from the x-axis which lies in the HOH plane perpendicular to the C_2 axis) and neglecting the asymmetry parameter $\bar\eta$, we obtain from equation (42) with $I = \frac{5}{2}$, a value $\tau_c = 0\cdot24 \times 10^{-11}$ sec in reasonable agreement with the results found from the ^1H and D resonances. Nothing is known about the coupling constant in the condensed state.[68]

4.1.2. *Other liquids*

We omit the discussion of "liquid gases" like CH_4, CD_4, C_2H_4, CF_4 and others. Spin-rotational interaction is an appreciable source of relaxation here, in some cases the behaviour of these liquids is not yet fully understood. Probably $CHCl_3$ represents an example of pure intermolecular proton relaxation. From $T_1 = 90$ sec,[69, 70] $D = 2\cdot7 \times 10^{-5}$ cm^2 sec^{-1}[71] at 25°C, and $\mathring{a}$ equal to the molecular diameter $5\cdot8 \times 10^{-8}$ cm we find that equation (34) yields $\tau_d = 1\cdot1 \times 10^{-10}$ sec. If we put $\mathring{a}$ equal to 20 per cent less than the molecular diameter ($4\cdot6 \times 10^{-8}$ cm), the result is $\tau_d = 5\cdot1 \times 10^{-11}$ sec. This and other results for τ_d will be discussed briefly below.

The most extensive measurement of the reorientation times of organic

---

† *Note added in proof.* J. G. Powles, M. Rhodes and J. H. Strange (*Mol. Phys.* **11**, 115, (1966)) find the same activation energy for the deuteron resonance as for the proton resonance.

molecules in pure liquids employing proton relaxation times is due to Zeidler.[35] The results are collected in Table 2. In all cases the dilution and extrapolation technique with fully deuterated compounds was used. Figs. 1 and 2 give some typical examples of the dependence of $1/T_1$ on the proton content of the liquid. Since on the right-hand side of equation (36) all terms are proportional to τ_c, an approximate generalization of equation (36) for protons in chemically equivalent and non-equivalent positions has been used[35]

$$\frac{1}{T_1} = F\tau_c$$

with

$$F = \gamma^4\hbar^2 \frac{1}{n} \sum_{\substack{i,j \\ i<j}} 2\alpha_{ij}b_{ij}^{-6},$$

b_{ij} is the distance between proton i and j, $\alpha_{ij} = \frac{3}{2}$ if i and j are chemically equivalent and $\alpha_{ij} = 1$ if they are not. F has been calculated from the geometry of the molecules; Table 2 presents some results. If one used $\alpha_{ij} = \frac{3}{2}$ for all proton pairs the difference in the resulting F is not great. In Table 2 the corresponding F and τ_c values are quoted in parentheses. It is likely that C_1 for organic molecules (see equation (20)) is not very different from unity. Thus we have $\tau_c = \frac{1}{3}\tau_r$. The reorientation times are also included in Table 2. A priori it is not known whether the CH_3 group in the respective molecules undergoes an internal rotation with time constants $\ll \tau_c$. If this should be the case, the reorientation times of the molecule would be greater by a factor as given by equations (24) and (20). The τ_d values in Table 2 are calculated from the measured intermolecular rates by equation (34), using the self-diffusion coefficients given in Table 2. The latter were measured by various authors and are in satisfactory agreement. The distance of closest approach $\mathring{a}$ has been put equal to the molecular diameter a, the latter was determined from:

$$\mathring{a} = a = 2a^*; \qquad \frac{4\pi a^{*3}}{3} N_2 = 0\cdot74V_M,$$

V_M = molar volume. The resulting τ_d values are longer than one would expect. Although we have certain indications that the translational jumping times are somewhat longer than the reorientation times, the reason for these surprising results is not yet known; but it might well be that the Torrey theory is too crude. For the example of $CHCl_3$ the effect of decreasing $\mathring{a}$ slightly has been demonstrated. The application of Hubbard's[21] correction does not alter the results appreciably.

In Table 2 are included the results of Bonera and Rigamonti,[72] of Eisner and Mitchell,[73] and of Powles and Figgins.[74] We are justified to assume that the longer T_1 results are the better ones since generally unsuitable equipment and impure substances have the tendency to shorten the measured

TABLE 2. ROTATIONAL CORRELATION AND REORIENTATION TIMES AND TRANSLATIONAL JUMPING TIMES FOR SOME PURE ORGANIC LIQUIDS.
$\theta^* = 25°C$

(All data taken from Zeidler[35] if not indicated otherwise.)

Compound	$10^2 \times (1/T_1)$ sec^{-1}	$10^2 \times (1/T_1)_{intra}$ sec^{-1}	$10^{-10} \times F$ sec^{-2}	$\tau_c \times 10^{11}$ sec	$\tau_r \times 10^{11}$ (approx.) sec	$\tau_d \times 10^{11}$ sec	$D \times 10^5$ cm^2 sec^{-1}
Benzene	5.3	0.9	0.84	0.11	0.32	7.6	2.5
Benzene[72,73]	5.5 ± 0.3	1.67	0.84	0.20	0.60	—	—
Benzene[74]	4.82	0.97	0.84	0.12	0.36	—	—
Pyridine	8.1	1.0	0.39 (0.60)	0.26 (0.17)	0.77 (0.50)	13.8	1.9
Cyclohexane	14.0	5.1	3.09 (4.23)	0.17 (0.12)	0.50 (0.36)	10.1	1.7
Methanol − d_1	11.1	4.7	4.11 (5.28)	0.11 (0.09)	0.34 (0.27)	4.1	2.2
Ethanol − d_1	19.9	12.3	3.93 (4.83)	0.31 (0.25)	0.94 (0.76)	1.0 (?)	1.0†
Acetic acid − d_1	20.7	9.8	4.11 (5.28)	0.24 (0.19)	0.72 (0.56)	19.0	1.3
Methyl iodide	8.0	6.4	5.28	0.12	0.36	1.4	3.5
Acetonitrile	6.2	4.4	5.28	0.08	0.25	1.9	5.4
Acetone	6.4	3.5	4.35 (5.55)	0.08 (0.06)	0.24 (0.19)	3.5	5.2
Acetone[72]	6.45 ± 0.3	5.3	(4.90)	(0.11)	(0.33)	—	—
Dimethyl sulphoxide	34.3	18.0	4.47 (5.67)	0.40 (0.32)	1.21 (0.95)	15.7	0.8

† Not determined by the spin echo method, (C_2H_5OH).

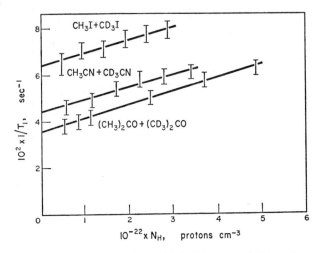

FIG. 1. Proton relaxation rates of methyl iodide, acetonitrile, and acetone in mixtures of the proton containing and deuterated compounds. $\theta^* = 25°C$. (Zeidler[35])

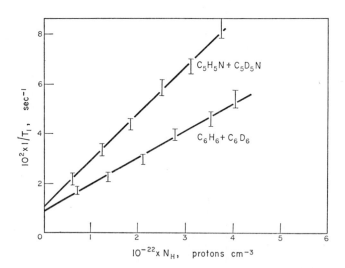

FIG. 2. Proton relaxation rates of benzene and pyridine in mixtures of the proton containing and deuterated compounds. From Figs. 1 and 2 the much smaller intramolecular relaxation rate of the aromatic and heterocyclic compounds as compared with the aliphatic compounds may be seen, $\theta^* = 25°C$. (Zeidler[35])

T_1 values. Bonera and Rigamonti have measured the temperature dependence of $1/T_1$ for acetone and benzene. These authors obtained the result that in both cases the relaxation mechanism is essentially dipolar. For benzene, Powles and Figgins[74] showed that at temperatures slightly above 25°C, spin-rotational interaction becomes important. According to the latter authors the activation energy for the rotational motion of the benzene molecule is 1·9 kcal mole^{-1} in good agreement with Woessner's C_6D_6 study.[57] Green and Powles[75] reported the following approximate correlation times: $\tau_c = 0\cdot19 \times 10^{-11}$, $\tau_c = 0\cdot26 \times 10^{-11}$, and $\tau_c = 0\cdot37 \times 10^{-11}$ sec at 25°C for fluorobenzene, chlorobenzene and bromobenzene, respectively, the intermolecular relaxation rate has been estimated.

Moniz and Gutowsky[76] have measured T_1 for ^{14}N resonances in a number of organic compounds. ^{14}N has spin $I = 1$ and it may be shown that the relaxation is wholly due to quadrupole interaction. The quadrupole coupling constant is known for several of the liquids measured by Moniz and Gutowsky, partly from solid state pure quadrupole resonances, partly from microwave studies in the gaseous state. The coupling constant obtained for condensed phase is to be preferred only when that from the condensed phase is not available; the value for the isolated molecule has been reduced by 10 per cent. In Table 3 the correlation times are summarized, they follow from the measured T_1 and the known coupling constants after insertion in equation

TABLE 3. ROTATIONAL CORRELATION TIMES AND ^{14}N RELAXATION TIMES FOR SOME NITROGEN-CONTAINING COMPOUNDS IN THE PURE LIQUID STATE
$\theta^* = 25°C$[76]

Compound	T_1 msec	$\tau_c \times 10^{11}$ sec	Compound	T_1 msec	$\tau_c \times 10^{11}$ sec
CH_3CN	5·0	0·098	$(C_2H_5)_3N$	1·4	0·18
$CH_2(CN)_2$	0·8	0·55	$(CH_2NH_2)_2$	1·5	0·29
$CH_2=CHCN$	3·5	0·135	Cyclohexylamine	1·5	0·42
CH_2ClCH_2CN	0·8	0·59	Pyridine	2·5	0·12

(42). It is again very likely that for the organic liquids considered here, a correlation factor $C_1 \approx 1$ (equation (20)) is appropriate, thus the τ_c values are essentially equal to one third of the tumbling times of the molecules. Again, the reorientation frequency about the C—N or C—C bonds is not known. It is remarkable that τ_c as determined from the proton resonance for CH_3CN (Table 2) agrees with the result of Table 3 as well as can be expected considering the accuracy of the T_1 measurements. Thus it follows that the rotational motion of acetonitrile is essentially isotropic, not anisotropic, as stated in reference 76. Fast reorientation about the C—C axis seems to be absent. The agreement between the τ_c's of Tables 2 and 3 for pyridine is worse. It is likely that the value determined by Moniz and Gutowsky is somewhat too low. This is suggested by the fact that line width measurements of the ^{14}N resonance also yield a longer correlation time ($\tau_c = 2 \times 10^{-12}$ sec[77]).

Similar experiments have been performed for molecules containing chlorine. If Cl, Br or I are covalently bound, the quadrupole interaction causes a very strong relaxation mechanism. Only ^{35}Cl and ^{37}Cl have sufficiently small quadrupole moments so as to be directly observable by NMR measurements in the covalently bound state. The relaxation time measurements are line width measurements here. The line width is of the order of 10 gauss. The relations between line width and relaxation times are given by equations (44) and (45). For highly fluid liquids the equality $T_1 = T_2$ holds. The ^{35}Cl quadrupole coupling constants have been measured by pure quadrupole resonance for a number of (solid) compounds. Thus, for the corresponding liquids application of equation (42) yields the correlation time $\tau_c(I = \frac{3}{2})$. Pertinent measurements with ^{35}Cl have been performed by Masuda[78] and by O'Reilly and Schacher.[79] The results are summarized in Table 4. C_1 is assumed to be about unity.

4.2. Mixtures

In a mixture of two components A and B the characterization of two composition ranges is particularly simple. These are the concentration ranges where one of the components is present to a relatively small amount, that is

TABLE 4. ROTATIONAL CORRELATION TIMES τ_c, MOLECULAR REORIENTATION TIMES τ_r, AND REORIENTATIONAL ACTIVATION ENERGIES E_r FOR SOME CHLORINE-CONTAINING COMPOUNDS. $\tau_r \approx 3\tau_c$

Compound	$\tau_c \times 10^{11}$ sec	$\tau_r \times 10^{11}$ sec	Temperature	E_r kcal mole^{-1}	Literature reference
TiCl$_4$	0·51	1·5	room temp.	—	78
	0·42	1·2	25°C	1·0 ± 0·1	79
	0·31	0·9	100°C	—	79
VOCl$_3$	0·32	1·0	room temp.	—	78
CrO$_2$Cl$_2$	0·26	0·8	room temp.	—	78
SiCl$_4$	0·29	0·9	room temp.	—	78
CCl$_4$	0·17	0·5	25°C	1·3 ± 0·1	79
	0·11	0·3	100°C		
CHCl$_3$	0·18	0·5	25°C	1·4 ± 0·1	79
	0·11	0·3	100°C		
C$_6$H$_5$Cl	0·41	1·2	25°C	1·6 ± 0·2	79
	0·23	0·6	100°C		

$x_B \ll 1$ or $x_A \ll 1$ (x_A, x_B mole fraction of molecule A and B respectively). If for instance $x_B \ll 1$, then we conventionally denote species B as the solute and molecule A as the solvent. Having at hand such a mixture with $x_B \ll 1$ the possible experiments immediately split up in two classes: either we study the relaxation behaviour of the solute or we investigate that of the solvent. Clearly, the first class of experiments yields information concerning the microdynamic behaviour of the solute, the second class reveals microdynamic details of the solvent. For the investigation of the solvent a special method or a model of interpretation is used in very many cases: the structural division of the bulk of A in solvation spheres around particles B and in essentially unperturbed or free liquid A. In some cases the kind of interaction of the spin is also different in the neighbourhood of particle B. This nature of the solvent as a "mixture" of structural non-equivalent regions introduces new formal features to the description of the relaxation rate which we have not yet presented. The necessary formulae will be given below.

We begin this section with the discussion of results obtained for the solute B. There are very few experiments yielding the reorientation time of a particle over the whole concentration range $0 < x_B \leqslant 1$. The results of these measurements will be reported here.

4.2.1. *Solute studies in non-aqueous solutions*

Mitchell and Eisner[80] measured the proton relaxation time of benzene, cyclohexane and chlorobenzene (denoted here as B) as a function of concentration in mixtures with CS_2 and CCl_4. Only the extrapolation values $x_B \to 0$ are of interest for us because all other relaxation rates contain a certain intermolecular contribution. The evaluation is again to be carried out by equations (36) and (18) and yields the results given in Table 5. Just as in all organic liquids without specific interactions, putting $C_1 \approx 1$ seems to be the most reasonable assumption and the reorientation times of the molecules should be about three times as long as the correlation times.

It is surprising that C_6H_{12} shows the shortest reorientation time in spite of the fact that cyclohexane is the biggest molecule as compared with the other

TABLE 5. ROTATIONAL CORRELATION TIMES FOR SOME ORGANIC
MOLECULES IN CS_2 AND CCl_4 IN THE LIMIT OF INFINITE DILUTION.
$\tau_r \approx 3\tau_c$, room temperature[80]

Compound	$\tau_c(CS_2) \times 10^{11}$ sec	$\tau_c(CCl_4) \times 10^{11}$ sec
C_6H_{12}	0·11	0·15
C_6H_6	0·15	0·23
C_6H_5Cl	0·22	0·38

two. This finding is not corroborated by the behaviour in the pure liquid (see Table 2).

Next we discuss the study of Pritchard and Richards[81] concerning the microdynamic behaviour of toluene, p-xylene, o-xylene and pyrene in the limit of infinite dilution in CS_2. Benzene has also been examined by these authors, their value for τ_c is about 20 per cent shorter than that reported by Mitchell and Eisner for CS_2. The correlation times as determined by Pritchard and Richards are summarized in Table 6. The most interesting

TABLE 6. ROTATIONAL CORRELATION TIMES OF DIFFERENT PROTONS IN SOME AROMATIC COMPOUNDS DISSOLVED IN CS_2 IN THE LIMIT $x_B \to 0$. ($\theta^* = 25°C$)[81]

Compound	$\tau_c \times 10^{11}$ sec.	
	ring	methyl
Benzene	0·12	—
Toluene	0·17	0·13
p-Xylene	0·24	0·18
o-Xylene	0·19	0·14
Pyrene	0·60	—

result in Pritchard and Richards' paper is the fact that in all three cases the correlation time found for the methyl group is smaller than the correlation time for the ring protons by a factor 0·75—the constancy of this factor of 0·75 is in itself rather surprising. Let us consider toluene first and assume for the moment that the toluene molecule is completely rigid. Then we would expect a slightly anisotropic rotational motion. The motion about a molecular axis which coincides with the C_3 axis of the methyl group will be faster than the motion about the two axes perpendicular to it. Now, all proton–proton vectors within the methyl group are perpendicular to the first mentioned axis of preferred molecular reorientation. None of the next neighbour proton–proton vectors in the ring is perpendicular to this axis; however, two of these vectors are parallel to it. The average angle of the proton–proton vectors in the ring relative to the axis of preferred rotation is thus <90°. Now, the contribution of a given proton–proton vector is affected the more by a facilitated rotation about an axis the more its angle relative to this axis approaches 90°. Thus the protons in the methyl group "report" the increased reorientation about the corresponding axis to a greater degree than the ring–protons do. The quantitative background of these arguments have been given by Woessner.[11] From Woessner's results with $D_{r_1}/D_{r_2} \approx 2$, one would estimate that the anisotropy and geometry effect reduces the relaxation rate of the methyl protons to about 90 per cent of the rate for the ring protons. The experimental value of reduction is 75 per cent: thus, assuming that the C_1 value is about unity for all axes of reorientation the remaining reduction of the methyl relaxation rate should come from the internal reorientation of the methyl group relative to the whole molecule (the aromatic skeleton). Woessner's formula given in equation (24) tells us that the time constant τ_0 of this internal motion must be relatively long; a numerical evaluation yields $\tau_{c_1}/\tau_0 \approx \frac{1}{3}$. Thus, on the average, after a time corresponding to one reorientation process of the whole molecule, one 120-degree jump of the methyl group occurs.

Two corrections have to be mentioned. In the treatment by Pritchard and Richards the average distance from the methyl to the ring protons has been used. If there exists an internal motion, then the contribution from these proton–proton pairs has to be reduced because the protons move relative to one another (Woessner[31]). This effect would increase the resulting τ_c for the methyl group in a more accurate evaluation. However, as shown by Green and Powles[75] a slight amount of spin-rotational interaction for the methyl protons in the pure liquid at 25°C is already present which should occur also in the CS_2 solution. Consequently, the dipolar relaxation rate of the methyl protons is actually somewhat smaller and will compensate approximately for the increase of τ_c resulting from the consideration of the partial motion of the intramolecular proton–proton vectors. It is interesting to note that the temperature dependence of the relaxation time of the methyl protons as measured by Green and Powles[75] for the pure liquid is about the same as that for the ring protons. Hence the activation energy of the internal motions should not be very different from that of the reorientation of the whole molecule. The intramolecular effect in Green and Powles' experiment is somewhat hidden by the intermolecular contribution. Measurements at high dilution would settle this question more clearly. Of course, the arguments just presented are somewhat tentative. No account has been taken of the fact that we have a molecule with non-equivalent protons studied by the method of adiabatic fast passage, i.e. the exact change in time of the magnetizations is not yet known theoretically. However, the discussion shows what further experiments could be performed in order to get the desired information. For the other homologues of toluene studied by Pritchard and Richards the situation should be similar.

4.2.2. Solutes in aqueous solutions

The microdynamic behaviour of a number of ions and of neutral particles has been studied by the present author together with Zeidler.[58] The respective particles were dissolved in D_2O. The proton relaxation time of the solute particles has been measured and the dilution and extrapolation technique applied. Models of the molecules were used to calculate the correlation times from equations (36) and (18). Possible effects of non-equivalent protons have not been taken into consideration. The results are summarized in Table 7.

Many of the solute particles are large, thus infinitesimal step rotational diffusion is very likely, i.e. $C_1 \approx 1$, and the reorientation time is about three

TABLE 7. ROTATIONAL CORRELATION TIMES OF SOME SOLUTE PARTICLES IN D_2O IN THE LIMIT OF INFINITE DILUTION ($\theta^* = 25°C$)[58, 86]

Solute particle	$(1/T_1)_{intra}$ sec^{-1}	$\tau_c \times 10^{11}$ sec	Solute particle	$(1/T_1)_{intra}$ sec^{-1}	$\tau_c \times 10^{11}$ sec
$(CH_3)_4N^+$	0·10	0·15	$C_2H_5COO^-$	0·15	0·33
$(C_2H_5)_4N^+$	0·35	0·53	C_2H_5COOH		
$(C_3H_7)_4N^+$	0·95	1·7	$C_3H_7COO^-$	0·22	0·50
$(C_4H_9)_4N^+$	2·10	4·0	C_3H_7COOH		
			$C_4H_9COO^-$	0·28	0·70
CH_3OH	0·048	0·09			
CH_3COCH_3	0·047	0·08			

times the measured correlation time. Whether this holds true also for the small particles like methanol and acetone is an open question. We would suppose that the residence time of a water molecule in a site of the first hydration sphere of the solute molecule is one to two times longer than in ordinary water (see below), that is, we have a residence time of 0·5 to $1·0 \times 10^{-11}$ sec. The solute particle will be surrounded by ten to twelve H_2O molecules. Thus we would find for τ in equation (20), $\tau \approx 0·08 \times 10^{-11}$ sec if the coupling between the water molecule and the solute molecule is strong enough to disturb the diffusional motion of the latter under each translational jump process of an H_2O molecule in the hydration sphere. It is hardly reasonable to assume $C_1 \ll \frac{1}{2}$. Hence with $C_1 = \frac{1}{2}$ we obtain from equation (20) that $\tau_r = 0·16 \times 10^{-11}$ sec for acetone and $\tau_r = 0·18 \times 10^{-11}$ sec for methanol. Summarizing, we get the result $0·16 \times 10^{-11} \leqslant \tau_r \leqslant 0·24 \times 10^{-11}$ sec and $0·18 \leqslant \tau_r \leqslant 0·27 \times 10^{-11}$ sec for acetone and methanol respectively. It is very surprising that these reorientation times within the water cage are not distinguishably different from the corresponding reorientation times in the pure liquids CH_3COCH_3 and CH_3OH. We note that the reorientation time of the surrounding water molecules is definitely greater by a factor of about two to three. Unfortunately, we do not yet know which role the internal motion plays for the correlation time under consideration. From the general experience so far available it is believed that the reorientation of the methyl or methylene group about the C—C bond is never very much faster than the reorientation of the whole molecule if the latter is of moderate size. So in lack of any evidence contradicting it we may put forward the statement that the measured reorientation times essentially describe the reorientation of the whole molecule. For very big molecules, the concept of a reorientation time gets meaningless if the molecule is not strictly rigid. Clearly, further experiments are necessary. Isotopic substitution techniques will be a very valuable aid. Severe discrepancies with dielectric relaxation time measurements in certain systems present appreciable difficulties to our understanding. The rather long dielectric relaxation times of pure aliphatic alcohols (methanol 79×10^{-11}, ethanol 144×10^{-11} sec,[82, 83])† would suggest that the OH group which carries the electric dipole moment undergoes very slow reorientations. In solutions of methanol and ethanol in inert solvents, the dielectric relaxation time is roughly the same as the reorientation time as determined in pure alcohols by NMR methods regarding the molecule as rigid.[84]

The observation that the rotational motion of organic molecules in aqueous solution is surprisingly fast has been confirmed by Clifford[85] who studied the reorientation behaviour of several alkyl sulphates and aliphatic alcohols in water in the limit of infinite dilution. Clifford used solutions of the alcohols (C_2 to C_6) in CCl_4 to compare their correlation times in this inert solvent with that in their cage in water and found that the limiting values for infinite dilution in D_2O and CCl_4 are practically the same. For C_2H_5OH the intramolecular relaxation rate in CCl_4 in very dilute solutions was found to be even higher than in D_2O. At 32°C Clifford reports for C_2H_5OH in D_2O $1/T_1 = 0.07$ sec^{-1} as a limiting value, in CCl_4 he finds 0.24 sec^{-1}. This exceptional result is not yet understood; it may be mentioned, however, that Zeidler has measured $(1/T_1)_{intra} = 0.123$ sec^{-1} at 25°C for pure ethanol (see Table 2) which, if an appropriate temperature correction is applied, is rather close to Clifford's D_2O value. The intramolecular relaxation rates in D_2O at infinite dilution of several alkyl sulphates (C_2 to C_8) are also very similar to those for the alcohols with the same number of C atoms in CCl_4 (again except C_2H_5OH). Clifford claims that his results show that both in D_2O and CCl_4 considerable, rapid, large amplitude internal motion must take place in the CH_2 chains. Unfortunately, the author does not give details concerning the way he calculated the expected relaxation rate caused by pure Brownian motion of the rigid molecule. Furthermore, these calculated relaxation rates are presented as a sole criterion for the existence of internal motions, no micro-viscosity factor or any other correction being mentioned.

The concentration dependence of the reorientation time of acetone in aqueous solution has been measured by Hertz and Zeidler.[58] Fig. 3 shows the results. The abscissa indicates the proton concentration in the respective solution. Curve (a) represents the proton relaxation rate in $(CH_3)_2CO + (CD_3)_2CO$ mixtures, curve (b) gives the proton relaxation rate of $(CH_3)_2CO +$

† However, there is a small dispersion step in the frequency range corresponding to the dielectric relaxation times of the alcohols in inert solvents.

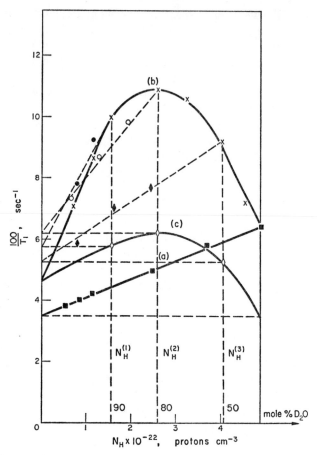

Fig. 3. Proton relaxation rates in mixtures of $(CH_3)_2CO$, $(CD_3)_2CO$ and D_2O, as a function of proton content. Details are explained in the text; $\theta^* = 25°C$. (Hertz and Zeidler.[58])

D_2O mixtures and the three straight lines connecting curve (b) with the ordinate show the dependence of the proton relaxation rate on the proton concentration in $(CH_3)_2CO + (CD_3)_2CO + D_2O$ mixtures. For each of these straight lines the total acetone content is constant but the ratio

$$\frac{\text{number of } (CH_3)_2CO \text{ molecules}}{\text{number of } (CH_3)_2CO \text{ molecules} + (CD_3)_2CO \text{ molecules}}$$

goes from one to zero. Finally, curve (c) shows the concentration dependence of the intramolecular relaxation rate which is proportional to the correlation time of the acetone molecule, with varying acetone concentration in the mixture with water. At a mole fraction of 0·8 of water molecules the correlation time of acetone takes on its maximum value of $\tau_c = 0.11 \times 10^{-11}$ sec compared with $\tau_c = 0.08 \times 10^{-11}$ sec for the limit of infinite dilution, and about the same ratio should apply for the reorientation times. Similar results were obtained for methanol in aqueous solution.[86] At infinite dilution the correlation time is $\tau_c = 0.09 \times 10^{-11}$ sec, and at the maximum value at about $x_{CH_3OH} = 0.45$ it is $\tau_c = 0.14 \times 10^{-11}$ sec. This is an interesting result because in water many liquid solutes with alkyl groups show remarkable anomalies at mole fractions of about 0·1 to 0·3. We mention but a few examples: viscosity, velocity of sound, sound absorption, and proton relaxation rate of water show a maximum at about this concentration range; the partial molar volume, the excess entropy of mixing and the excess enthalpy of mixing have a minimum at about the same composition. Thus one sees that even the reorientational motion of the solute molecules is affected by the general structural change occurring at about this critical concentration range; see, for example, references 2 and 87. We generally assume that an increase in water structure is occurring if solutes with alkyl groups are added to water (see below).

Finally, we quote the activation energy of the reorientation of the $N(C_2H_5)_4^+$ ion in aqueous solution in the limit of infinite dilution, which is 3.3 ± 1 kcal mole^{-1}. This is less than the appropriate activation energy for the reorientation of the surrounding water molecules[58] in the same temperature range. The correlation time of the chlorate ion in aqueous solution is $\tau_c = 0.45 \times 10^{-11}$ and 0.24×10^{-11} sec at 25 and 100°C, respectively, $E_r = 1.5 \pm 0.4$ kcal mole^{-1}.[79]

4.2.3. *The micro dynamic behaviour of the solvent*

As was mentioned above the great majority of NMR relaxation time studies performed on the nuclei of the solvent had the purpose of obtaining information on the microdynamic behaviour of the solvation spheres. Generally, the structural properties of the solvation sphere are distinct from those of the remainder of the solvent farther apart from the solute particles. Thus if we were able to keep a solvent molecule fixed in a given solvation sphere it would show a relaxation time different from the pure unperturbed solvent liquid. Interpretation of this relaxation time as obtained for the particle in the solvation sphere in terms of the formulae of Section 2 would yield the desired microdynamic parameters. Usually, however, an individual solvent molecule diffuses from one solvation sphere to the free solvent, then enters into another solvation sphere and so forth. As a consequence of this process, the measured relaxation rate contains contributions from all microscopic regions of the solution. The way in which the various contributions enter into the expression for the total relaxation rate depends mainly on the mean life-times of the solvent molecules in the various regions. We denote as τ_a the mean life-time in the free or unperturbed solvent, the life-times in the various solvation spheres b, $c \ldots$ may be called τ_b, $\tau_c \ldots$. The relaxation times in the respective regions are called T_{ia}, T_{ib}, $T_{ic} \ldots$, $i = 1, 2$ (generally $T_{i\nu}$). If we add a prime to the symbol $1/T_\nu$ the intramolecular relaxation rate is concerned. The mole fractions of molecules in the various regions of the liquid are denoted as x_a, x_b, $x_c \ldots$,

$$x_a + x_b + x_c + \ldots = 1.$$

Let us begin with the case of very fast exchange. This means that the life-time of the solvent molecule in a given microscopic region is of the same order of magnitude as the correlation time, i.e. the correlation function itself depends on the details of the exchange of the particle between the different microscopic regions. The resulting problem for dipolar interaction in a two-spin molecule (e.g. water) has been treated by Beckert and Pfeifer[88] and subsequently by the present author,[7] using the model of rotational diffusion as given in equations (11), (13) and (14). Detailed calculations have been performed only for a system consisting of two "sub-liquids". Sub-liquid b comprises all microscopic regions formed by the solvation spheres of one dissolved (ionic) species and sub-liquid a is the remainder of the system, i.e. the unperturbed solvent. The result for conditions of extreme narrowing is the following: if the general type of the diffusive rotational motion is the same everywhere in the solution, i.e. if the rotation-step constant C_1 (see equation (11)) for sub-liquid a (C_{1a}), for sub-liquid b (C_{1b}), and for the transition from a to b and vice versa (C_{1ab}) is the same,† then the intra-molecular relaxation rate of the solution is given to a good approximation by

$$\left(\frac{1}{T_i}\right)_{\text{intra}} = x_a \frac{1}{T'_{ia}} + x_b \frac{1}{T'_{ib}} \quad i = 1, 2 \tag{46}$$

even for τ_b, $\tau_a \lesssim \tau_{ca}$, τ_{cb} (τ_{ca}, τ_{cb} rotational correlation times in a and b respectively). T'_{ia} and T'_{ib} (where $i = 1, 2$) are the intramolecular relaxation times in the sub-liquids a and b, respectively, if we were able to keep the particle fixed in the corresponding sub-liquid. They are given by equation (18) (e.g. for water) with $\tau_c = \tau_{ca}$ and $\tau_c = \tau_{cb}$ respectively. If the $C_{1a, b, ab}$ approach zero or if τ_a and $\tau_b \gg \tau_{ca}$, τ_{cb}, then equation (46) holds exactly. However, if C_{1a}, C_{1b}, and C_{1ab} have markedly different values and τ_a and $\tau_b \lesssim \tau_{ca}$, then $(1/T_1)_{\text{intra}}$ is a very complicated non-linear function of x_b, its behaviour is completely different from that given by equation (46).[7] Of course, for equation (46) to hold with constant $1/T'_{ia, b}$ it is necessary that the $\tau_{ca, b}$ and the C_{1i} values are independent of the composition of the mixture.

---

† An additional condition is that $\left| 1 - \frac{\tau_{ca}}{\tau_{cb}} \right| \lesssim 1$, which, however, should not be critical.

Equation (46) with the associated conditions is also valid if the relaxation is caused by intramolecular quadrupole interaction.

Now we turn to the intermolecular relaxation rate. For a diamagnetic solution of only one solute species with spin S in the centre of the solute particle, the intermolecular relaxation rate (of the solvent protons) under conditions of extreme narrowing has been calculated to be approximately:[7]

$$
\left(\frac{1}{T_{1,2}}\right)_{\text{inter}}
$$

$$
= \frac{3\pi\gamma_I^4\hbar^2 C_I \tilde{\tau}_d^{(I)}}{\mathring{a}^3}\left(\frac{1}{3} + \frac{2}{15}\frac{\mathring{a}^2}{\tilde{D}_I\tilde{\tau}_d^{(I)}}\right) + \frac{3}{2}\frac{\gamma^4\hbar^2 n_c x_b \zeta \tau_2'(\tau_{b'} + \tilde{\tau}_d^{(I)})}{n_h(R')^6(2\tau_2' + \tau_{b'})}
$$

$$
+ \left(\frac{1}{T_{1,2}}\right)_{\text{capt}} + \frac{8\pi\gamma_I^2\gamma_S^2\hbar^2 C_S S(S+1)\tilde{\tau}_d^{(IS)}}{3R_{IS}^3}\left(\frac{1}{3} + \frac{2}{15}\frac{R_{IS}^2}{\tilde{D}_{IS}\tau_d^{(IS)}}\right)
$$

$$
+ \frac{4\gamma_I^2\gamma_S^2\hbar^2 S(S+1)n_c x_b \tau_2'(\tau_b' + \tilde{\tau}_d^{(IS)})}{3(R_{IS}')^6 n_h(\tau_2' + \tau_{b'})} + \left(\frac{1}{T_{1,2}}\right)_{\text{capt}}^{(IS)}, \quad x_b \ll 1. \quad (47)
$$

$C_I =$ concentration of spins $I = \frac{1}{2}$ of the solvent,

$\tilde{\tau}_d^{(I)} =$ concentration dependent average translational jumping time of solvent molecules,

$\tilde{D}_I =$ concentration dependent self-diffusion coefficient of the solvent,

$\mathring{a} =$ closest distance of approach of solvent spins,

$n_c =$ number of solvent molecules rigidly bound in the first solvation sphere (rigidly bound means that at the corresponding site the molecule does not perform any motion which changes the distance from any intramolecular point of the molecule to the centre of the ion),

$n_h =$ total number of solvent molecules in the solvation shell (e.g. defined geometrically),

$R' =$ shortest intermolecular proton–proton distance within the rigid first solvation sphere,

$\zeta =$ number of neighbours at distance R' in the rigid first solvation sphere,

$\tau_2' =$ correlation time for the rotational motion of the rigid first hydration sphere, rotational jump diffusion is excluded,

$\tau_{b'} =$ lifetime of a molecule in the rigid first solvation sphere,

$\left(\frac{1}{T_{1,2}}\right)_{\text{capt}}, \left(\frac{1}{T_{1,2}}\right)_{\text{capt}}^{(IS)} =$ contributions to the relaxation rate due to solvent molecules captured by the solvation shell during a time of order of the correlation time,

C_S = concentration of solute spins,

R'_{IS} = closest distance of approach between solute and solvent spin within the first solvation sphere,

$R_{IS} = R'_{IS}$ if $n_c = 0$, otherwise = closest distance of approach between solute and solvent spin in the second solvation sphere,

$$\tilde{\tau}_d^{(IS)} = \tfrac{1}{2}(\tilde{\tau}_d^{(I)} + \tilde{\tau}_d^{(S)}),$$
$$\tilde{D}_{IS} = \tfrac{1}{2}(\tilde{D}_I + \tilde{D}_S),$$

$\tilde{D}_S$ = concentration dependent self-diffusion coefficient of spin S,

$\tilde{\tau}_d^{(S)}$ = concentration dependent translational jumping time for spin S.

Detailed expressions for $(1/T_{1,2})_{capt}$ and $(1/T_{1,2})_{capt}^{(IS)}$ are given in reference 7. Furthermore, it is shown in reference 7 that these terms may be neglected in the majority of applications. On the one hand, in diamagnetic electrolyte solutions very often $n_c = 0$ holds, and in this case we see that equation (47) is merely the sum of equations (34) and (41) with concentration dependent parameters $\tau_d^{(I)}$, $\tau_d^{(IS)}$, D_I and D_{IS}. On the other hand, if $n_c \neq 0$ (n_c should be equal to 4 or 6) and if $\tau_{b'} \gg \tau_2'$ and $\tau_{b'} \gg \tilde{\tau}_d^{(I)}$, $\tilde{\tau}_d^{(IS)}$, then, apart from constant factors we discover equation (18) and equation (38c) as the second and the fifth term on the right-hand side of equation (47), i.e. we find the expressions for the intramolecular relaxation rate of a pair of like and unlike spins. Now the notation "intramolecular" refers to the stable and rigid hydration complex as molecular species. Equation (47) concerns magnetic dipole–dipole interaction, a corresponding formula for quadrupole interaction is not yet available.

Next we consider solutions with paramagnetic particles, i.e. S is an electronic spin. In this case the condition of extreme narrowing is generally not fulfilled. Thus we have to introduce the frequency dependence of the relaxation rate for the unlike spin terms in equation (47) (the fourth and fifth terms). The translational term $(1/T_{1,2})_{transl}^{(IS)}$ replacing the fourth term on the right-hand side of equation (47) has been calculated by Pfeifer.[33] As stated in Section 2.5, Pfeifer's formula includes the effect of the electron spin relaxation time τ_s. The general result is complicated and will not be reproduced here. It is only valid for the limiting case $R_{IS}^2/\langle r^2 \rangle \gg 1$. Thus only an effective diffusion coefficient characterizes the motion of the particles and (apart from τ_s) no true microdynamical parameters appear. Qualitatively, the most important result of Pfeifer's treatment is the fact that the curve representing the relaxation rate as a function of frequency forms a flatter, more extended dispersion step in the case of translational motion than it does for rotational motion.

Furthermore, if we neglect for simplicity $\tilde{\tau}_d^{(IS)}$ in the fifth member of equation (47) the latter may be written

$$\frac{4}{3}\frac{\gamma_I^2\gamma_S^2\hbar^2 S(S+1)n_c x_b}{n_h(R'_{IS})^6}\tau'_c$$

with

$$\frac{1}{\tau'_c}=\frac{1}{\tau_{b'}}+\frac{1}{\tau'_2}.$$

In order to replace this expression in the correct way for paramagnetic particles, according to the foregoing we simply substitute for it the frequency dependent "intramolecular" relaxation rate given by equation (38a) and after dropping (for reasons of simplicity) $(1/T_{1,\,2})_{\text{capt}}$ and $(1/T_{1,\,2})_{\text{capt}}^{(IS)}$ we obtain:

$$\left(\frac{1}{T_1}\right)_{\text{inter}}$$

$$=\frac{3\pi\gamma_I^4\hbar^2 C_I\tilde{\tau}_d^{(I)}}{\mathring{a}^3}\left(\frac{1}{3}+\frac{2}{15}\frac{\mathring{a}^2}{\tilde{D}_I\tilde{\tau}_d}\right)$$

$$+\frac{3}{2}\frac{\gamma_I^4\hbar^2 n_c x_b\zeta\tau'_2(\tau'_b+\tilde{\tau}_d^{(I)})}{n_h(R')^6(2\tau'_2+\tau_{b'})}+\left(\frac{1}{T_1}\right)_{\text{transl}}^{(IS)}+\frac{2}{5}\frac{\gamma_I^2\gamma_S^2\hbar^2 S(S+1)x_b n_c}{n_h(R'_{IS})^6}$$

$$\times\left\{\frac{1}{3}\frac{\tau'_c}{1+(\omega_I-\omega_S)^2\tau'^2_c}+\frac{\tau'_c}{1+\omega_I^2\tau'^2_c}+\frac{2\tau'_c}{1+(\omega_I+\omega_S)^2\tau'^2_c}\right\}\qquad(48a)$$

$$\left(\frac{1}{T_2}\right)_{\text{inter}}$$

$$=\frac{3\pi\gamma_I^4\hbar^2 C_I\tilde{\tau}_d^{(I)}}{\mathring{a}^3}\left(\frac{1}{3}+\frac{2}{15}\frac{\mathring{a}^2}{\tilde{D}\tilde{\tau}_d}\right)$$

$$+\frac{3}{2}\frac{\gamma_I^4\hbar^2 n_c x_b\zeta\tau'_2(\tau_{b'}+\tilde{\tau}_d^{(I)})}{n_h(R')^6(2\tau'_2+\tau_{b'})}+\left(\frac{1}{T_2}\right)_{\text{transl}}^{(IS)}+\frac{4\gamma_I^2\gamma_S^2\hbar^2 S(S+1)n_c x_b}{15\,n_h(R'_{IS})^6}$$

$$\times\left\{\tau'_c+\frac{1}{4}\frac{\tau'_c}{1+(\omega_I+\omega_S)^2\tau'^2_c}+\frac{3}{4}\frac{\tau'_c}{1+\omega_I^2\tau'^2_c}\right.$$

$$\left.+\frac{3}{2}\frac{\tau'_c}{1+\omega_S^2\tau'^2_c}+\frac{3}{2}\frac{\tau'_c}{1+(\omega_I+\omega_S)^2\tau'^2_c}\right\}.\qquad(48b)$$

Here we have assumed that the electron spin relaxation time τ_s is much longer than τ'_c. The term τ_s could be incorporated into τ'_c ($1/\tau''_c=1/\tau'_c+1/\tau_s^{(89,\,90)}$), however, this case is of little practical importance, because when $\tau_s \ll \tau'_c$ equations (48a) and (48b) usually no longer hold for other physical reasons as mentioned in Section 2.5 (see, for instance, references 91, 92).

If the paramagnetic particle is uncharged (free radical) or if it is charged but complexed appropriately by non-solvent ligands, then n_c in equation (48a, b) is practically zero (see for instance refs. 93–97; note that the time constants τ_d reported in the literature as determined from the relaxation rate of purely translational motion in paramagnetic systems usually are only the reciprocal of an effective diffusion coefficient multiplied by a constant factor and by R_{IS}^2). For the transition metal cations, if non-solvent particles do not penetrate into the first coordination sphere, we have $\tau'_b \gg \tau'_2$ and we may put $n_c = n_h$ and consequently $\tau'_2 = \tau_{cb}$ if rotational jump diffusion is excluded. For ions partially or stepwise complexed by non-solvent ligands we would expect the intermediate case $n_c = 5, 4, 3 \ldots$, and $\tau_{b'} \approx \tau'_2$. Examples for the latter case may be given by some Cr(III) complexes for which Günther and Pfeifer[94] were unable to find the correct dispersion behaviour.

Now we multiply all purely translational terms in equations (47), (48a) and (48b) (i.e. the first and fourth terms in equation (47), the first and third terms in equations (48a) and (48b)) by $1 = x_a + x_b$. If we neglect $(1/T_{1,\,2})_{\text{capt}}$ and $(1/T_{1,\,2})_{\text{capt}}^{(IS)}$ and add equations (46) and (47) or equations (46) and (48a) or (48b) the result may be written:

$$\frac{1}{T_i} = x_a \frac{1}{T_{ia}} + x_b \frac{1}{T_{ib}}, \quad i = 1, 2. \tag{49}$$

$1/T_{ia}$ and $1/T_{ib}$ are the factors of x_a and x_b in the resulting sum, they are (to a fair approximation) the total relaxation rates of the proton in the sub-liquids a and b respectively at the composition x_a, x_b. We recall that for equation (49) to hold it is necessary that the C_{1i}/τ_j values are essentially equal for all regions of the liquid or that the times τ_a, τ_b or $\tau_{b'}$ are sufficiently long compared with the correlation times τ_{ca}, τ_{cb}, τ'_2. For the bare (paramagnetic) transition metal ions the latter condition is always fulfilled.

Equation (49) as it stands may be used—and has been used—for the study of neutral solute particles. Examples dealing with hydration spheres of aliphatic alcohols and (non-dissociated) fatty acids will be reported below. Studies on free radicals will not be mentioned again because, apart from effective diffusion coefficients, they do not yield relevant microdynamical information.

For electrolyte solutions, however, equation (49) is not yet satisfactory because these systems contain at least two different solutes, namely the anion and cation pair necessary for electrical neutrality. A calculation corresponding to that given by Beckert and Pfeifer[88] and by Hertz[7] for a system of three sub-liquids has not yet been performed. However, Zimmerman and Brittin[98] using the solution of the Chapman–Kolmogoroff equation have shown that under very general conditions the following relation holds

$$\frac{1}{T_i} = x_a \frac{1}{T_{ia}} + x_b \frac{1}{T_{ib}} + x_c \frac{1}{T_{ic}} + \ldots + x_p \frac{1}{T_{ip}}, \quad i = 1, 2 \qquad (50)$$

if the exchange is fast in a time scale given by the relaxation times, i.e. $T_{ia}, T_{ib} \ldots \gg \tau_a, \tau_b \ldots$, where $T_{ia}, T_{ib}, \ldots$ are the relaxation times of the sub-liquids $a, b, \ldots, p$ (sometimes called "phases") at the composition $x_a, x_b, \ldots, x_p$. On the other hand, the exchange has to be slow enough for the $T_{ia, b} \ldots$, to be well-defined quantities, i.e. the correlation functions must not depend on the exchange process. Now, the detailed microdynamical analysis[7, 88] has revealed the conditions under which equation (49) holds, and from the identity of equations (49) and (50) for a system of two sub-liquids we may conclude with a high degree of certainty that equation (50) holds generally—at least approximately—for a system composed of sub-liquids $a, b, c, d, \ldots, p$ if the same microdynamic conditions are fulfilled, i.e. all C_{1i}/τ_j equal ($i = a, b, ab, ac \ldots, j = a, b, c, \ldots$), the validity being the better the smaller the C_{1i} values and the smaller the values $\tau_{ca}/\tau_a, \tau_{cb}/\tau_b, \tau_{cc}/\tau_c, \ldots, \tau_{cp}/\tau_p$. The observed approximate linear dependence of $(1/T_{1, 2})$ on the composition and the additivity of the terms $1/T'_{ib}, 1/T'_{ic} \ldots i = 1, 2$, shows that the necessary conditions are satisfied in real systems, where their fulfilment may be doubtful a priori.

Before summarizing the final expressions for the relaxation rates of a solvent particle in solution we have to introduce another completely new relaxation mechanism which is important only in mixtures. This is the scalar relaxation of the first kind.

This additional relaxation is caused by scalar interaction between the spins I and S, the interaction energy is given by $A\mathbf{S} \cdot \mathbf{I}$, where A is the coupling constant. Suppose that I is the spin the relaxation process of which we are studying (the solvent spin), S may be any other spin in the solution, but the relaxation due to scalar interaction is of particular interest if the solute is a paramagnetic ion and thus S is the spin of its unpaired electrons. If the solvent molecule is a member of the first coordination sphere—in coordination spheres further apart there may also be a slight effect—the spin I sitting on this molecule undergoes scalar interaction $A\mathbf{S} \cdot \mathbf{I}$ with the unpaired electrons of the ion. A is proportional to the probability of finding an unpaired electron at the position of the nucleus with spin I. The scalar interaction is independent of the orientation of the vector connecting spin I with the centre of the ion. Thus tumbling of the solvation sphere does not affect this interaction. However, if the molecule under consideration leaves the first solvation sphere and diffuses into the free solvent or to another spin S its scalar interaction with the first spin is from then on practically zero. Thus the coupling constant of the scalar interaction is a fluctuating quantity, fluctuating between the two values A and zero. The correlation function of the scalar coupling, i.e. of the coupling constant is

$$\overline{A(0)A(t)} \sim e^{-t/\tau_b}. \tag{51}$$

τ_b is the average time the solvent molecule resides in a site near the ion where the scalar interaction is effective. It is most reasonable for further applications to identify τ_b with $\tau_{b'}$ the mean life-time in the first (rigid) solvation sphere. Application of the general theory of relaxation yields the results[6]

$$\left(\frac{1}{T_1}\right)_{sc} = \frac{2A^2}{3} x_b S(S+1) \frac{\tau_b}{1 + (\omega_I - \omega_S)^2 \tau_b^2} \tag{52}$$

$$\left(\frac{1}{T_2}\right)_{sc} = \frac{A^2}{3} x_b S(S+1) \left\{ \tau_b + \frac{\tau_b}{1 + (\omega_I - \omega_S)^2 \tau_b^2} \right\}. \tag{53}$$

We thus see that the time τ_b which characterizes the microdynamic process of solvent molecule exchange in the first solvation sphere may be obtained from T_1 or T_2 measurements if the coupling constant A is known.

It can happen, however, that the electron spin relaxation time τ_s is of the same order of magnitude or shorter than τ_b. In this case we have to substitute for τ_b the time constant τ_e with[91]

$$\frac{1}{\tau_e} = \frac{1}{\tau_s} + \frac{1}{\tau_b}.$$

If $\tau_s \ll \tau_b$, then we have relaxation similar to scalar relaxation of the second kind.[6]

Since $(1/T_{1,\,2})_{sc}$ is an independent relaxation rate the total relaxation rate for a solution of paramagnetic ions is found by adding $(1/T_{1,\,2})_{sc}$ to the rate as given by equation (50). Now the final results for the relaxation rate of the solvent spin may be collected.

(a) *Diamagnetic solutions* (*extreme narrowing*). In the case of relaxation by magnetic dipole–dipole interaction, experimental results are only available for aqueous solutions. Thus we specialize the following formulae for water as the solvent (for any molecule containing two spins the same formulae are valid, for other molecules the extension is obvious). We identify sub-liquid a with the free water, sub-liquid b and c with the sum of all cationic and anionic hydration shells respectively, i.e.

$$T_{1a} \equiv T_1^0, \quad T_{1b} \equiv T_1^+, \quad T_{1c} \equiv T_1^-, \quad x_b \equiv x^+, \quad x_c \equiv x^-,$$
$$\tau_{ca} \equiv \tau_c^0, \quad \tau_{cb} \equiv \tau_c^+, \quad \tau_{cc} \equiv \tau_c^-.$$

The nuclear spin on the cation is S, the nuclear spin on the anion S'.

Thus far we have no practical example of a diamagnetic electrolyte solution for which the second and fourth term on the right-hand side of equation (47) are of importance. The same holds true for the terms $(1/T_1)_{\text{capt}}$ and $(1/T_1)_{\text{capt}}^{(IS)}$ or $(1/T_1)_{\text{capt}}^{(IS')}$. Hence, we may neglect these terms and from equations (50), (47) and (18) we obtain:

$$
\frac{1}{T_1} = \frac{3}{2} \frac{\gamma_I^4 \hbar^2}{b^6} \{(1 - x^+ - x^-)\tau_c^0 + x^+\tau_c^+ + x^-\tau_c^-\}
$$

$$
+ \frac{3\pi\gamma_I^4\hbar^2 C_I \tilde{\tau}_d^{(I)}}{a^3} \left(\frac{1}{3} + \frac{2}{15} \frac{\mathring{a}^2}{\tilde{D}_I \tilde{\tau}_d^{(I)}}\right)
$$

$$
+ \frac{8\pi\gamma_I^2\gamma_S^2\hbar^2 S(S+1) C_S \tilde{\tau}_d^{(IS)}}{3R_{IS}^3} \left(\frac{1}{3} + \frac{2}{15} \frac{R_{IS}^2}{\tilde{D}_{IS}\tilde{\tau}_d^{(IS)}}\right)
$$

$$
+ \frac{8\pi\gamma_I^2\gamma_{S'}^2\hbar^2 S'(S'+1) C_{S'} \tilde{\tau}_d^{(IS')}}{3R_3^{IS'}} \left(\frac{1}{3} + \frac{2}{15} \frac{R_{(IS)'}^2}{\tilde{D}_{IS'}\tilde{\tau}_d^{(IS')}}\right) \tag{54}
$$

with

$$
x^\pm = \frac{n_{\bar{h}}^\pm m}{55\cdot 5} \nu^\pm.
$$

$\nu^\pm$ are the stoichiometric numbers of the salt in solution, $\tilde{D}_I$ and $\tilde{\tau}_d^{(I)}$ are the self-diffusion coefficient and the average translational jumping time of the water molecule in the electrolyte solutions under consideration. $\tilde{\tau}_d^{(I)} = \langle r^2 \rangle/6\tilde{D}_I$, the concentration dependence of $\langle r^2 \rangle$ has been discussed elsewhere.[7, 100] The symbol m represents the molality of the dissolved salt. The meaning of the symbols in equation (54) not indicated before are obvious by analogy. In order to find the intramolecular relaxation rate the three translational terms in equation (54) have to be calculated from measured or estimated data. Very often the third and fourth terms may be neglected, since $\gamma_{S,S'}^2 \ll \gamma_I^2$ or the nucleus concerned has no magnetic moment. The grouping of terms in equations (54) and (50) has sometimes been performed in a somewhat different way which, however, is not essential.[99] If the self-diffusion coefficient is not known, the following approximate empirical relation may be used:[99, 100]

$$
\frac{1}{\tilde{D}_I} \approx \alpha \frac{1}{T_1} \qquad \alpha \approx 1\cdot 7 \times 10^5 \text{ sec}^2 \text{ cm}^{-2}. \tag{55}
$$

Suppose the solute particles have small γ_S and $\gamma_{S'}$ or have no magnetic moment so that the two last terms of equation (54) may be dropped, then, if in the parenthesis of the second term on the right-hand side $\frac{1}{3}$ is negligible as compared with $\frac{2}{15} \mathring{a}^2/\tilde{D}_I\tilde{\tau}_d^{(I)}$ (the substitution of $\tilde{\tau}_d^{(I)}$ by $\langle r^2 \rangle/6\tilde{D}_I$ is another possibility) we have from equations (54) and (55)

$$\frac{1}{T_1} = \frac{3}{2}\frac{\gamma_I^4\hbar^2}{b^6}\{(1 - x^+ - x^-)\tau_c^0 + x^+\tau_c^+ + x^-\tau_c^-\} + \frac{2\pi}{5}\frac{\gamma_I^4\hbar^2 C_I}{\mathring{a}} \propto \frac{1}{T_1} \quad (56)$$

or

$$\left(\frac{1}{T_1}\right)_{\text{intra}} = \frac{3}{2}\frac{\gamma_I^4\hbar^2}{b^6}\{(1 - x^+ - x^-)\tau_c^0 + x^+\tau_c^+ + x^-\tau_c^-\}$$

$$= \frac{1}{T_1}\left(1 - \frac{2\pi}{5}\frac{C_I\gamma_I^4\hbar^2}{\mathring{a}}\propto\right) \quad (57)$$

as an approximate relation to find the intramolecular relaxation rate from measured T_1 data. The term $\mathring{a}$ varies slightly with concentration, since with increasing concentration the ions more and more prevent the water spins from approaching to one another, i.e. in the average $\mathring{a}$ increases, C_I is practically constant or decreases slightly.

As mentioned above, the correctness of the term for the intramolecular contribution in equation (54) (the first term on the right-hand side) depends on the magnitudes of the rotation-step coefficients C_{1i} and the life-times τ_j, $i = a, b, ab, \ldots, j = a, b, \ldots$. Generally, the only way to decide whether the necessary conditions for equation (54) to hold are fulfilled is to study $1/T_1$ over a relatively extended concentration range and to examine whether indeed $(1/T_1)_{\text{intra}}$ does depend linearly on the concentration. However, it may happen as a consequence of the structure of the electrolyte solution that a linear concentration dependence cannot occur even if the above-mentioned microdynamic conditions are satisfied. The origin of this behaviour in our model may be ascribed to the overlap of hydration spheres at higher concentration. In this event, however, it is possible to use certain corrections to make linear the concentration dependence of $(1/T_1)_{\text{intra}}$ at least to some degree: details have been described elsewhere.[7, 100] If this is not feasible the evaluation of $(1/T_1)_{\text{intra}}$ cannot be performed without using information from other sources. However, actual examples of such behaviour are unknown.

A further check on the validity of the relation for the intramolecular relaxation rate in equation (54) is to examine whether or not the additivity of the ionic contributions as predicted by equation (54) is experimentally verified. Again this is so for the majority of examples.

For the evaluation of the intramolecular relaxation rate, n_h^+ and n_h^- have to be known. These quantities are generally not available. Fabricand and co-workers have tried to determine hydration numbers as well as correlation times from measured proton relaxation rates in diamagnetic electrolyte solutions.[101] Their results have been criticized by several authors who suggest that they are not physically meaningful.[7, 102] Probably the best procedure is to choose reasonable numbers for $n_h^\pm$ without relying on any other physico-chemical sources and to interpret the resulting parameters as properties of an adequate model. For Al^{3+}, Ga^{3+}, and Be^{2+} true first co-ordination numbers are available from ^{17}O NMR measurements;[103-5] however, proton relaxation time measurements of τ_c^+ have not yet been reported for these systems. For Mg^{2+}, $n_h^+ = 6$ seems to be a good figure.[106] But even in cases where true first coordination numbers are available, one never knows how many water molecules are affected dynamically. We have still two unknown quantities in the intramolecular contribution of equations (54) or (56), namely τ_c^+ and τ_c^-. In order to reduce to one the number of unknown time constants, a possible procedure is to put $\tau^{K+} = \tau^{cl-}$ [99, 107] since the transference numbers of both these ions are almost exactly $\frac{1}{2}$. Of course, other proposals to ascribe a definite value to one $\tau_c^\pm$ will be found in the literature.[101, 108] The possibility of determining one of the terms in the intramolecular rate directly, for instance $x^-\tau_c^-$ by measuring the ^{19}F nuclear relaxation of the F^- ion is discussed in reference 100. In order to find the $\tau_r^\pm$'s, the reorientation times of the water molecule in the hydration sphere C_{1b} and C_{1c} and the respective τ's have to be known (see equation (20)). Here, of course, only reasonable estimates are available, e.g. $\frac{1}{4} \leqslant C_1 \leqslant \frac{3}{4}$.[7, 100] If $\tau_c^\pm$ is found to be $\ll 8\tau_c^0$, then the life-time τ_b of the water molecule in the hydration shell can be estimated to be about the same as the reorientation time τ_r.[99, 100] Figure 4 shows some typical proton relaxation rates in diamagnetic electrolyte solutions as a function of the concentration. Table 8 summarizes some relative correlation times $\tau_c^\pm/\tau_c^0$ and reorientation times $\tau_r^\pm/\tau_r^0$ as obtained from equation (54).[100] Other data reported in the literature[101, 107-8] are of similar magnitude if proper transformations from other hydration numbers and other splittings into single ion contributions are performed. In some cases the intermolecular relaxation rate has not been taken into consideration. One sees from Table 8 that for the so-called structure-breaking ions (I^-, Br^-, Rb^+, Cs^+) the correlation and reorientation times are smaller than in the undisturbed water, i.e. in the hydration sphere the local fluidity is greater. The question of the localization of the more fluid water molecules relative to the surface of the ion has been examined by using the concentration dependence of the relaxation rate of the ionic nuclear resonances.[99] The activation energies of the tumbling process of water molecules in the hydration sphere as determined from the measured temperature dependence of the water and the ionic resonances are reported in Table 9. The approximate equality of the numerical values as determined from the water protons and from the nucleus in the centre of the ion serves as a check that in both cases the molecules on the surface of the ions are concerned. For details the literature should be consulted.[100] In the case of Na^+ there seems to be a discrepancy which is not yet explained.

So far all explicit formulae presented for diamagnetic ions consider T_1.

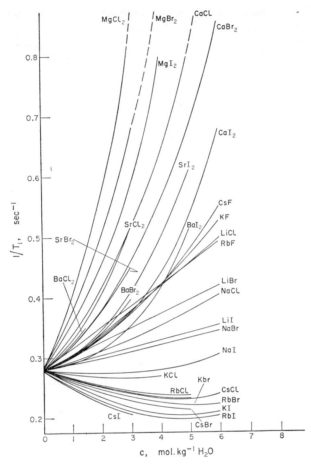

Fig. 4. Proton relaxation rates in some diamagnetic electrolyte solutions. The different behaviour of "structure-forming" and "structure-breaking" ions may be seen. $\theta^* = 25°C$.[100]

In pure (neutral) water T_2 is slightly shorter than T_1 which is due to the rather slow exchange of protons between different water molecules in the presence of electron coupled spin–spin interaction with ^{17}O.[110] Apart from this effect $T_1 = T_2$ for a great number of diamagnetic electrolyte solutions.[111] However, for Al^{3+} solutions T_2 has been found to be less than T_1 and frequency dependent. From the difference of $1/T_2$ and $1/T_1$ and the measured chemical shift of the water protons in Al^{3+} solutions the life-time of the proton in the hydration shell of Al^{3+} has been determined by Lohmann[111] to be $\approx 10^{-5}$ sec; the exact figure depending somewhat on concentration and pH value. The

TABLE 8. RATIO OF THE CORRELATION TIME (AND REORIENTATION TIME) OF THE WATER MOLECULE IN THE HYDRATION SHELL OF SOME DIAMAGNETIC IONS TO THE CORRESPONDING TIMES IN PURE WATER. $n_h^\pm$ has been assumed to have a value of 6, $\tau_c(K^+)$ has been put equal to $\tau_c(Cl^-)$, $\theta^* = 25°C$[100]

Ion	$\tau_c^+/\tau^0 \approx \tau_r^+/\tau^0$†	Ion	$\tau^\mp/\tau_0 \approx \tau^\pm/\tau_r^0$
Li^+	2·3	Sr^{++}	3·1
Na^+	1·6	Ba^{++}	2·7
K^+	0·9	F^-	2·3
Rb^+	0·6	Cl^-	0·9
Cs^+	0·5	Br^-	0·6
Mg^{++}	5·2	I^-	0·3
Ca^{++}	3·5		

† The ratio $\tau_r^\pm/\tau_r^0$ is not very sensitive to the choice of parameters C_1 and τ.

TABLE 9. ACTIVATION ENERGIES FOR THE REORIENTATION PROCESS OF WATER MOLECULES IN THE HYDRATION SHELL OF SOME DIAMAGNETIC IONS. The underlined symbols indicate that the reported activation energy is determined from the relaxation time of the corresponding ionic nucleus. All data from reference 100 except that for NaCl which is taken from reference 109

Ion	E_r kcal mole^{-1}	Salt	E_r kcal mole^{-1}
Li^+	4·3	Li$\underline{Cl}$	2·6
Na^+	4·2	Li$\underline{Br}$	2·9
K^+	3·3	Na$\underline{Br}$	2·9
Rb^+	2·9	Cs$\underline{Br}$	2·3
Cs^+	2·7	Li$\underline{I}$	2·5
Mg^{++}	5·2	Na$\underline{I}$	2·7
Ca^{++}	4·6	K$\underline{I}$	2·7
Sr^{++}	3·7	Cs$\underline{I}$	2·5
Ba^{++}	3·7	$\underline{Na}$Cl	2·5
		$\underline{Rb}$Cl	2·5
F^-	4·3		
Cl^-	3·3	H_2O	3·3
Br^-	2·9		
I^-	2·7		

formula to be used is similar to equation (68) given below. Since from ^{17}O chemical shift measurements[103] we know that the life-time of the ^{17}O nucleus in the Al^{3+} hydration sphere is much greater than 10^{-4} sec, we conclude that here the water molecule does not exchange as a stable molecule. For some other strongly hydrated cations similar experiments should be possible.

We now discuss the case that the solvent nuclei are relaxed by quadrupole interaction (D_2O, $H_2^{17}O$). It is not known as yet whether the ionic electrostatic field produces an intermolecular contribution to the relaxation rate. This effect seems to be not very important and we will neglect the intermolecular terms in equation (50). Thus from equations (50) and (42) we obtain:

$$\frac{1}{T_1} = \frac{3}{40} \frac{2I + 3}{I^2(2I - 1)} \left(\frac{eQ}{\hbar}\right)^2 \left\{ (1 - x^+ - x^-)\left(1 + \frac{\tilde{\eta}_0^2}{3}\right)\left(\frac{\partial^2 V^0}{\partial z'^2}\right)^2 \tau_c^0 \right.$$
$$\left. + x^+\left(1 + \frac{\tilde{\eta}_+^2}{3}\right)\left(\frac{\partial^2 V^+}{\partial z'^2}\right)^2 \tau_c^+ + x^-\left(1 - \frac{\tilde{\eta}_-^2}{3}\right)\left(\frac{\partial^2 V^-}{\partial z'^2}\right)^2 \tau_c^- \right\}. \quad (58)$$

The meaning of the symbols should be obvious from the fore-going. If neutral solutes are considered $x^- = 0$ and the subscript $^+$ has to be replaced

TABLE 10. RATIO OF THE CORRELATION TIME (AND REORIENTATION TIME) OF THE WATER MOLECULE IN THE HYDRATION SHELLS OF PARTICLES WITH ALKYL GROUPS TO THE CORRESPONDING TIMES IN PURE WATER

$n_h^{\pm}$ and n_h^b are assumed numbers, $\theta^* = 25°C^{(58)}$

Solute particle	Water molecule	$n_h^{\pm}$	$\tau_c^{\pm}/\tau_c^0 \approx \tau_r^{\pm}/\tau_r^0$
$(CH_3)_4N^+$	D_2O	10	1·6
$(C_2H_5)_4N^+$	D_2O	15	2·1
$(C_3H_7)_4N^+$	D_2O	20	3·1
$(C_4H_9)_4N^+$	D_2O	25	2·9
CH_3COO^-	D_2O	12	1·7
$C_2H_5COO^-$	D_2O	15	2·0
$C_3H_7COO^-$	D_2O	18	2·1
$C_4H_9COO^-$	D_2O	21	2·0
$(CD_3)_4N^+$	H_2O	10	2·0
CD_3COO^-	H_2O	12	1·8
		n_h^b	$\tau_{cb}/\tau_c^0 \approx \tau_{rb}/\tau_r^0$
CH_3COOH	D_2O	12	1·4
C_2H_5COOH	D_2O	15	1·4
C_3H_7COOH	D_2O	18	1·6
CH_3OH	D_2O	10	1·6
C_2H_5OH	D_2O	13	1·8
C_3H_7OH	D_2O	16	2·0
CH_3COCH_3	D_2O	13	1·5
$C_2H_5COCH_3$	D_2O	16	1·5
CH_3CN	D_2O	10	1·2
C_2H_5CN	D_2O	13	1·3
$(CD_3)_2SO$	H_2O	15	1·5

by the subscript b. Using equation (58) in order to determine the $\tau_c^{\pm}$ we have to assume that the quadrupole coupling constants $(eQ/h)(\partial^2 V^{\pm}/\partial z'^2)$ are equal to that of water, which might not be strictly true. Equation (58) has been used for the study of ordinary electrolyte solutions. The results are in fair agreement with data obtained from proton T_1 measurements.[99, 63] For ^{17}O the ion dependence of the quadrupole coupling constant seems to be greater.[63]

The hydration properties of tetra-alkyl ammonium ions, carboxylate ions, fatty acids, alcohols, ketones and nitriles have also been studied with help of equation (58).[58] These solutes were also found to be hydrated (see Table 10), this hydration effect of inert alkyl groups is the "iceberg effect" or the hydration of the second kind which has been discussed very much recently (see, for example, reference 2). The lengthening of the reorientation times is considered to be a consequence of the "increase of structure" in the aqueous solvent. The experiments just mentioned have been confirmed by proton relaxation time measurements in aqueous solutions of tetra-alkyl ammonium salts[58, 112] and alkyl sulphates.[113] Figure 5 represents a picture of the

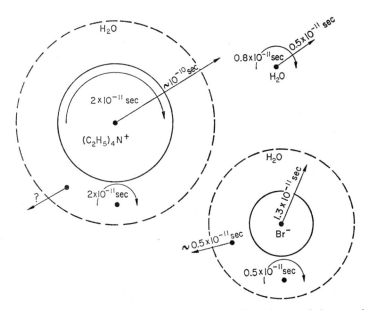

FIG. 5. "Microdynamic structure" of an aqueous solution of tetra-ethyl ammonium bromide. The regions occupied by the solute particles are surrounded by the hydration spheres. Time constants with circular arrows indicate reorientation times, those with straight arrows indicate translational jumping times. The term $(\langle r^2 \rangle)^{1/2}$ is indicated by the assumed length $(= a)$ of the straight arrows. The reorientation times are upper limits, C_1 has been put ≈ 1 everywhere. $\theta^* = 25°C$. (After Hertz and Zeidler,[58] but, modified.)

H

"microdynamic structure" of tetraethyl ammonium bromide in the limit of infinite dilution.[58] The ionic translational jumping times have been determined from conductivity data, self-diffusion measurements by spin echo techniques would be another source in favourable cases.

(b) *Paramagnetic solutions.* The paramagnetic ion (spin S) is supposed to be the cation, its solvation sphere is labelled by the letter b, a denotes the free solvent and c the solvation sphere of the anions,

$$\tau_s \gg \tau_c^+ \equiv \tau_{cb}, \quad \tau_b \gg \tau_c^+ .$$

For all systems considered here the relation holds:

$$\omega_I \ll \omega_S. \tag{59}$$

Then from equations (50), (52) and (53) we have:

$$\frac{1}{T_1} = \frac{x_a}{T_1^0} + \frac{x_b}{T_{1b}} + \frac{x_c}{T_{1c}} + \left(\frac{1}{T_1}\right)_{sc}$$

$$= \frac{x_a}{T_1^0} + \frac{x^+}{T_1^+} + \frac{x^-}{T_1^-} + \tfrac{2}{3}A^2 x^+ S(S+1) \frac{\tau_e}{1 + \omega_S^2 \tau_e^2}, \tag{60}$$

$$\frac{1}{T_2} = \frac{x_a}{T_2^0} + \frac{x_b}{T_{2b}} + \frac{x_c}{T_{2c}} + \left(\frac{1}{T_2}\right)_{sc}$$

$$= \frac{x_a}{T_2^0} + \frac{x^+}{T_2^+} + \frac{x^-}{T_2^-} + \tfrac{1}{3}A^2 x^+ S(S+1) \left\{ \tau_e + \frac{\tau_e}{1 + \omega_S^2 \tau_e^2} \right\}. \tag{61}$$

Now it may easily be shown that if we have $I = \frac{1}{2}$, i.e. there is no quadrupole interaction, for paramagnetic ions as solutes and x^+ not too small we can neglect all intramolecular contributions (relative to the solvent molecule) and from the dipolar intermolecular contribution we have only to retain the terms concerning the I–S interaction. Thus with equations (60), (61), (59), (48a), and (48b) we obtain:

$$\frac{1}{T_1} = \left(\frac{1}{T_1}\right)_{transl}^{(IS)} + \frac{2}{5} \frac{\gamma_I^2 \gamma_S^2 \hbar^2 x^+ S(S+1)}{(R_{IS}^{'6})} \left(\tau_c^+ + \frac{(7/3)\tau_c^+}{1 + \omega_S^2 (\tau_c^+)^2} \right)$$

$$+ \tfrac{2}{3} A^2 x^+ S(S+1) \frac{\tau_e}{1 + \omega_S^2 \tau_e^2} \tag{62a}$$

$$\frac{1}{T_2} = \left(\frac{1}{T_2}\right)_{transl}^{(IS)} + \frac{1}{15} \frac{\gamma_I^2 \gamma_S^2 \hbar^2 x^+ S(S+1)}{(R_{IS}^{'6})} \left(7\tau_c^+ + \frac{13\tau_c^+}{1 + \omega_S^2 (\tau_c^+)^2} \right)$$

$$+ \tfrac{1}{3} A^2 x^+ S(S+1) \left\{ \tau_e + \frac{\tau_e}{1 + \omega_S^2 \tau_e^2} \right\}, \tag{63a}$$

$$x^+ = \frac{n_h^+ m M_0}{1000}, \quad n_h^+ = 6 = n_c, \quad \frac{1}{\tau_e} = \frac{1}{\tau_s} + \frac{1}{\tau_b},$$

m = molality (mole kg^{-1} solvent), M_0 = molar mass of solvent.

Equations (62a) and (63a) contain the two microdynamic parameters τ_c^+ and τ_b. Here τ_c^+ is the correlation time for the rotational motion of the rigid solvation complex. It is essentially one-third of the reorientation time of the solvation complex. Of course, at the same time $3\tau_c^+$ is the reorientation time τ_r^+ of a solvent molecule in the rigid solvation complex. $\tau_b = \tau_b'$ is the mean life-time of a solvent molecule in the (rigid) solvation complex.

The measurement of the frequency and temperature dependence of the relaxation rates $1/T_{1,\,2}$ allows the determination of the two desired times τ_c^+ and τ_b and of the coupling constant A. Also R'_{IS} may be obtained. The temperature dependence of the two microdynamical parameters is:

$$\tau_c^+ = \tau_c^{+0} \exp\left(E_r/RT\right)$$

$$\tau_b = \tau_b^0 \exp\left(E_b/RT\right).$$

The electron spin relaxation time τ_s may be measured directly by electron spin resonance methods, τ_s is also frequency dependent; its temperature dependence is given roughly by

$$\tau_s = \tau_s^0 \exp\left(-E_s/RT\right)$$

with E_s having a value of about 1 to 4 kcal mole^{-1} (for details see the literature references 91, 114 and 115 and the references cited therein).

The terms $(1/T_{1,\,2})_{\text{transl}}^{(IS)}$ amount to about 10–30 per cent of the other dipolar terms in equations (62a) and (63a); in a number of studies for the analysis of the experimental results $(1/T_{1,\,2})_{\text{transl}}^{(IS)}$ has been neglected (see the work of Pfeifer and coworker[116, 115] where it has been used).

The solvent nuclei may have $I > \frac{1}{2}$ and thus relax by quadrupole interaction in the pure solvent. Now in general we are not allowed to drop the intramolecular contribution to the relaxation rate. However, the intermolecular contribution from all diamagnetic particles may safely be ignored. On the other hand it may happen that the contribution from the magnetic dipole–dipole interaction is small compared with the scalar and the quadrupolar relaxation rate. If the covalent binding in the solvation sphere is not much stronger than in the free solvent we have simply (for $x^+ \ll 1$):

$$\frac{1}{T_1} = \frac{1}{T_1^0} + \tfrac{2}{3}A^2 x^+ S(S+1)\frac{\tau_e}{1 + \omega_s^2 \tau_e^2} \tag{62b}$$

$$\frac{1}{T_2} = \frac{1}{T_2^0} + \tfrac{1}{3}A^2 x^+ S(S+1)\left\{\tau_e + \frac{\tau_e}{1 + \omega_s^2 \tau_e^2}\right\} \tag{63b}$$

where

$$\frac{1}{T_i^0}\frac{\tau_c^{\pm}}{\tau_c^0} \ll A^2 S(S+1)\frac{\tau_e}{1 + \omega_s^2 \tau_e^2}$$

is supposed to hold.

These equations apply under proper conditions for the ^{17}O resonance[114] in aqueous solutions of paramagnetic ions (regarding an additional contribution to $1/T_2$ see below).

For $\tau_s \ll \tau_b$ we may write equations (62a) and (63a) in the form:

$$\frac{1}{T_1} = (1 - x^+)\left(\frac{1}{T_1}\right)^{(IS)}_{\text{transl}} + \frac{x^+}{(T_1^+)^{\text{ion}}} \tag{62c}$$

$$\frac{1}{T_2} = (1 - x^+)\left(\frac{1}{T_2}\right)^{(IS)}_{\text{transl}} + \frac{x^+}{(T_2^+)_{\text{ion}}}. \tag{63c}$$

In the same way we write for equations (62b) and (63b) with $\tau_s \ll \tau_b$

$$\frac{1}{T_1} = \frac{1}{T_1^0} + \frac{x^+}{[T_1^+]_{\text{ion}}}, \quad x^+ \ll 1 \tag{62d}$$

$$\frac{1}{T_2} = \frac{1}{T_2^0} + \frac{x^+}{[T_2^+]_{\text{ion}}}, \quad x^+ \ll 1. \tag{63d}$$

$(1/T_{1,2}^+)_{\text{ion}}$ and $[1/T_{1,2}^+]_{\text{ion}}$ are the true relaxation rates in the solvation sphere of the paramagnetic cation, whereas the quantities $1/T_{1,\,2b}$ or $1/T_{1,\,2}^+$ appearing in equations (50) or (60) and equation (61) represent the relaxation rates in the solvation sphere apart from the scalar contribution. As mentioned above, in equations (62c) and (63c) the assignment of $(1 - x^+)(1/T_{1,\,2})^{(IS)}_{\text{transl}}$ and $x^+(1/T_{1,\,2})^{(IS)}_{\text{transl}}$ to the free water and the hydration shell respectively is only an approximation. However, this procedure is of no practical consequence.

If the condition holds that $\tau_b \gtrsim T_{1,\,2}^+$ then equations (62a) and (63a) are no longer valid. In this case for $\tau_b \gg \tau_s$, $x^+ \ll 1$ and $(T_{1,\,2})^{(IS)}_{\text{transl}} \gg (T_{1,\,2}^+)_{\text{ion}}$ it may be shown from the formulae of Zimmerman and Brittin[98] that the relation holds

$$\frac{1}{T_i} = \left(\frac{1}{T_i}\right)^{(IS)}_{\text{transl}} + \frac{x^+}{(T_i^+)_{\text{ion}} + \tau_b} \quad i = 1, 2. \tag{64a}$$

In the limit $\tau_b \gg (T_i^+)_{\text{ion}}$ we have

$$\frac{1}{T_i} = \left(\frac{1}{T_i}\right)^{(IS)}_{\text{transl}} + \frac{x^+}{\tau_b} = \left(\frac{1}{T_i}\right)^{(IS)}_{\text{transl}} + \frac{1}{\tau_a^0}\dagger \quad i = 1, 2 \tag{65a}$$

with

$$\frac{\tau_b}{\tau_a^0} = x^+,$$

† In the literature the grouping of the terms in equations (60) and (61) has sometimes been performed in a somewhat different form, particularly the scalar term in its general form has been included in $(1/T_i^+)_{\text{ion}}$.[33, 115-16] However, since in all cases where τ_b becomes comparable with T_i^+ the electron spin relaxation time is much shorter than τ_b, no trouble arises with equation (64a).

τ_a^0 is the mean life-time of the solvent particle in the liquid not bound to the cation.

In the same way for $\tau_b \gtrsim [T_{1,2}^+]_{\text{ion}} \gg \tau_s$ equations (62b) and (63b) or equations (62d) and (63d) break down. Now again we have with $x^+ \ll 1$, $T_1^0 \gg [T_i^+]_{\text{ion}}$ $i = 1, 2$

$$\frac{1}{T_i} = \frac{1}{T_i^0} + \frac{x^+}{[T_i^+]_{\text{ion}} + \tau_b} \quad i = 1, 2 \tag{64b}$$

and

$$\frac{1}{T_i} = \frac{1}{T_i^0} + \frac{1}{\tau_a^0} \quad \text{for} \quad \tau_b \gg [T_i^+]_{\text{ion}} \quad i = 1, 2. \tag{65b}$$

Furthermore, even if the explicit expressions for the relaxation times in the free and bound state are not known, formulae like equations (64) and (65) may be applied. The last equations offer another independent way of determining τ_b from measured relaxation rates. In the case of equation (65a, b) the increase of the relaxation rate (line broadening) is only determined by the velocity of the (chemical) particle exchange.

So far we have not yet mentioned the possibility that the chemical shift of the nucleus under consideration in the solvent is different in the different environments formed by the sub-liquids a, b, c . . . . In this case the rapid exchange of the spin between the various environments causes an additional contribution to the transversal relaxation rate which offers another possibility of determining the life-time of a particle in the respective environment, a solvation sphere, for instance. The theoretical approach to this problem is based on the inclusion of exchange terms into the Bloch equations and is due to Gutowsky, McCall and Slichter[117] and to McConnell in a somewhat modified but analogous form.[118] Swift and Connick[114] have worked out detailed formulae to be used here and we quote their results for a system of two different environments. Thus the formulae given below may be applied again to a solution of a paramagnetic cation. The solvation spheres of the cation form one environment, the free solvent together with the solvation spheres of the diamagnetic anions forms the other. An additional condition for the validity of the formulae to be given is that $x^+ \ll 1$, i.e. small concentration of the paramagnetic ions. The effect of the diamagnetic anions is again neglected. Swift and Connick's result for the transversal relaxation rate of the solvent spins is:

$$\frac{1}{T_2} = \left(\frac{1}{T_2}\right)_{\text{transl}}^{(IS)} + \frac{1}{\tau_a^0} \frac{((1/T_2^+)_{\text{ion}})^2 + 1/\tau_b(1/T_2^+)_{\text{ion}} + \Delta\omega_b^2}{\{(1/T_2^+)_{\text{ion}} + 1/\tau_b\}^2 + \Delta\omega_b^2}. \tag{66}$$

This formula concerns nuclei where magnetic dipole–dipole interaction causes the relaxation in the pure solvent. For nuclei relaxing by quadrupole interaction of the type as discussed in connection with equations (62b) and

(63b) $(1/T_2)^{(IS)}_{transl}$ and $(1/T_2^+)_{ion}$ have to be replaced by $1/T_2^0$ and $[1/T_2^+]_{ion}$. In the following all formulae given for the dipolar case hold analogously for this special quadrupole case after replacing $(1/T_2)^{(IS)}_{transl}$ and $(1/T_2^+)_{ion}$ by $1/T_2^0$ and $[1/T_2^+]_{ion}$, respectively.

Again, equations like (66) and the following formulae may be used with any other relaxation times in the bound and free states. Proton relaxation studies in solutions of C_0^{2+} and N_i^{2+} are examples of this situation. The chemical shift $\Delta\omega$ of the solvent spin in solution is given by[114]

$$\Delta\omega = x^+ \frac{\Delta\omega_b}{\tau_b^2[((1/T_2^+)_{ion} + 1/\tau_b)^2 + \Delta\omega_b^2]} \tag{67a}$$

$\Delta\omega_b$ is the chemical shift of the solvent spin in the solvation sphere of the cation. If it is caused by scalar interaction we have

$$\Delta\omega_b = \frac{S(S+1)\gamma_s A}{3kT\gamma_I} \omega_I. \tag{67b}$$

If $\tau_b^{-1}(1/T_2^+)_{ion} \gg ((1/T_2^+)_{ion})^2, \Delta\omega_b^2$, equation (66) yields a limiting law which is identical with equation (63) with $\tau_s \ll \tau_b$; if $\Delta\omega_b^2 \gg ((1/T_2^+)_{ion})^2, (1/\tau_b)^2$ or $((1/T_2^+)_{ion})^2 \gg \Delta\omega_b^2 \gg \tau_b^{-1}(1/T_2^+)_{ion}$ we obtain equation (65) as limiting law. Equation (64) may also be derived from equation (66). However, if $(1/\tau_b)^2 \gg \Delta\omega_b^2 \gg \tau_b^{-1}(1/T_2^+)_{ion}$ a new result appears, namely

$$\frac{1}{T_2} = \left(\frac{1}{T_2}\right)^{(IS)}_{transl} + x^+\tau_b\Delta\omega_b^2. \tag{68}$$

Under these conditions the measured chemical shift is

$$\Delta\omega = x^+\Delta\omega_b. \tag{69}$$

Thus τ_b can be determined from equation (68) together with equation (69). Finally if the exchange is sufficiently slow $\tau_b^{-1} \ll \Delta\omega_b$, $(1/T_2^+)_{ion}$ sufficiently small and x^+ sufficiently great to make possible a separate investigation of the NMR signal of the spins in the solvation sphere (with observed relaxation rate $(1/T_2^+)_b$) we have

$$\left(\frac{1}{T_2^+}\right)_b = \left(\frac{1}{T_2^+}\right)_{ion} + \frac{1}{\tau_b}, \tag{70}$$

which is again equivalent to equation (65) and allows a direct determination of the life-time τ_b if by lowering of the temperature one is able to measure $(1/T_2^+)_{ion}$.

The existence of the chemical shift $\Delta\omega_b$ does not cause additional effects for $1/T_1$, equations (62), (63), (64) and (65) remain unchanged.

In Table 11 are collected some results obtained by the formulae just presented.

Figures 6, 7 and 8 give examples of the observed behaviour of the proton relaxation in paramagnetic solutions.

TABLE 11. LIFE-TIMES τ_b OF SOLVENT MOLECULES IN THE SOLVATION SPHERE OF PARAMAGNETIC IONS, CORRELATION TIMES τ_c^+ FOR THE REORIENTATION OF THE SOLVATION SPHERE (REORIENTATION TIME $\tau_r^+ \approx 3\tau^+$) AND ACTIVATION ENERGIES FOR THESE TWO PROCESSES. $\theta^* = 25°C$ if not indicated otherwise.

Solute	Solvent	τ_b sec	E_b kcal mole⁻¹	$\tau_c^+ \times 10^{11}$ sec	$\tau_r^+ \times 10^{11}$ sec	E_r kcal mole⁻¹	Determined by Equation	Literature reference
Mn^{2+}	H_2O	$2\cdot7\times10^{-8}$	$8\cdot4$	$3\cdot1$	—	$5\cdot5$	63a, 62a	119
Mn^{2+}	H_2O	$2\cdot5\times10^{-8}$	$8\cdot1$	$3\cdot1$	$9\cdot3$	$4\cdot5$	63a, 62a	91
Mn^{2+}	H_2O	—	—	$5\cdot6$[a]	$9\cdot3$	$4\cdot3$	62a	116
Mn^{2+}	H_2O	$2\cdot6\times10^{-8}$[a] / $2\cdot0\times10^{-8}$[b]	—	$3\cdot2$	17[a] / $9\cdot6$	—	63a, 62a	95
Mn^{2+}	$H_2^{17}O$	$1\cdot4\times10^{-8}$[b] / $3\cdot2\times10^{-8}$	$8\cdot1$[c]	$1\cdot6$[b]	$4\cdot8$[b]	$4\cdot3$	63b, 65b	114
V^{2+}	H_2O	—	—	$2\cdot1$	$6\cdot3$	—	62a	91
Cr^{3+}	H_2O	$4\cdot5\times10^{-6}$	10	$8\cdot3$	25	—	64a, 62a	91
Cr^{3+}	H_2O	$1\cdot8\times10^{-6}$ / 9×10^{-7}[b]	—	$4\cdot5$ / $2\cdot45$[b]	$13\cdot5$ / $7\cdot5$	—	64a, 62a	95
Cr^{3+}	$H_2^{17}O$	very long	—	—	—	—	—	120
Cu^{2+}	H_2O	—	—	~2	~6	—	62a	91
Cu^{2+}	H_2O	—	—	$4\cdot2$[a] / $2\cdot6$ / $1\cdot15$[b]	12[a] / 8 / $3\cdot5$[b]	$2\cdot8$ (?)	62a	119
Cu^{2+}	H_2O	—	—	—	—	—	62a	95
Cu^{2+}	$H_2^{17}O$	5×10^{-9}	5[c]	—	—	—	63b	114, 125
Ni^{2+}	$H_2^{17}O$	$3\cdot7\times10^{-5}$	$11\cdot6$[c]	—	—	—	65b, caused by $\Delta\omega_b$(68)	114
Co^{2+}	$H_2^{17}O$	$7\cdot4\times10^{-7}$	$8\cdot0$[c]	—	—	—	68	114
Fe^{2+}	$H_2^{17}O$	$3\cdot1\times10^{-7}$	$7\cdot7$[c]	—	—	—	68	114
Gd^{3+}	H_2O	—	—	2	6	$2\cdot4$ (?)	62a	91
Mn^{2+}	CH_3OH	~5×10^{-6}	$5\cdot4{+2\cdot5 \atop -1}$	$2\cdot4$	$7\cdot2$	—	62a	119
Mn^{2+}	HCOOH	3×10^{-8}	$10\cdot2^{-1}$	$5\cdot1 \pm 0\cdot8$	~15	$2\cdot8 \pm 0\cdot3$	65a, 62a	115
Co^{2+}	CH_3OH	$5\cdot5\times10^{-5}$	$13\cdot8$ ([c]?)	—	—	—	63a, 62a	121
Co^{2+}	CH_3OH	$5\cdot5\times10^{-5}$	$13\cdot2$[c]	—	—	—	70	122
Ni^{2+}	CH_3OH	9×10^{-4}	$10\cdot6$[c]	—	—	—	66	123
Ni^{2+}	CH_3OH	1×10^{-3}	$16\cdot4$[c]	—	—	—	70	124
Co^{2+}	CH_3OH	2×10^{-6}	—	—	—	—	66	123
Co^{2+}	$CH_3OH + H_2O$	~3×10^{-5}[d]	—	—	—	—	70	122

(a) $\theta^* = 5°C$.　(b) $\theta^* = 65°C$.
(c) The activation energy is defined by the relation $\tau_b = \text{const } T^{-1} \exp(E_b/RT)$.
(d) τ_b is the life-time of the CH_3OH molecule.

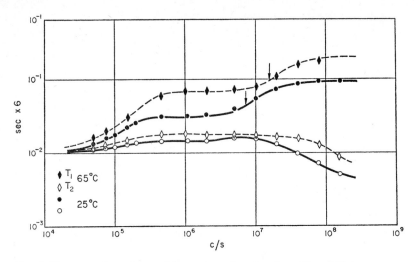

FIG. 6. Frequency dependence of the proton relaxation times T_1 and T_2 in aqueous MnCl$_2$ solution. The arrows indicate the dispersion step $\omega_s \tau_c^+ = 1$ according to equation (62a). $m = 1 \cdot 53 \times 10^{-3}$ mole kg^{-1} H$_2$O. (Hausser and Noack.[95])

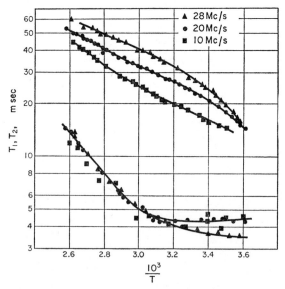

FIG. 7. Temperature dependence of the proton T_1 (upper curves) and T_2 (lower curves) observed at different resonance frequencies for aqueous solutions of Mn^{2+} of molality 5×10^{-3} mole kg^{-1} H$_2$O. The typical deviations from straight lines for the T_1 curves are due to the presence of the dispersion step. (Gutowsky, McCall and Slichter.[119])

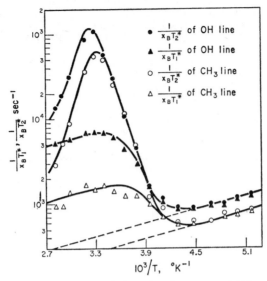

FIG. 8. Temperature dependence of the relaxation rates of different protons in methanol caused by the presence of Co^{2+} $(1/T^*_{1,2})$ are normalized to unit mole fraction x_B of $Co(ClO_4)_2$. Apart from the normalization factor, $(1/T^*_1)$ and $(1/T^*_2)$ are given by equations of type (64a) and (66) respectively. (Luz and Meiboom.[123])

REFERENCES

1. J. D. BERNAL, in *Liquids, Structure, Properties, Solid Interactions*, Ed. T. J. HUGHEL, p. 25, Amsterdam, London, New York, 1965.
2. H. G. HERTZ, *Ber. Bunsenges. physik. Chem.* **68**, 907 (1964).
3. E. R. ANDREW, *Nuclear Magnetic Resonance*, Cambridge 1956.
4. J. A. POPLE, W. G. SCHNEIDER and H. J. BERNSTEIN, *High-Resolution Nuclear Magnetic Resonance*, McGraw-Hill, New York, 1959.
5. C. P. SLICHTER, *Principles of Magnetic Resonance*, New York, Evanston, and London, 1963.
6. A. ABRAGAM, *The Principles of Nuclear Magnetism*, Oxford, 1961.
7. H. G. HERTZ, *Ber. Bunsenges. physik. Chem.* in press.
8. W. A. STEELE, *J. Chem. Phys.* **38**, 2404 (1963).
9. W. A. STEELE, *J. Chem. Phys.* **38**, 2411 (1963).
10. W. B. MONIZ, W. A. STEELE and J. A. DIXON, *J. Chem. Phys.* **38**, 2418 (1963).
11. D. E. WOESSNER, *J. Chem. Phys.* **37**, 647 (1962).
12. (a) H. SHIMIZU, *J. Chem. Phys.* **37**, 765 (1962); (b) *J. Chem. Phys.* **40**, 754 (1964).
13. F. PERRIN, *J. Phys. Radium* **5**, 497 (1934).
14. L. D. FAVRO, *Phys. Rev.* **119**, 53 (1960); see also *Fluctuation Phenomena in Solids*, Ed. R. E. BURGESS, p. 79, New York and London, 1965.
15. W. H. FURRY, *Phys. Rev.* **107**, 7 (1957).
16. D. E. WOESSNER, *J. Chem. Phys.* **36**, 1 (1962).
17. H. VERSMOLD, Diplomarbeit, University of Münster, Germany.
18. H. C. TORREY, *Phys. Rev.* **92**, 962 (1953).
19. S. CHANDRASEKHAR, *Revs. Modern Phys.* **15**, 1 (1943); or *Noise and Stochastic Processes*, Ed. N. WAX, p. 3, New York, 1954.
20. I. OPPENHEIM and M. BLOOM, *Can. J. Phys.* **39**, 845 (1961).

21. P. S. Hubbard, *Phys. Rev.* **131**, 275 (1963).
22. H. Schneider, *Ann. Physik* 7. Flg. **16**, 135 (1965).
23. P. S. Hubbard, *Phys. Rev.* **109**, 1153 (1958); **111**, 1746 (E) (1958).
24. P. S. Hubbard, *Phys. Rev.* **128**, 650 (1962).
25. G. W. Kattawar and M. Eisner, *Phys. Rev.* **126**, 1054 (1962).
26. H. Schneider, *Ann. Physik* 7. Flg. **13**, 313 (1964).
27. H. Schneider, *Z. Naturforsch.* **19a**, 510 (1964).
28. L. K. Runnels, *Phys. Rev.* **134**, A28 (1964).
29. R. L. Hilt and P. S. Hubbard, *Phys. Rev.* **134**, A 392 (1964).
30. D. Fenzke, *Ann. Physik* 7. Flg. **16**, 281 (1965).
31. D. E. Woessner, *J. Chem. Phys.* **42**, 1855 (1965).
32. I. Solomon, *Phys. Rev.* **99**, 559 (1955).
32a. H. Pfeifer, D. Michel, D. Sames and H. Sprint, *Mol. Phys.* **11**, 591 (1966).
33. H. Pfeifer, *Ann. Physik* 7. Flg. **8**, 1 (1961).
34. See, for example, M. E. Rose, *Elementary Theory of Angular Momentum*, New York, London, 1957.
35. M. D. Zeidler, *Ber. Bunsenges. physik. Chem.* **69**, 659 (1965).
36. P. S. Hubbard, *Phys. Rev.* **131**, 1155 (1963).
37. See for example, E. L. Hahn, *Physics Today* **6**, 11, 4 (1953).
38. See, for example, J. G. Powles, *Ber. Bunsenges. physik. Chem.* **67**, 328 (1963).
39. H. Y. Carr and E. M. Purcell, *Phys. Rev.* **94**, 630 (1954).
40. S. Meiboom and D. Gill, *Rev. Sci. Instr.* **29**, 688 (1958).
41. E. R. Hunt and H. Y. Carr, *Phys. Rev.* **130**, 2302 (1963).
42. N. Bloembergen, E. M. Purcell and R. V. Pound, *Phys. Rev.* **73**, 679 (1948).
43. H. C. Torrey, *Phys. Rev.* **130**, 2306 (1963).
44. I. Oppenheim, M. Bloom and H. C. Torrey, *Can. J. Phys.* **42**, 70 (1964).
45. W. A. Anderson and J. T. Arnold, *Phys. Rev.* **101**, 511 (1956).
46. R. Hausser, *Z. Naturforsch.* **18a**, 1143 (1963).
47. J. G. Powles and D. W. G. Smith, *Physics Letters* **9**, 239 (1964).
48. D. W. G. Smith and J. G. Powles, *Mol. Phys.* **10**, 457 (1966).
49. K. Krynicki, *Physica* **32**, 167 (1966).
50. See for instance, A. Loewenstein and T. M. Connor, *Ber. Bunsenges. physik. Chem.* **67**, 280 (1963).
51. D. W. McCall and D. C. Douglass *J. Phys. Chem.* **69**, 2001 (1965).
52. J. H. Wang, *J. Phys. Chem.* **69**, 4412 (1965).
53. See, for instance, K. E. Larsson in *Thermal Neutron Scattering*, Ed. P. A. Egelstaff, Chap. 8, London and New York, 1965.
54. P. K. Sharma and S. K. Joshi, *Phys. Rev.* **132**, 1431 (1963); P. K. Sharma and R. P. Gupta, *Phys. Rev.* **138**, A 1045 (1965).
55. J. W. Hennel, *Acta Phys. Polon.* **18**, 387 (1959).
56. J. H. Simpson and H. Y. Carr, *Phys. Rev.* **111**, 1201 (1958).
57. D. E. Woessner, *J. Chem. Phys.* **40**, 2341 (1964).
58. H. G. Hertz and M. D. Zeidler, *Ber. Bunsenges. physik. Chem.* **68**, 821 (1964).
59. D. W. Posener, *Australian J. Phys.* **13**, 168 (1960).
60. E. B. Treacy and Y. Beers, *J. Chem. Phys.* **36**, 1473 (1962).
61. P. Waldstein, S. W. Rabideau and J. A. Jackson, *J. Chem. Phys.* **41** 3407 (1964).
62. M. Weissmann, *J. Chem. Phys.* **44**, 422 (1966).
63. F. Fister and H. G. Hertz, to be published.
64. Z. Luz and S. Meiboom, *J. Chem. Phys.* **39**, 366 (1963).
65. R. E. Connick and E. D. Stover, *J. Phys. Chem.* **75**, 2075 (1961).
66. J. A. Glasel, *Proc. Nat. Ac. Sci.* **55**, 479 (1966).
67. M. J. Stevenson and C. H. Townes, *Phys. Rev.* **107**, 635 (1957).
68. S. W. Rabideau and J. A. Jackson, *J. Chem. Phys.* **41**, 3405 (1964).
69. J. Blicharski, J. W. Hennel, K. Krynicki, J. Mikulski, T. Waluga and G. Zapalski, *Compte rendu du 9e Colloque Ampère*, 452 (1960).
70. A. M. Pritchard and R. E. Richards, *Trans. Faraday Soc.* **62**, 2014 (1966).
71. E. v. Goldammer, private communication.

72. G. Bonera and A. Rigamonti, *J. Chem. Phys.* **42**, 171 (1965).
73. M. Eisner and R. Mitchell, *Bull. Amer. Phys. Soc.* **6**, 363 (1961).
74. J. G. Powles and R. Figgins, *Mol. Phys.* **10**, 155 (1966).
75. D. K. Green and J. G. Powles, *Proc. Phys. Soc.* **85**, 87 (1965).
76. W. B. Moniz and H. S. Gutowsky, *J. Chem. Phys.* **38**, 1155 (1963).
77. D. Herbison-Evans and R. E. Richards, *Mol. Phys.* **7**, 515 (1964).
78. Y. Masuda, J., *Phys. Soc. Japan* **11**, 670 (1956).
79. D. E. O'Reilly and G. E. Schacher, *J. Chem. Phys.* **39**, 1768 (1963).
80. R. W. Mitchell and M. Eisner, *J. Chem. Phys.* **33**, 86 (1960).
81. A. M. Pritchard and R. E. Richards, *Trans. Faraday Soc.* **62**, 1388 (1966).
82. R. W. Rampolla, R. C. Miller and C. P. Smyth, *J. Chem. Phys.* **30**, 566 (1959).
83. J. A. Saxton, R. A. Bond, G. T. Coats and R. M. Dickinson, *J. Chem. Phys.* **37**, 2132 (1962).
84. F. Buckley and A. A. Maryott, *Tables of Dielectric Dispersion Data for Pure Liquids and Dilute Solutions*, Nat. Bur. Stand. Circ. 589 (1963).
85. J. Clifford, *Trans. Faraday Soc.* **61**, 1276 (1965).
86. H. G. Hertz and M. D. Zeidler, unpublished data.
87. F. Franks and D. J. G. Ives, *Quart. Rev.* **20**, 1 (1966).
88. D. Beckert and H. Pfeifer, *Ann. Physik.* 7. Flg. **16**, 262 (1965).
89. J. S. Dohnanyi, *Phys. Rev.* **125**, 1824 (1962).
90. D. Sames, *Ann. Phys.* 7. Flg. **15**, 363 (1965).
91. N. Bloembergen and L. O. Morgan, *J. Chem. Phys.* **34**, 842 (1961).
92. U. Lindner, *Ann. Phys.* 7. Flg. **16**, 319 (1965).
93. H. S. Gutowsky and J. Chow Tai, *J. Chem. Phys.*, **39**, 208 (1963).
94. K. Günther and H. Pfeifer, *Zhurn. strukt. Khimi* **5**, 193 (1964); English translation, *J. Struct. Chem.* **5**, 177 (1964).
95. R. Hausser and F. Noack, *Z. Physik* **182**, 93 (1964).
96. K. H. Hausser, G. J. Krüger and F. Noack, *Z. Naturforsch.* **20a**, 91 (1965).
97. H. E. Heinze and H. Pfeifer, *Z. Physik* **192**, 329 (1965).
98. J. R. Zimmerman and W. E. Brittin, *J. Phys. Chem.*, **61**, 1328 (1957).
99. H. G. Hertz and M. D. Zeidler, *Ber. Bunsenges. physik. Chem.* **67**, 774 (1963).
100. L. Endom, H. G. Hertz, B. Thül and M. D. Zeidler, *Ber. Bunsenges. physik. Chem.* (in press).
101. B. P. Fabricand, S. S. Goldberg, R. Leifer and S. G. Ungar, *Mol. Phys.* **7**, 425 (1964); **9**, 399 (1965).
102. G. T. Jones and J. G. Powles, *Mol. Phys.* **8**, 607 (1964).
103. R. E. Connick and D. N. Fiat, *J. Chem. Phys.* **39**, 1349 (1963).
104. M. Alei and J. A. Jackson, *J. Chem. Phys.* **41**, 3402 (1964).
105. T. J. Swift, O. G. Fritz and T. A. Stephenson, *J. Chem. Phys.* **46**, 406 (1967).
106. J. H. Swinehart and H. Taube, *J. Chem. Phys.* **37**, 1579 (1962).
107. H. G. Hertz, *Ber. Bunsenges. physik. Chem.* **67**, 311 (1963).
108. V. M. Vdovenko and V. A. Shcherbakov, *Zhurn. struktur. Khimi* **1**, 28, 122 (1960); English translation, *J. Struct. Chem.* **1**, 25, 111 (1960).
109. M. Eisenstadt and H. L. Friedman, to be published.
110. See, for instance, the most recent paper on this subject: R. E. Glick and K. C. Tewari, *J. Chem. Phys.* **44**, 546 (1966), and literature cited therein.
111. W. Lohmann, *Z. Naturforsch.* **19a**, 814 (1964).
112. S. Danyluk and E. S. Gore, *Nature* **203**, 748 (1964).
113. J. Clifford and B. A. Pethica, *Trans. Faraday Soc.* **61**, 1 (1965).
114. T. J. Swift and R. E. Connick, *J. Chem. Phys.* **37**, 307 (1962); **41**, 2553 (E) (1964).
115. R. Sperling and H. Pfeifer, *Z. Naturforsch* **19a**, 1342 (1964).
116. H. Pfeifer, *Z. Naturforsch* **17a**, 279 (1962).
117. H. S. Gutowsky, D. W. McCall and C. P. Slichter, *J. Chem. Phys.* **21**, 279 (1953).
118. H. M. McConnell, *J. Chem. Phys.* **28**, 430 (1958).
119. R. E. Bernheim, T. H. Brown, H. S. Gutowsky and D. E. Woessner, *J. Chem. Phys.* **30**, 950 (1959).

120. R. E. CONNICK and R. E. POULSON, *J. Chem. Phys.* **30,** 759 (1959).
121. D. GESCHKE and H. PFEIFER, *Zhurn. strukt. Khimi* **5,** 201 (1964); English translation, *J. Struct. Chem.* **5,** 184 (1964).
122. Z. LUZ and S. MEIBOOM, *J. Chem. Phys.* **40,** 1058 (1964).
123. Z. LUZ and S. MEIBOOM, *J. Chem. Phys.* **40,** 2686 (1964).
124. Z. LUZ and S. MEIBOOM, *J. Chem. Phys.* **40,** 1066 (1964).
125. R. E. CONNICK and D. FIAT, *J. Chem. Phys.* **44,** 4103 (1966).

CHAPTER 6

SOLVENT EFFECTS AND NUCLEAR MAGNETIC RESONANCE

Pierre Laszlo

Institut de Chimie des Substances Naturelles, Gif-sur-Yvette, France,
and Department of Chemistry, Princeton University, Princeton,
New Jersey, 08540, U.S.A.

CONTENTS

SUMMARY

Solvent effects are defined in a very general way as consequences of intermolecular forces, and their relevance to NMR arises from the relatively high concentrations necessary, compared with other spectroscopic methods. The purpose of this review is to encompass solvent effects on the NMR spectra of various nuclei, to present the characteristics of particular solvents, and to suggest means for disposing of unwanted solvent shifts, and to show how the deliberate use of solvent shifts, conversely, may help in the determination of molecular conformations and structures. Once the nature of the reference compounds and inactive solvents is defined, the concept of the reaction field will be critically examined, and shown to be in fact partly non-specific; one then proceeds to a consideration of specific effects, for polar solutes, choosing to discuss theoretical treatments of the equilibria involved in hydrogen bonding solvents, while the crucial underlying assumptions are considered with respect to the special case of benzene. Considerations of liquid crystals, conformational equilibria, exchange phenomena are included.

INTRODUCTION

The proposed title for this chapter was to be "Solvent Effects *in* Nuclear Magnetic Resonance". We have chosen instead to discuss here "Solvent Effects *and* Nuclear Magnetic Resonance", since NMR is a quite exceptional potential tool for the study of molecular interactions in liquids, so that our aim will be threefold:

(a) to provide a *description* of these solvent effects, that is to present a critical review of the modifications to the shieldings and coupling constants for various nuclei incurred when the solvent for the studied molecule is changed from a reference solvent;

(b) to *discuss* the origin of these solvent effects in terms of the principal intermolecular forces occurring, whether "non-specific" like the reaction field or the van der Waals shifts, or "specific" like the hydrogen bond or the orientation effects encountered between aromatic solvents and polar solutes;

(c) to point out the use that can be made of these effects: both inside NMR spectroscopy (to help with spectral analysis, the determination of signs of coupling constants, etc.) and outside NMR spectroscopy (to help in the precise determination of organic structures when special orientation effects, in a nematic phase or more commonly in an aromatic solvent, are used to study the geometry of the solute and its deformations).

It is impossible to provide a full coverage of this subject with all pertinent references, since these are overwhelmingly numerous. The author has instead chosen to concentrate on a few points, and to express his personal views, if not the catholicity of his interests, rather freely. The works chosen as examples for discussion are a few recent leading references, to which the reader is urged to address himself whenever the coverage of the present review is inadequate.

As far as notation is concerned, we have not superimposed any uniformity, and we have instead attempted to respect the symbolisms that the various authors have considered most appropriate for the description of their ideas.

LIST OF ABBREVIATIONS

TMS   tetramethylsilane.

ASIS   Aromatic Solvent(s) Induced Shift(s).

Δ_Y^X positive sign if a diamagnetic shift (towards high field), going from solvent X to solvent Y.

Before proceeding further, it may be useful to give very brief classifications of the more usual solvents, these examples are chosen in the realm of proton NMR, and of the principal intermolecular forces (see Tables 1 and 2).

1. GAS-TO-LIQUID PHASE SHIFTS; DISPERSION FORCES

A downfield shift always results for protons in changing substances from the gaseous to the liquid state, and is due to the non-specific dispersion forces. We shall not consider the intermolecular shifts in the gas phase

TABLE 1. LIST OF THE MORE USUAL SOLVENTS FOR ^{1}H NMR

Solvent	Dielectric Constant ϵ^a	Perdeuterated[b]	Low temp.[c]
n-Hexane	1·90	−	+
Me$_4$C		−	+
TMS	1·89	−	+
C$_6$H$_{12}$	2·02	+	−
C$_6$H$_{11}$Me	1·99	+	+
CCl$_4$	2·24	0	−
CHCl$_3$	4·81	+	+
Dioxane	2·22	+	−
Et$_2$O	4·34	+	+
CS$_2$	2·64	0	+
C$_6$H$_6$	2·26	+	−
PhMe (toluene)	2·35	+	+
C$_5$H$_5$N	12·3	+	
C$_6$H$_{10}$O (cyclohexanone)	18·3	−	−
Me$_2$CO	20·7	+	+
MeOH	33·6	+	−
MeCN	35·1	+	+
MeNO$_2$	35·0	−	−
Me$_2$NCHO	37·6	+	+
Me$_2$SO	46	+	−
AcOH	6·19	+	−
H$_2$O	80·4	+	−
H$_2$C⇒CHCl		−	+
CH$_2$Cl$_2$	9·08	+	+
Sulpholane	44	+	− *

^{a} Dielectric constant at ambient temperatures (30–40°C).
^{b} Commercially available at the time of writing.
^{c} Down to at least −50°C; (*) suitable for high temperatures (< 250°C).

TABLE 2. THE MAIN INTERMOLECULAR FORCES

Nature	Strength[a]	Additivity	Temperature effect
Dipole–dipole	1 kcal mole^{-1}	YES	YES also variations with the solvent dielectric constant
Polarization (dipole induced–dipole)	10^{-2} kcal mole^{-1}	NO	±
Dispersion (van der Waals)[b]	10^{-1} kcal mole^{-1}	± (partially specific)	NO

^{a} Order of magnitude.
^{b} Induced dipole–induced dipole (or, more generally, the interaction of two transition densities), their fundamental property is the variation with distance, in r^{-6}–r^{-12} (the so-called 6–12 potential).

caused by the effect of pressure, perturbation from introduced foreign gases, or temperature changes. We refer the reader to the leading references of Raynes, Buckingham and Bernstein,[1] Petrakis and Bernstein,[2] Widenlocher, Dayan and Vodar,[3] Widenlocher and Dayan,[4, 5] Rummens and Bernstein,[6] Gordon and Dailey,[7] Buckingham,[8] Petrakis and Sederholm.[9] In gases, the chemical shifts can be separated into van der Waals dispersion terms σ_W, polar electric terms σ_E, and a contribution from the magnetic anisotropy σ_A of the other molecules in the medium. Rummens and Bernstein[6] have shown that inclusion of a repulsive part $1 - (r_0/r)^6$ in the dominating σ_W markedly improves the agreement with experiment. Buckingham[8, 10] has provided a quite adequate description of the σ_E term. Schug[11] has refined the anterior treatments of σ_A, to consider not only nearest neighbours, but *all* molecules in the solution; his assumptions are (i) cylindrical symmetry for the magnetically anisotropic molecules, (ii) complete randomness in the liquid state, (iii) the σ_A term being proportional to the volume fractions of the anisotropic molecules. The good fit obtained for benzene in n-hexane, carbon tetrachloride, and carbon disulphide, gave an empirical set of σ_W values, but the same procedure did not work with toluene, which does not approach cylindrical symmetry, as benzene does.

As for the gas-to-liquid shifts, Bothner-By[12] has devised an empirical method to separate the solute and the solvent contributions. Using β_i^j to denote the excess shift in ppm from gas to liquid solvent j, for solute i:

$$\beta_i^j = \delta_j^0 - \delta_i^j + \frac{2\pi}{3} \chi_i \tag{1}$$

where δ_i^0 and δ_i^j are respectively the gas and liquid phase shifts, referred to some external standard, and $(2\pi/3)\chi_i$ the usual bulk susceptibility correction for the solute. The values of the β_i^j's found by Bothner-By[12] are shown in Table 3. Below each solute and solvent in Table 3 is given their increments x_i or y_j, according to equation (2) which predicts rather

$$-\beta_i^j = x_i y_j \tag{2}$$

accurately a large number of experimental shifts. It is observed that the y_i terms are increased by the existence of permanent dipole moments, while the x_i's depend upon the acidic character of the C—H bond which is being polarized by the solvent, and upon the ease of approach of the solvent.

Howard, Linder and Emerson[13] have used a more theoretical approach to evaluate the dispersion shift of a solute at infinite dilution in an isotropic solvent. This shift σ_W had been shown by Stephen,[14] and by Marshall and Pople[15] to be proportional to E^2, the square of the electrostatic field acting upon the molecule, the best values of the proportionality constant being taken either as -0.74×10^{-12} [15] or as -1×10^{-12} ppm cm^2 (e.s.u.)$^{-1}$. [16, 17]

TABLE 3. GAS-TO-LIQUID CHEMICAL SHIFTS (ppm)

Solvent (y_j)	Me_2CO (0·017)	CMe_4 (0·027)	C_5H_{10} (0·030)	C_6H_{12} (0·033)	CH_2Cl_2 (0·053)	CCl_4 (0·052)
Solute (x_i)						
C_5H_{10} (5·1)	−0·08	−0·13	−0·15	−0·17	−0·30	−0·29
C_2H_6 (5·5)	−0·08		−0·16	−0·20	−0·30	
CMe_4 (6·0)	−0·10	−0·16	−0·18	−0·20	−0·30	−0·33
Me_2O (6·0)	−0·13	−0·13	−0·15	−0·18	−0·39	
$H_2C=CH_2$ (6·7)	−0·11	−0·18	−0·20	−0·22		
TMS (7·0)	−0·15			−0·25	−0·32	−0·36
CH_2Cl_2 (9·0)	−0·61	−0·20	−0·24	−0·30	−0·57	−0·53
MeCl (9·1)	−0·38	−0·20		−0·28	−0·55	−0·53

This static electric field is equivalent to the free energy F for the dispersion of the solute:

$$F = -\tfrac{1}{4}\langle m^2 \rangle_2 g \, \frac{\nu_1}{\nu_1 + \nu_2} \tag{3}$$

where subscripts 1 and 2 refer to solvent and solute, respectively, ν is the mean frequency of absorption, $\langle m^2 \rangle_2$ being the non-zero quadratic mean of the solute's dipole oscillating in a medium described by the parameter g:

$$g = \frac{2n^2 - 2}{2n^2 + 1} \frac{1}{a^3} \tag{4}$$

where n is the refractive index of the medium, and a the radius of the free solute molecule, assimilated to a sphere.

$$F = -\alpha_2 E^2/2 \tag{5}$$

where α_2 is the solute polarizability. Hence,

$$E^2 = \tfrac{1}{2}\langle m^2 \rangle_2 \frac{g}{\alpha_2} \frac{\nu_1}{\nu_1 + \nu_2} \tag{6}$$

and the only problem left is that of the evaluation of ν_1 and ν_2. There are two distinct procedures for their estimation, either through the London approximation, namely: $\nu = I/h$, or through the proportionality of $\langle m^2 \rangle_2$ to the molar magnetic susceptibility χ_M[13]:

$$\langle m^2 \rangle_2 = -6 \frac{m_e c^2}{N} \chi_M \tag{7}$$

which, taken together with the London expression[18]

$$\langle m^2 \rangle_2 = \tfrac{3}{2} h \nu_2 \alpha_2 \tag{8}$$

gives:

$$\nu_2 = -4 \frac{m_e c^2 \chi_M}{h N \alpha_2} \tag{9}$$

and

$$\sigma_W = -\tfrac{3}{4} \times 10^{-12} \, hg \, \frac{\nu_1 \nu_2}{\nu_1 + \nu_2}. \tag{10}$$

2. THE PROBLEM OF REFERENCE STANDARDS FOR THE MAGNETIZED SAMPLE

2.1. *External Reference and Bulk Susceptibility Corrections*

The solvent effect on the chemical shift is, by definition, the change in the resonance positions of some given proton of the solute at infinite dilution between solvents A and B, measured from some common reference. For this purpose, two distinct techniques, which are introduced in the book of Pople, Schneider, and Bernstein,[19] can be used. In one method a reference compound, separated from the sample and serving as an "absolute" standard is used and a correction is applied to account for the difference in magnetic susceptibility between the reference and the sample. In the other method, the reference compound is dissolved in the solution studied and thus there is no need to apply a susceptibility correction. However, the molecule may not be totally inert, chemically or physically, towards the other molecules of the medium, so that the observed solvent effect would become the difference between the *large* solvent shift, e.g. for a polar solute, and the *small* solvent effect for a non-polar reference such as a hydrocarbon reference like cyclohexane or tetramethylsilane. An evaluation of the respective merits of these approaches follows.

The technique used by Fratiello and Douglass[20] to measure volume magnetic susceptibilities, following an earlier suggestion of Reilly, McConnell and Meisenheimer,[21] uses a pair of stationary coaxial cylindrical tubes in the probe. Since the sample does not spin, a broad resonance is obtained. The reference is enclosed in the inner tube, and the sample is in the surrounding annular region, two peaks separated by $\Delta \nu$ are observed, such that:

$$\frac{\Delta \nu}{4 \pi \nu_0} = \chi_{\text{ref}} \frac{a^2}{r^2} - \chi_g \left(\frac{a^2 - b^2}{r^2} \right) - \chi_{\text{sol}} \frac{b^2}{r^2} \tag{11}$$

a and b being, respectively, the inner and outer radii of the central tube, r the mean radius of the peripheral torus; χ_{ret}, χ_g, and χ_{sol} are, respectively, the magnetic susceptibilities for the reference (for which benzene or cyclohexane are the best choice), the glass walls, and the sample of unknown susceptibility. Fratiello and Douglass[20] found a value for $\chi_2 = -0\cdot805 \pm 0\cdot005 \times 10^{-6}$. Frost and Hall[22] find that such a non-spinning cylindrical external reference is difficult to make without introducing a large geometrical asymmetry, or tilting of one tube relative to the other. They proposed instead to have the external reference in a spherical cavity, for which no susceptibility correction is needed in principle, and they carefully examine the correspondence between the two types of geometries for the standard container, cylindrical or spherical. Writing the relation:

$$\delta = \Delta\nu + g(\chi_{\text{ret}} - \chi_{\text{sol}}) \qquad (12)$$

between the true chemical shift δ and the observed shift with an external reference $\Delta\nu$, g being the shape factor for the interface of contact, the difference between the cylindrical and the spherical-referred shifts is:

$$\Delta\nu_{\text{cyl}} - \Delta\nu_{\text{sph}} = (g_{\text{cyl}} - g_{\text{sph}})(\chi_{\text{ret}} - \chi_{\text{sol}}). \qquad (13)$$

A plot of $\Delta\nu_{\text{cyl}} - \Delta\nu_{\text{sph}}$ against $\chi_{\text{ret}} - \chi_{\text{sol}}$, for dioxane–water mixtures, is indeed a straight line passing through the origin. Its slope is $2\cdot080$, somewhat inferior to the theoretical prediction of $2\cdot095$, which may be due to a non-zero value of g_{sph} or to a value of g_{cyl} smaller than $2\pi/3$.

Frei and Bernstein[23] have directly compared external references in a cylindrical and a spherical cavity in the same solution, and they also reported a shape factor $g = 2\cdot058$ (smaller than $2\pi/3 = 2\cdot095$) attributing this to a departure from sphericity of the blown sphere. Lussan[24] has used an elaborate method to check that the true shape factor is identical to $2\pi/3$. He remarked that the failure of the measurements of Bothner-By and Glick[25-9] to lead to this value, but to much higher values of g between $2\cdot33$ and $3\cdot00$, is due to the specific interactions existing in the binary samples studied, such as the methylene chloride–methylene bromide system. In order to approximate ideal solution behaviour, Lussan[24] dissolved small amounts of a paramagnetic compound ($CoCl_2 . 6H_2O$, with a maximum concentration of $0\cdot1$ M) in ethanol, since it has been shown by Venkarteswarlu and Sriraman[30] that hydrogen bonding does not affect the magnetic susceptibility; because the solution, which is referenced to external water, remains dilute it can be considered ideal and one can vary the susceptibility by large amounts using variable quantities of the Co(II) salt. When $\Delta\chi$ remains smaller than $0\cdot5 \times 10^{-6}$, the experimental and theoretical lines $\Delta\delta = (2\pi/3)\Delta\chi$ are identical to within $\pm10^{-8}$.

Frost and Hall[31] have extended the theory of Pople, Schneider and

Bernstein[19] to spherical cavities, inside a cylindrical sample container. In the general case:

$$\delta = \Delta\nu - \left[\left(\frac{4\pi}{3} - \alpha_1\right)\chi_{ref} + (\alpha_1 - \alpha_2)\chi_g + \left(\alpha_2 - \frac{4\pi}{3}\right)\chi_{sol}\right] \quad (14)$$

which reduces to:

$$\delta = \Delta\nu + \left(\alpha - \frac{4\pi}{3}\right)(\chi_{ref} - \chi_{sol}) \quad (15)$$

when $\alpha_1 = \alpha_2 = \alpha$; $(\alpha - 4\pi/3)$ is the g factor of above.

For a cylindrical reference, with $\alpha = (2\pi/3)$, the well-known susceptibility correction is:

$$\delta = \Delta\nu - \frac{2\pi}{3}(\chi_{sol} - \chi_{ref}). \quad (16)$$

For a perfect sphere, $\alpha = 4\pi/3$ and the observed shift is the true separation. However, it is an experimental fact that, for the actual spheres that can be blown, α_1 differs from α_2. This is found simply by interchanging, in two successive measurements, the contents of the inner sphere and of the outer volume: the difference between both measurements is non-zero and depends upon the real shape of the imperfect sphere *and* upon the material from which it is made. The method of the cylindrical reference is therefore capable of better accuracy than the other method.[32] From equation (11) above, it follows that for a series of measurements with the same set of coaxial tubes, and an identical reference:

$$\Delta\nu = A\chi_{sol} - B - C. \quad (17)$$

A calibration curve can be established relating the unknown χ_{sol} with the measured $\Delta\nu$. To get the largest separations, acetone was chosen by Lauwers and van der Kelen[32] as the external reference. The excellent linearity obtained proves the validity of equation (17) and of the method, the constant slope depending only upon the sample tube parameters. However, there is a vertical translation of this calibration line when the position of the probe insert is varied, which accounts for the Fratiello and Christie[34] statement that recalibration of the coaxial tubes has to be frequent. This observation has no obvious explanation. Contrasting statements, by Douglass and Fratiello[35] that $\Delta\nu$ depends upon the X and Z homogeneity settings, and by Lauwers and van der Kelen[32] that it does not depend upon these, nevertheless point out to the necessity of a symmetrical peak shape for such susceptibility measurements. The precision obtained by Lauwers and van der Kelen[32] is about 2%. Nash and Maciel[33] have used in their ^{13}C studies two concentric, thin-walled spherical bulbs about 0·2 and 1·4 ml in volume, the small inner bulb containing the external reference, a saturated solution of sodium acetate enriched to 55% ^{13}C at the carbonyl carbon.

2.2. *Relative Merits of the Internal and the External Reference*

Let us restate the problem. Use of an external reference, enclosed in a special container separated by a glass wall from the sample investigated avoids interactions involving the reference compound. On the other hand, whereas an internal reference allows for direct measurement of chemical shifts, a susceptibility correction, which depends upon the shape of the cavity, is needed with an external reference. Lussan[24] has pointed out that the magnetic susceptibilities of organic compounds are of the order of -10^{-6}; if chemical shifts are desired within 10^{-2} ppm, the values of the susceptibilities ought to be determined with a comparable absolute precision, which implies better than 0·25% relative accuracy. The various NMR susceptibility determinations have accuracies no better than 2%, and are to be rejected on that account. Lussan[24] himself uses a Pascal balance[36] at exactly the same temperature as the NMR probe and obtains results reproducible to within 1–3%. Because of the tediousness of such corrections which are necessary with the external referencing procedure we consider that the better method is that of internal referencing if one *carefully checks* in preliminary experiments that the chosen reference does not interact with the solvent. Even if there is a slight interaction, the implicit phenomenological definition of "solvent effects" for most people incorporates it, so that no great harm is done, if the discussion is confined to an empirical level. In Table 4 we present the data of Lussan,[24] to the effect that an internal reference is exactly equivalent to the *correct* external reference.

However, no reference is perfect; for the common organic compounds, tetramethylsilane (TMS) is reasonably inert, but still some variations may occur, as displayed by the chemical shift between the singlet resonances of cyclohexane and tetramethylsilane in dilute solution in various solvents, measured accurately by Reisse and Ottinger[37] (see Table 5). These measurements emphasize that caution is necessary in the interpretation and discussion of solvent shift data particularly in deciding when small observed differences are really significant.

2.3. *The Choice of the Internal Reference*

For the regular proton NMR work, tetramethylsilane (TMS) is by far the best internal reference, cyclohexane the next best; hexamethyl disiloxane has some disadvantages but is also useful. As water-soluble internal references, Jones *et al.*[38] have suggested acetonitrile, dioxane and tert-butanol; however, these substances are not chemically inert nor isotropic, further they

TABLE 4. CHEMICAL SHIFTS OF CHLOROFORM IN CHLOROFORM–
CYCLOHEXANE MIXTURES

Mole fraction CHCl$_3$	Internal ref.[a]	External ref.[b]
0·171	5·695	5·715
0·305	5·715	5·715
0·500	5·740	5·740
0·575	5·745	5·745
0·619	5·775	5·775
0·802	5·805	5·805
0·904	5·815	5·815
1·000	5·845	5·845

[a] Relative to cyclohexane (ppm).
[b] Converted to external cyclohexane, and corrected for the bulk susceptibility (ppm).

TABLE 5. VARIATIONS OF THE CYCLOHEXANE–TMS CHEMICAL SHIFT
IN VARIOUS SOLVENTS[a]

Solvent	60 Mc/s shift (c/s, ±0·2)
CCl$_4$	0·0
CDCl$_3$	0
C$_5$H$_5$N	+4·9
PhNH$_2$	+2
C$_6$H$_6$	+1·5
PhNO$_2$	+4·3
Dioxane	0
CS$_2$	+0·7
Me$_2$CO	0
MeOH	−1·3
MeCN	−0·3
AcOH	−0·1

[a] Relative to the value in carbon tetrachloride.

are not sufficiently volatile to be disposed of easily for recovery of the pure sample.

As a suitable internal reference for ^{13}C studies, Maciel and Natterstad[39] suggest a 1:1 ratio of dimethyl carbonate to sample, since:

(a) it has a conveniently intense reference signal;
(b) it is a relatively good solvent;
(c) the small variations encountered are of the order of the usual experimental uncertainties in comparing the results of the *external* referencing technique from different investigators;
(d) the maximum solvent shift of the carbonyl resonance is ∼1·1 ppm, and is for the methyl resonance ∼1·8 ppm, (see Table 6).

TABLE 6. REFERENCE SIGNALS OF DIMETHYL CARBONATE IN VARIOUS
SOLVENTS[a]

Solvent	$^{13}C=O$	^{13}Me	Δ
Dimethyl carbonate	$-28 \cdot 0$	$+74 \cdot 9$	$102 \cdot 9$
C_6H_6	$-27 \cdot 2$	$+74 \cdot 6$	$101 \cdot 8$
$PhNMe_2$	$-28 \cdot 1$	$+74 \cdot 6$	$102 \cdot 7$
CCl_4	$-27 \cdot 9$	$+75 \cdot 4$	$103 \cdot 3$
MeCN	$-28 \cdot 3$	$+73 \cdot 9$	$102 \cdot 2$
Dioxane	$-27 \cdot 9$	$+75 \cdot 7$	$103 \cdot 6$
THF[b]	$-28 \cdot 0$	$+75 \cdot 0$	$103 \cdot 0$
AcOH	$-28 \cdot 2$	$+75 \cdot 1$	$103 \cdot 3$
MeI	$-27 \cdot 8$	$+74 \cdot 4$	$102 \cdot 2$
Me_2CO	$-28 \cdot 0$	$+74 \cdot 9$	$102 \cdot 9$
m-C_6H_4–Me, OH	$-28 \cdot 3$	$+73 \cdot 9$	$102 \cdot 2$

[a] ppm from external benzene.
[b] Tetrahydrofuran, $(CH_2)_4O$.

For ^{31}P studies it has been proposed that phosphorus oxide, P_4O_6, should be used as a reference in preference to phosphoric acid or trimethyl phosphite.[40] P_4O_6 has a high phosphorus content per unit volume and an extremely narrow (0·3 c/s) ^{31}P resonance line which appears in a spectral region (112·5 ppm below the resonance of phosphoric acid H_3PO_4) convenient for referencing the majority of phosphorus compounds. In many fluorine studies $CFCl_3$ has been used as both the solvent and internal reference, the chemical shift at infinite dilution in this weakly associative solvent being the standard procedure recommended by Filipovich and Tiers.[41] Taft et al.[42] in their study of the fluorobenzenes, found it convenient to employ 1,1,2,2-tetrachloro-3,3,4,4-tetrafluorocyclobutane as internal reference, on account of the high fluorine content, so that a 2% concentration per volume is sufficient. They also used an external reference, and showed the additivity to an excellent approximation, of the shifts referred to the above internal reference, and of that of this reference itself with respect to the external reference, to provide the required externally referred chemical shifts. The internal referencing problem for fluorine resonance is not great, due to the variety and inertness of the existing Freons.

2.4. The Problem of the Inert Reference Solvent

Even more important than the problem of the reference compound, which anyway is present in a small amount, is that of the reference solvent in which shifts are measured to serve as *intra*molecular constants to be compared with the chemical shifts in another active solvent, where we observe a combination of these intramolecular, and intermolecular factors. The correct procedure is obviously to measure the gas phase shift of the pure solute, instead of using a solvent believed to be inactive, but this is seldom practical. The ideal inert solvent must possess a dielectric constant as near unity as possible since, as discussed in Section 4, polar solutes are submitted to an electric field shift which increases with the dielectric constant of the medium. We defer to another section the discussion of the disadvantages of chloroform, widely used due to its good solubility properties, but which forms hydrogen bonds with most polar solutes, thus creating extraneous solvent shifts.

The special case of carbon tetrachloride will now be discussed. This solvent has been extensively utilized in NMR studies, for historical reasons essentially. In 1958 Tiers[43] defined the *tau scale*, and his definition has sufficient merit to be still used. It refers to the chemical shift for a molecule compared with

TABLE 7. COMPARISON OF CHEMICAL SHIFTS[a] IN CARBON TETRACHLORIDE AND CYCLO-HEXANE SOLUTIONS

Solute protons	CCl_4	C_6H_{12}	Δ	Ref
MeCl	186	172	+14	(44)
CH_2Cl_2	318	309	+9	(44)
$CHCl_3$	436	426	+10	(44)
CH_3CH_2Cl	90	83	+7	(44)
CH_3CH_2Cl	211	205	+6	(44)
CH_3CHCl_2	352	346	+6	(44)
CH_3CHCl_2	134	118	+16	(44)
CH_2ClCH_2Cl	222	215	+7	(44)
CH_3CCl_3	165	158	+7	(44)
$CH_2ClCHCl_2$	236	229	+7	(44)
$CH_2ClCHCl_2$	344	337	+7	(44)
CH_2ClCCl_3	257	250	+7	(44)
$CHCl_2CHCl_2$	356	347	+9	(44)
$CHCl_2CCl_3$	367	359	+8	(44)
H_A in I[b]	414·4	409·9	+4·5	(45)
H_B in I[b]	487·1	485·8	+1·3	(45)
cis-CHCl=CHCl	298·9	288·2	+10·7	(46)
trans-CHCl=CHCl	294·5	287·1	+7·4	(46)
$Me_2CO \; \delta(^{13}C=O)$	+1·3	+2·4	−1·1	(47)
$Me_2CO \; \delta(C=^{17}O)$	−5	−8	+3	(48)
$Me_2CO \; \nu(C=O)$ stretch	1719 cm^{-1}	1723 cm^{-1}		(49)

[a] At 60 Mc/s, in c/s from internal TMS for ^1H resonance.

b

I

internal TMS *at infinite dilution in carbon tetrachloride* which obviously was chosen because it is aprotic, isotropic, and was believed to be completely inert towards all solutes. In fact, the latter is not true, and the use of carbon tetrachloride may also lead to solvent shifts when compared with the hydrocarbon reference solvents, n-hexane, cyclohexane, neopentane (CMe_4), and tetramethylsilane (see Table 7). These $\Delta_{CCl_4}^{C_6H_{12}}$ shifts are in the same direction as the much larger shifts due to hydrogen bonding, e.g. with $CHCl_3$, MeOH or CF_3COOH, and can be interpreted (cf. reference 48) in terms of a higher pi-bond polarity in the more polar CCl_4 (for example, in the case of acetone increased importance of canonical structure III) or in the nearly

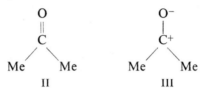

II   III

equivalent formulation of the reaction field and van der Waals interactions.[47] These effects in carbon tetrachloride solution can be attributed to:

(a) The reaction field (see Section 4) since carbon tetrachloride ($\epsilon = 2\cdot24$) is a slightly more polar solvent than cyclohexane ($\epsilon = 2\cdot02$). If one takes $(\epsilon - 1)/(\epsilon + 1)$ as the simplest form for the solvent dielectric function, it gives values of $0\cdot383$ (CCl_4) and $0\cdot338$ (C_6H_{12}), which are significantly different, and would lead to a $\sim10\%$ change in the shielding of a proton arising from the reaction field.

(b) Dispersion effects, due to the four very polarizable chlorine atoms of the solvent interacting with the solute are also to be taken into consideration. These would be less with chlorine than with iodine.

(c) The occurrence of charge transfer complexes between the halogen atoms of the solvent (electron-acceptor) and an aromatic or an heteroatomic solute (electron-donor). The interatomic distances and stabilities for these complexes are comparable with those of the hydrogen bonded complexes.[50] Such charge transfer complexes between carbon tetrachloride and oxygenated molecules have been demonstrated by a variety of methods.[50]

(d) Possibly weak hydrogen bonding of the solute protons on to the chlorine atoms of CCl_4.

One can conclude therefore that whenever solubility allows the solvents n-hexane, neopentane, cyclohexane, decalin, tetramethylsilane, etc., ought to be preferred to carbon tetrachloride. However, despite the anomalies it is preferred to the anisotropic carbon disulphide, which in turn is far superior to the ubiquitous chloroform or deuterochloroform.

The problem with low solubilities in a good "inactive" solvent is basically that of low NMR sensitivity, which can be markedly increased (by a factor of 10–50) by recourse to the CAT technique.[51] Use of a flowing liquid sample also increases the signal-to-noise ratio, since the continuous supply into the probe of unsaturated nuclei lowers the saturation parameter.[52] A review of this field is available.[53]

3. SOLVENT EFFECTS AND CONFORMATIONAL EQUILIBRIA

Before the advent of NMR, infra-red and dipole moment evidence had shown that the relative populations of conformers and rotamers can change with the polarity of solvent, and that in some cases the direction of such changes can be predicted by simple electrostatic considerations. The originality of the NMR technique, and its limitations, compared with optical spectroscopy, arises from measurement of a single spectrum, at a given temperature, which is the weighted average of the spectra for the different conformations. The classical analytical procedure is therefore to change the temperature, plot the observed coupling constants and/or chemical shifts as a function of temperature. By assigning a temperature independent set of chemical shifts and coupling constants to each conformer, one can derive the respective conformer populations at each temperature, and consequently determine the thermodynamic characteristics of the equilibrium. A recent review by Reeves[54] can be consulted. The weakness of these methods comes from the underlying assumptions: the ΔE terms, representing the enthalpy differences between the conformers, are postulated to be temperature independent; the coupling constants, which in fact are vibrational and torsional averages at any temperature for a given isomer, are expected to show some temperature dependence but are assumed constant. The same is also true for chemical shifts, which like coupling constants, are affected by association effects with the solvent, which are themselves temperature dependent. Gutowsky, Belford and McMahon[55] have pointed out, in their extensive study of the ethanic rotator, how much the adverse effects of molecular association influence the chemical shift data which, despite their much greater intrinsic accuracy compared with coupling constants, are rendered useless for the precise determination of thermodynamic and molecular parameters. Finegold[56] has studied 1,2-di-substituted propanes, using the alternative approach of varying the solvent rather than varying the temperature. The main advantages of the former are that the observed variations are larger, and that, since temperature is kept constant, the populations of the vibrational levels do not change appreciably. Each type of proton is labelled, according to the Finegold[56] nomenclature, in the pertinent *trans* (*t*) and *gauche* (*g* and *g'*) rotational isomers:

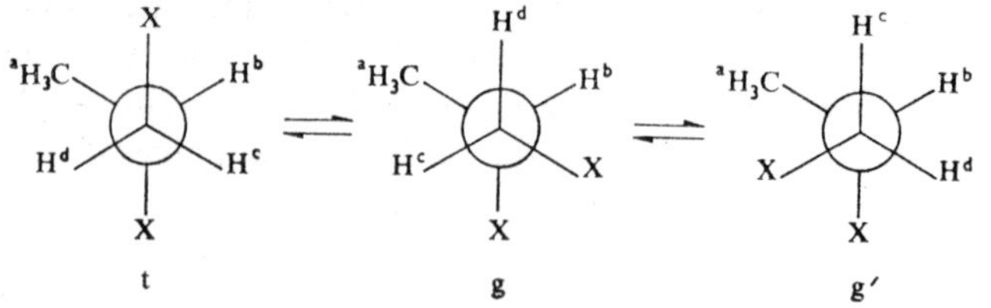

While the vicinal coupling $^3J_{ab}$ associated with the methyl group does not vary appreciably (the corresponding potential function having three equal minima) when the solvent is changed from carbon tetrachloride to dimethyl sulphoxide, the vicinal coupling constants $^3J_{bc}$ and $^3J_{bd}$ for 1,2-dichloro- and 1,2-dibromopropane (X = Cl or Br) change by 20–40%, suggesting that the ΔE terms indeed depend upon the solvent. Finegold[56] demonstrates that the original Karplus equation[57] is not an adequate analytical expression to deduce the average dihedral angle in each solvent, and hence the populations of *t*, *g* and *g'*. Instead, Finegold[56] uses the known ΔE and $\Delta E'$ free energy differences between the *trans* (*t*) and preferred *gauche* (*g*), and the *trans* (*t*) and the highly hindered *gauche* (*g'*) rotamer respectively. These differences obtained from electron diffraction, dielectric constant and infra-red measurements in the gas phase or for the pure liquid, are used to derive approximate values of the ^3J_g and ^3J_t coupling constants for the frozen conformations of a 1,2-disubstituted propane. Finegold further assumes, perhaps incorrectly, that the sums $^3J_{bc} + {}^3J_{bd}$ are the same for the *trans* (*t*) and preferred *gauche* (*g*) diastereomers. He gets $^3J_g = -1.5$ c/s and $^3J_t = +15.5$ c/s, in very good agreement with the findings of Gutowsky, Belford, and McMahon[55] for other ethanic rotators studied by variable temperature techniques. The energy differences obtained for the different solvents are summarized in Table 8, from which it appears that ΔE decreases when the dielectric constant of the solvent is increased. This "is fully consistent with the accepted view that the *gauche* form, which is the more polar, will predominate in the more polar media".[56] Similar results characterize the 1,2-dibromopropane system. For the *meso*-2,3-dibromobutane studied by Anet,[58] the vicinal coupling constant $J_{AA'}$ is greater by ~1 c/s in carbon tetrachloride than in the polar pure liquid, again suggesting a greater concentration of the *trans* form *t* in the less polar solvent:

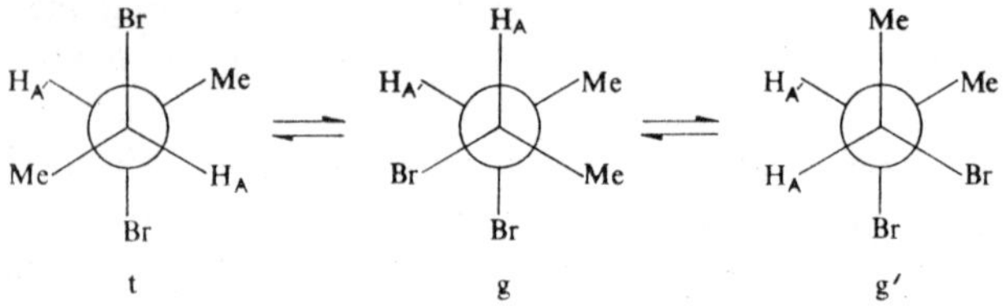

TABLE 8. DECREASING ORDER OF FREE ENERGIES FOR
1,2-DIHALOPROPANES

Solvent	Mole fraction solute	Approx. ΔE (kcal mole^{-1})
1,2-Dichloropropane		
CCl$_4$	0·141	1·0
Neat	—	0·2
C$_6$H$_6$	0·101	0·2
CHCl$_3$	0·116	0·2
MeCN	0·0766	(0·14–0·4)
Me$_2$CO	0·0849	(0·12–0·3)
Me$_2$SO	0·104	0
1,2-Dibromopropane		
CCl$_4$	0·231	1·4–1·7
CHCl$_3$	0·190	1·2–1·3
C$_6$H$_6$	0·0948	1·2
Neat	—	1·0
MeCN	0·126	0·6
Me$_2$SO	0·170	0·2

Abraham, Cavalli and Pachler,[59] working with 1-chloro-2-bromoethane, have also analysed the change in coupling constant with the solvent. Let $N = J + J'$, and $L = J - J'$, according to the classical notation for AA′BB′ systems. They found that N and L change from 16·55 and $-5·43$ c/s respectively, in carbon tetrachloride solution to 13·80 and $-1·09$ c/s in formamide, whilst the corresponding values for temperature-caused variations over 150°C in the pure liquid are smaller by an order of magnitude, 15·12 and $-3·15$ at $-10°$C to 14·78 and $-2·75$ c/s at 140°C. In a series of solvents, the values of N and L, when plotted against the dielectric function of the medium $x = (\epsilon - 1)/(2\epsilon + 1)$ give curves with little scatter. These authors have conclusively shown that a dipolar theory does not account for the observations, but that the inclusion of non-polarizable quadrupole terms is sufficient to bring quantitative agreement with the energy difference ΔE^s between the *gauche* and *trans* rotamers in a solvent of dielectric constant ϵ. They also demonstrated that the ΔE term is indeed not temperature independent, but increases with temperature, since both the density and the dielectric constant decrease with increasing temperature. The solvent dependence of N and L is used to define the relevant parameters for the *gauche* and *trans* rotamers: $N_t = 18·4$, and $N_g = 10·3$ c/s; $L_t = -8·3$, and $L_g = 4·2$ c/s. A similar treatment applied to CH$_2$BrCH$_2$F and CH$_2$ClCH$_2$F showed that the values of J_g vary widely between the isomers, 5·1–5·5 c/s for the *trans* and 2·2–2·7 for the *gauche*, as already noted by Williams and Bhacca in the cyclohexane series.[60] The theory of Abraham, Cavalli and

Pachler[59] leads to the following results:

$$\varDelta E^s = E_g^s - E_t^s = \varDelta E^v - \frac{kx}{1 - \dfrac{2\alpha x}{a^3}} + \frac{3hx}{5 - x} \tag{18}$$

with $\varDelta E^v = E_g^v - E_t^v$ (in the vacuum); α the molecular polarizability, and a the solute molecular radius; the factors k and h are constants which depend on the differences in the dipole (μ) and quadrupole moments (q) of the isomers, e.g. for each isomer:

$$k_i = \frac{\mu^2}{a^3} \tag{19}$$

$$h_i = \frac{3}{2a^5} [4(q_{xx}^2 + q_{zz}^2 + q_{xx}q_{zz}) + 3(q_{xz} + q_{zx})^2]. \tag{20}$$

The calculated and observed $\varDelta E$ terms are in fair agreement: 1·62 and 1·20 kcal mole^{-1} (CH$_2$ClCH$_2$Cl); 1·28 and 0·98 kcal mole^{-1} (CH$_2$ClCH$_2$Br); and, 0·97 vs. 0·86 kcal mole^{-1} (CH$_2$BrCH$_2$Br). A related study of Snyder[61] is also available.

In the alicyclic series, evidence derived from coupling constants was also used by Trager, Nist and Huitric[62] to measure the conformational preference in 4-hydroxycyclohexanone. The ^3J's, for a given conformation, are assumed to be solvent independent, and the signal width method of Garbisch,[63] Feltkamp and Franklin,[64] applied to the X protons for the weighted average of I and II, and for the anchored models III and IV, provides the results summarized in Table 9.
The poor accuracy of this modified Eliel method can be attributed to several factors:

(a) For a given molecule there is a slight dependence of the vicinal coupling constants on the solvent; for the *trans*-4-tert-butylcyclohexanol (III), the peak width of H$_X$ is 30·6 c/s in deuterochloroform and 30·0 c/s in pyridine; for the *cis*-4-tert-butyl cyclohexanol, (IV) the peak width of H$_X$ is constant at 11·0 c/s in both chloroform and pyridine; in 4-tert-butyl-*cis*-4-hydroxycyclohexanol-3,3,5,5-d$_4$ the same parameter varies only by 0·5 c/s between pyridine and 25% acetic acid.
(b) The model compounds (III) and (IV), could not be examined in D$_2$O because of their insolubility.
(c) Molecules (III) and (IV) are in fact inadequate models for the system studied, with a C—1 trigonal carbon atom.

The authors[62] indicate that an intramolecular hydrogen bonded boat form, in chloroform or pyridine solution, is ruled out by the observation that the

TABLE 9. SOLVENT VARIATION OF THE CONFORMATIONAL
EQUILIBRIUM FOR THE 4-HYDROXYCYCLOHEXANONE SYSTEM

Solvent	Percentage of the equatorial structure I
$CDCl_3$	$54 \pm 4\%$
C_5H_5N	$53 \pm 4\%$
D_2O	$61 \pm 4\%$

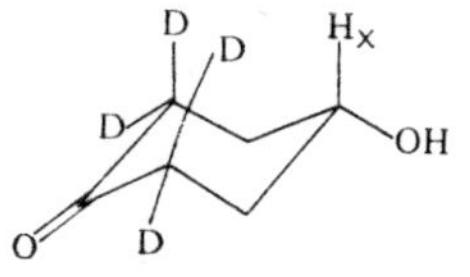

I

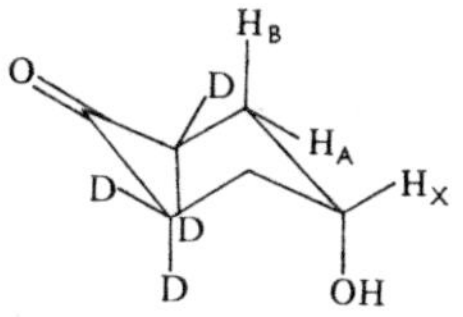

II

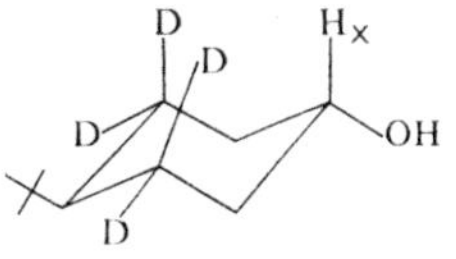

III

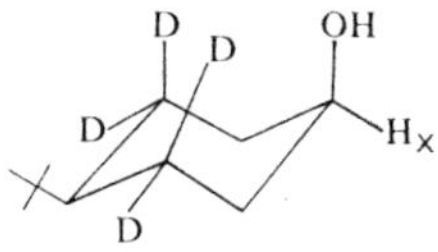

IV

benzoate ester of 4-hydroxycyclohexanone-2,2,6,6-d_4 conserves a constant signal width of 19·2 c/s, whether in pyridine or chloroform, compared with the 20·0 c/s of the parent alcohol. No speculation about the origin of the small variations observed between pyridine, chloroform and D_2O, could be attempted. In this work, Trager, Nist and Huitric[62] witnessed solvent effects upon the chemical shifts (these are reported in Section 10), and upon the conformational equilibria, the latter being assessed through the solvent independent coupling constants. A similar approach, using coupling constants, was applied by Lemieux and Lown[65] to the *trans*-1,2-dihalocyclohexanes (see Table 10). It is seen that, whether with chlorine or bromine substituents, the more polar diequatorial form VB predominates in the more polar solvents. The variation with the medium dielectric constant is rather smooth, except for the bromoform solvent, where the authors'[65] assumption that the vicinal coupling constants ^3J are independent of the nature of the solvent may be grossly invalid. It should be noted that bromoform is known to be a much better proton donor than chloroform,[66] and by preferential solvation of the less hindered diequatorial conformation VB would stabilize it energetically.

Takahashi[67] has measured the concentration effect of the long range proton spin–spin coupling constant in aliphatic ketones (see Table 11).

Takahashi[67] assumed a $\cos^2$ law for the angular dependence of these

TABLE 10. EFFECT OF SOLVENT ON THE CONFORMATIONAL EQUILIBRIA FOR *trans*-CYCLOHEXENE DIHALIDES

Y	Solvent	ϵ	Fraction[a] of the diequatorial structure V_B
Cl	CCl$_4$	2·24	0·58
Cl	MeCN	37·5	0·84
Br	CCl$_4$	2·24	0·24
Br	CS$_2$	2·64	0·25
Br	CHCl$_3$	4·81	0·26
Br	CHBr$_3$	4·39	0·45
Br	Me$_2$CO	20·7	0·57
Br	MeCN	37·5	0·62

[a] Calculated from: $J_{BX} = 10 \cdot 2n + 2 \cdot 04 (1 - n)$ c/s.

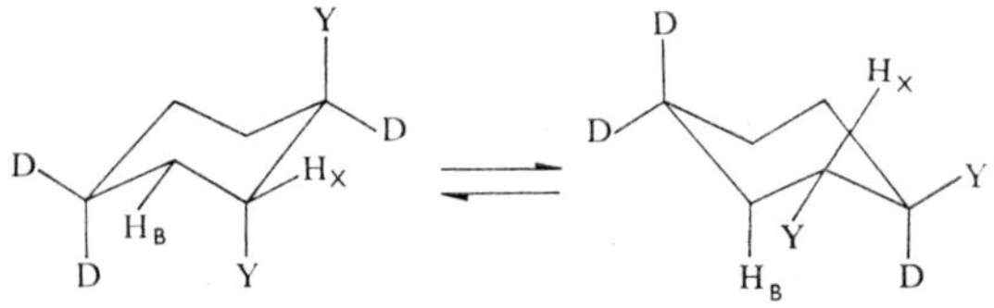

$$V_A \rightleftharpoons V_B$$

TABLE 11. LONG RANGE ^4J'S (c/s) FOR A NUMBER OF METHYL-KETONES, AT 23°C

Ketone	^4J:	
	Pure liquid	5 mole % in CCl$_4$
MeCOiPr	0·45 ± 0·05	0·44 ± 0·03
MeCOCH$_2$Br	0·40 ± 0·01	0·37 ± 0·01
MeCOCH$_2$Cl	0·33 ± 0·03	0·22 ± 0·03
MeCOCHCl$_2$	0·24 ± 0·03	0·30 ± 0·01

^4J's, and was able to calculate the following distribution of the rotamers for chloroacetone in the pure liquid:

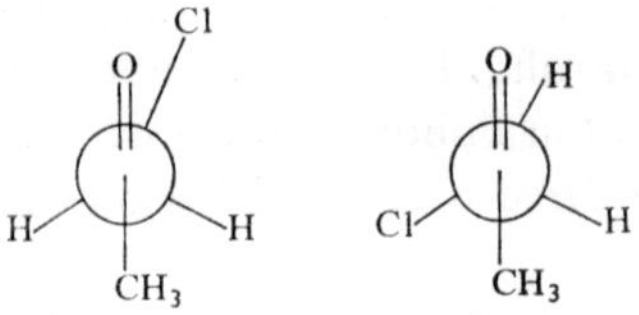

VI
66%

VII
33%

These values are in agreement with independent spectroscopic evidence; the less polar isomer (VII) increases in population in the less polar carbon tetrachloride solution, according to expectation.

A very astute study by Reisse, Celotti and Ottinger[68] shows how the effects of solvent upon the chemical shifts, *alone* can be separated into components: an intrinsic contribution of the solvent to the chemical shift of a proton in a given conformation, plus the solvent effect on the distribution of conformers. This new quantitative method of conformational analysis was prompted by the observation that when the chemical shift of the CHBr proton in 2-bromopropane (VIII) in various solvents is plotted against the chemical shift for the CHBr protons in 1,2-dibromoethane in the same solvents (carbon tetrachloride serving as the reference solvent) (see Fig. 1), a proportionality relationship is evident. This means, if the relevant conformations IXt and IXg are considered, that the solvent shift in 1,2-dibromoethane

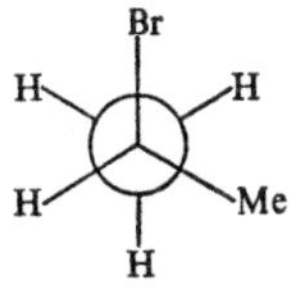

| VIII | IXt | IXg |

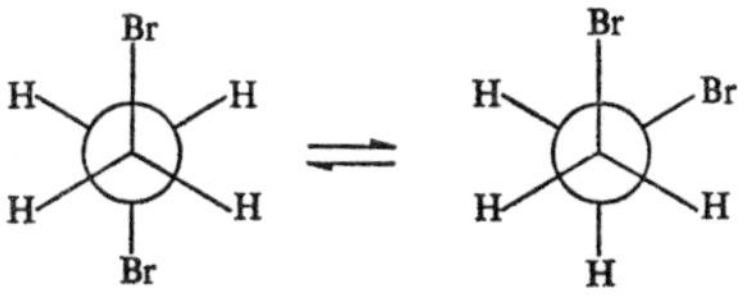

(IX) is conformation independent, since it is not influenced by the *gauche* or *trans* relationship of both bromine atoms, or by the solvent changes in the populations of the *t* and *g* rotamer. One possible generalization would be that the solvent shift uniquely characterizes the geminal relationship of the affected proton with the halogen on the same carbon atom. A drastic test of this empirical rule is to examine whether there is any sort of a correlation between the solvent shifts (again relative to carbon tetrachloride) established by Finegold[56] for the H_b, H_c and H_d protons of 1,2-dichloropropane (X) and 1,2-dibromopropane (XI), for which the nature of the substituents, the molecular dipole moments, and the rotational preference, all differ: the positive result, as plotted in Fig. 2, is indeed surprising. Even more remarkable is the observation that CHBr protons in the alicyclic series are subject, not only to intrinsic solvent shifts similar to those of the aliphatic series (which

$e-e\,(\delta_e)$ XII $a-a\,(\delta_a)$

I

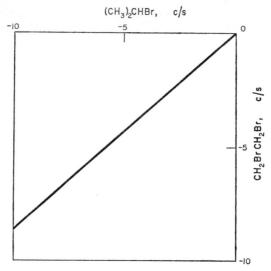

FIG. 1. Correlation between the solvent shifts for the CHBr protons of 2-bromo-propane and 1,2-dibromoethane. (Reisse, Celotti and Ottinger[68].)

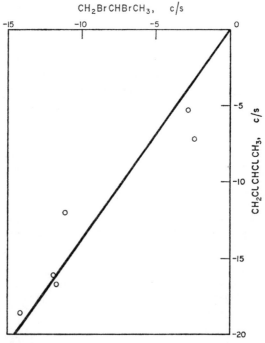

FIG. 2. Correlation between the solvent shifts for 1,2-dichloropropane and 1,2-dibromopropane.

therefore could serve as models) but also to the solvent shifts arising from conformational changes. For instance, the experimental points descriptive of the chemical shift variation of H—1 in (XII), between carbon tetrachloride and the same solvents as used for the correlation shown in Fig. 1, *all appear much above this line* (see Table 12) and it is concluded[68] that the ratio of the diequatorial to the diaxial conformer varies with the change of solvent, compared with its value in the reference solvent, carbon tetrachloride. It is then a logical suggestion that the magnitude of the deviation establishes the position of the conformational equilibrium. The quantitative treatment can be outlined as follows. It is postulated that H—1 appears at a lower field in the diaxial conformation (notation: δ_a), than for the diequatorial conformation (notation: δ_e), by analogy with the quasi totality of saturated cyclohexane systems. A solvent more polar than carbon tetrachloride is expected to shift the conformational equilibrium towards the more polar diequatorial conformer, and therefore the resonance of H—1 towards δ_e. This shift is superimposed upon the normal solvent effect, which is approximately given by the line depicted in Fig. 1. In any solvent S, one may write the classical Eliel relationship[143]:

$$\delta_{\mathrm{XII}}^{\mathrm{S}} = N_e^{\mathrm{S}}\delta_e^{\mathrm{S}} + N_a^{\mathrm{S}}\delta_a^{\mathrm{S}}. \tag{21}$$

If the hypothesis is made that the *intrinsic* solvent effect is similar for δ_e, δ_a, and δ_{XII}, the straightforward deduction is for the points characteristic of the system $\Delta\delta_{\mathrm{XII}} = f(\Delta\delta_{\mathrm{IX}})$ to lie on the same line of Fig. 1. Let us define as Δ the vertical deviation. One obtains:

$$\delta^{\mathrm{S}} - \Delta = N_e^{\mathrm{CCl_4}}\delta_e^{\mathrm{S}} - N_a^{\mathrm{CCl_4}}\delta_a^{\mathrm{S}}. \tag{22}$$

From relationships (21) and (22), it follows:

$$N_e^{\mathrm{S}} = N_e^{\mathrm{CCl_4}} + \frac{\Delta}{\delta_e^{\mathrm{S}} - \delta_a^{\mathrm{S}}}. \tag{23}$$

If an independent experimental method yields $N_e^{\mathrm{CCl_4}}$ and if the value for the difference $\delta_e^{\mathrm{S}} - \delta_a^{\mathrm{S}}$ is taken from the measurements on *cis-* and *trans-*4-tertbutylbromocyclohexane, as 43·6 c/s in carbon tetrachloride, it becomes possible to calculate N_e^{S} in any solvent S for which the resonance peak of H—1 can be observed. This procedure postulates the quantity $\delta_e^{\mathrm{S}} - \delta_a^{\mathrm{S}}$ to be solvent independent, and is equivalent to the approximate relationship:

$$N_e^{\mathrm{S}} = N_e^{\mathrm{CCl_4}} + \Delta/43\cdot6. \tag{24}$$

For the sake of simplicity, we shall restrain the argument to the case where the value of $N_e^{\mathrm{CCl_4}}$ is given by application of Eliel's method. It is emphasized here that the procedure being described is indeed different from Eliel's method, as it will be evident from the comparison between their respective results, in Table 12. The chemical shifts for H—1 in *cis*-3-*trans*-4-dibromo-tert-butylcyclohexane-3,5,5-d₃ (XIII) and *trans*-3-*cis*-4-dibromo-tert-butyl cyclohexane-3,5,5-d$_3$ (XIV) are respectively δ_e and δ_a and these are given together with δ_{XII} in Table 12.

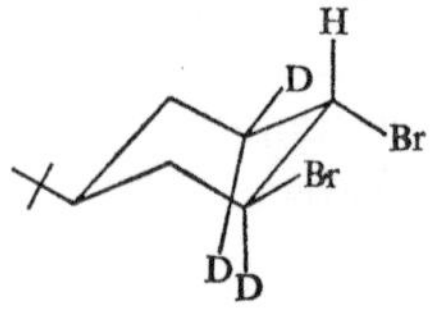

XIII 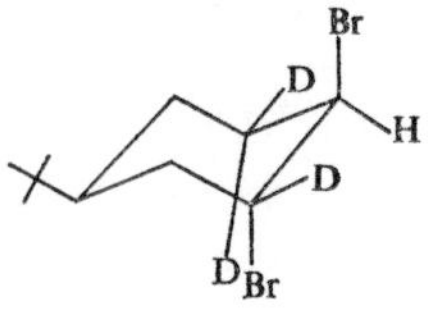   **XIV**

TABLE 12. SOLVENT SHIFTS OF *vic*-DIBROMOCYCLOHEXANES[a]

Solvent	$\Delta\delta$(XII)	$\Delta\delta_e$(XIII)	$\Delta\delta_a$(XIV)
CCl$_4$	0·0	0·0	0·0
C$_5$H$_5$N	+1·4	−14·3	−6·6
Dioxane	−2·5	—	−8·6
Me$_2$CO	+1·2	−14·8	−9·7
MeCN	+3·9	−12·8	−7·1
AcOH	+2·6	−8·4	−3·1
Me$_2$SO	−1·7	−20·8	−11·6
Me$_2$NCHO	+0·2	−20·3	−12·3

[a] In c/s at 60 Mc/s relative to the chemical shift in carbon tetrachloride.

Application to these values of relation (24) yields the values of the conformational constant N_e^S (first column of Table 13), compared with the values resulting from the application of the conventional Eliel method to the absolute chemical shifts for molecules XII, XIII and XIV, in any solvent S (last column of Table 13): the agreement is rather good. The method can be further improved by the use of better models than the aliphatics for the intrinsic solvent effects. Consideration of Table 12 shows that, if the solvent effect follows the same sequence for XIII and XIV, it is more pronounced on δ_e than on δ_a. Furthermore, it is quantitatively different from the solvent effect previously observed on VIII and IX. A possible new choice for the reference line is the system $0 \cdot 18 \Delta\delta_e + 0 \cdot 82 \Delta\delta_a$, as a function of $\Delta\delta_e$ (see Figure 3.) This is equivalent to the postulate that the solvent effect upon an average signal is equal to the weighted mean of the solvent effects for either the axial or the equatorial term. Also it is no longer possible to rely upon the invariance of $\delta_e^S - \delta_a^S$ with the solvent, and it is preferable to use the relationship:

$$N_e^S = N_e^{CCl_4} + \frac{\Delta}{\delta_e^S - \delta_a^S} \tag{25}$$

which yields the values of N_e^s in the middle column of Table 13.

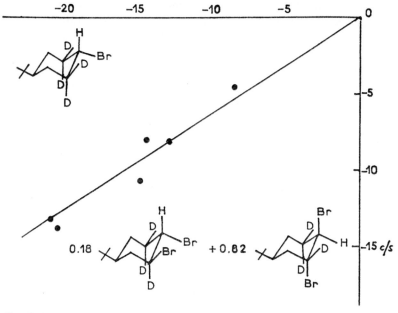

FIG. 3. Improved correlation for the solvent shifts in the 1,2-dibromocyclohexane series. Reisse Celotti and Ottinger[68]).

TABLE 13. COMPARISON OF THE N_e VALUES FOR *trans*-1,2-DIBROMOCYCLOHEXANE COMPUTED BY DIFFERENT METHODS

Solvent	N_e		
	Aliphatic models	Alicyclic models	Eliel
CCl$_4$	0·18[a]	0·18[a]	0·18
C$_5$H$_5$N	0·24	0·49	0·46
Dioxane	0·24	—	—
Me$_2$CO	0·40	0·46	0·49
MeCN	0·46	0·49	0·50
AcOH	0·34	0·38	0·37
Me$_2$SO	0·40	0·50	0·52
Me$_2$NCHO	0·45	0·55	0·54

[a] Assumed.

Obviously, the method based upon the anchored cyclohexane models for the solvent effects, and the Eliel method are equivalent, and the slight discrepancy observed between the middle and the last column of Table 13 is due to the small scatter of the experimental points. The important features to be stressed are that the elementary method based upon the aliphatic models provides coherent results (with the exception of pyridine) since the proportion of the more polar diequatorial conformer increases with the solvent polarity, and that there is only a small deviation compared with the Eliel method. This new method is applicable when that of Eliel fails, either through lack of the pertinent cis- and trans-substituted-4-tert-butylcyclo-hexanes or due to overlap of the corresponding resonances with solvent lines (as is the case here for dioxane). If an alternative value of $N_e^{CCl_4} = 0.32$ (which is equally probable as the above value of 0.18 from the available evidence) is assumed, then the N_e values calculated for the various solvents are in very good agreement with the determination of Lemieux and Lown[65] based upon coupling constants (see Table 14).

TABLE 14. COMPARISON OF SOLVENT EFFECTS[68] AND COUPLING CONSTANTS[65] DATA FOR THE vic-DIBROMOCYCLOHEXANES

Solvent	N_e	
	from solvent effects	from coupling constants
CCl_4	0.32[a]	0.24
C_5H_5N	0.38	—
Dioxane	0.38	—
Me_2CO	0.54	0.57
MeCN	0.60	0.62
AcOH	0.48	—
Me_2SO	0.54	—
Me_2NCHO	0.59	—

[a] Assumed (see text).

Dudek and Dudek[69] have remarked upon the influence of polar solvents upon the enthalpy of tautomeric interchange between enolimines and keto-amines (see Table 15); they state that the hydrogen bonding capability of the medium is a more dominant factor than its polarity in determining the position of the equilibrium, since the non-aromatic tautomer predominates in the less polar chloroform more than in acetonitrile. The equilibrium constants were determined from the values of the ^{15}NH coupling constants, postulating solvent independence for a given tautomer.

From the temperature dependence of the ^1H resonance spectra of N,N-disubstituted amides, Woodbrey and Rogers[70] have determined the solvent effects on the height of the energy barrier for hindered internal rotation about the C—N bond (see Table 16). This solvent dependence is explained[70] in terms of stabilization of the polar (and planar) excited state XVIa with increasing dilution in the non-polar solvent carbon tetrachloride. Woodbrey and

Ph   Ph

$$\text{XVK} \quad \rightleftharpoons \quad \text{XVE}$$

XVK   XVE

TABLE 15. KETO-ENOL EQUILIBRIUM FOR
[15]N-SUBSTITUTED ANILIDES (XV)

Solvent:	CDCl$_3$	CCl$_4$	MeCN
$-\Delta H$: kcal mole^{-1}	1·5	0·4	0·08

Rogers[70] also performed measurements for intermediate values of the concentration in methylene bromide and carbon tetrachloride, which are not given here. Fryer, Conti and Franconi[71] have provided a critical discussion of the NMR experimental methods used for the determination of barrier

TABLE 16. SOLVENT DEPENDENCE OF THE BARRIER TO ROTATION ABOUT THE C—N BOND IN N,N-DISUBSTITUTED AMIDES

Solvent	Concn.[a]	E_a (kcal mole^{-1})[b]	Log A[c]	$\Delta F^{\ddagger}_{298.2}$ [d]	T_c(°K)[e]
		N,N-dibenzylacetamide			
CH$_2$Br$_2$	38·0$_9$	7·3 ± 0·5	6·1 ± 0·3	16·5	343·7
CCl$_4$	38·2$_9$	6·4 ± 0·6	5·5 ± 0·4	16·3	334·5
		N,N-dimethylpropionamide			
Neat	100	9·2 ± 0·7	7·3 ± 0·5	16·7	334·4
CH$_2$Br$_2$	84·7$_1$	10·1 ± 0·8	7·9 ± 0·5	16·7	333·2
CH$_2$Br$_2$	10·1$_5$	6·7 ± 0·7	5·4 ± 0·5	16·7	338·3
CCl$_4$	69·3$_4$	8·6 ± 0·6	7·0 ± 0·4	16·5	330·3
CCl$_4$	11·0$_7$	6·3 ± 0·4	5·7 ± 0·3	16·0	316·4
		N,N-dimethylcarbamyl chloride			
Neat	100	7·3 ± 0·5	6·1 ± 0·3	16·5	326·0
CH$_2$Br$_2$	90·0$_3$	8·2 ± 0·3	6·6 ± 0·2	16·6	325·6
CH$_2$Br$_2$	10·7$_2$	7·3 ± 0·6	6·0 ± 0·4	16·5	325·9
CCl$_4$	71·3$_4$	6·9 ± 0·5	5·8 ± 0·4	16·5	326·5
CCl$_4$	10·9$_8$	6·8 ± 0·2	5·9 ± 0·2	16·2	317·4

[a] Mole %.
[b] Enthalpy of activation $\Delta H^* = E_a - RT$.
[c] Frequency factor.
[d] Free energy of activation.
[e] Coalescence temperature for the two N—CH$_2$R peaks.

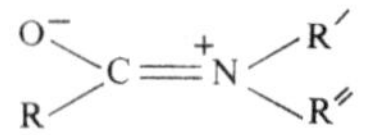

XVIa   XVIb

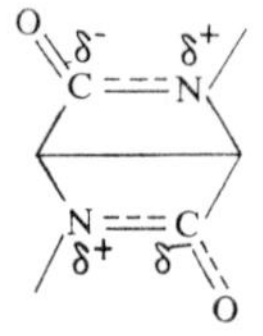

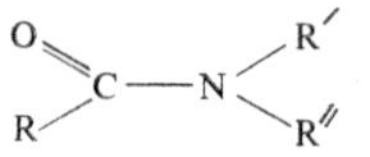

XVI

heights for amides, and they have indicated clearly the precision to be expected from such measurements.

4. THE REACTION FIELD

4.1. *Definition*

For polar solutes, large solvent shifts are due to the effect of the *reaction field*. This secondary electric field, arising in the polar or polarizable solvent under the influence of the permanent dipole moment of the solute, modifies in turn the electron distribution of the immersed molecule, to an extent depending upon its polarizability. Onsager[72] introduced this concept and used a particular physical model to calculate the magnitude of the reaction field. The entire molecule is reduced to a point dipole in the centre of a cavity (originally considered as a sphere, but later as a prolate or oblate spheroid of revolution) surrounded by an isotropic, homogeneous, polarizable continuous medium of dielectric constant ϵ. By polarizable, it is meant that the resultant field in the surrounding medium no longer vanishes as it would in the absence of an applied field. These calculations[72, 73] gave the reaction field R at the centre of the cavity as

$$R = \rho(\epsilon)\mu$$

where μ is the dipole moment of a free solute molecule, and where $\rho(\epsilon)$ depends on the shape of the cavity.

Lovell[74] has shown that this useful concept is meaningful only in an entirely macroscopic context, excluding the consideration of molecular interactions. One of the main shortcomings of this model is that no molecule, even diatomic, can be adequately represented by a point dipole (especially in a liquid), and that no liquid or gas can be considered microscopically homogeneous, especially in the immediate neighbourhood of the spherical cavity. Its validity, whether qualitative or quantitative, has been intensely discussed[74] and cannot really be tested, since the appropriate experiments are not easy to envisage.

Buckingham and co-workers[10, 75] have pointed out that the magnetic shielding of the nuclei in a chemical bond should depend upon the component along the bond of an electric field E

$$\Delta\sigma_E = -k_E \cdot E \cdot \cos\theta - k'_E \cdot E^2 \tag{26}$$

where k_E is a constant characteristic of an A—X covalent bond, along which $E \cos\theta$ is the projection of the local electric field E, and where σ_E is the variation in the screening of nucleus X. The minus sign ensures that the magnetic shielding of X be reduced if the direction of E is such as to reduce the electron density at the nucleus X. The theoretical interpretations of Buckingham[10] and Musher[76] also agree upon the negligible character of the quadratic term for protons, since $k_E \gg k'_E$ (for protons, $k_E \sim 2 \times 10^{-12}$ e.s.u., and $k'_E \sim 10^{-18}$ e.s.u.[10, 75, 76]). The dependence of $\Delta\sigma_E$ on E should thus be linear to a good first approximation, except when E is perpendicular to the A—X bond. The effect of the reaction field R, which should have a time average local strength considerably greater than that obtainable by ordinary laboratory procedures, could provide a suitable experimental test of this linear dependence.

4.2. Detailed Formulation

The reaction field R for a dipole μ in the centre of a spherical cavity of radius r is given by:[10, 72-6]

$$R = \frac{(n^2 - 1)}{3\alpha} \frac{(\epsilon - 1)}{\epsilon + n^2/2} \mu \tag{27}$$

where n is the refractive index, and

$$\alpha = \frac{(n^2 - 1)r^3}{n^2 + 2} \tag{28}$$

the polarizability of the solute. The value of r can be derived from the empirical equation:

$$r^3 = \frac{3M}{4\pi\rho N} \tag{29}$$

M and σ being the molecular weight and density of the solute respectively, and N the Avogadro number.

The more complex equation for R in an ellipsoidal cavity has been employed by Diehl and Freeman:[77]

$$R = \frac{3}{abc} \xi_a[1 + (n^2 - 1)\xi_a] \frac{(\epsilon - 1)}{(\epsilon + \beta)} \mu \qquad (30)$$

where a, b, c are the principal axes of the ellipsoid,

$$\beta = \frac{n^2 \xi_a}{(1 - \xi_a)}, \qquad (31)$$

ξ_a being a shape factor, which has been calculated by Ross and Sack[78] The case $\beta = 1\cdot0$ is that of a sphere with $n^2 = 2$, while the values $\beta = 2\cdot86$ and $\beta = 0\cdot5$ refer respectively to the *oblate* and *prolate* examples studied by Diehl and Freeman.[77]

From equations (26) and (30) it follows that the variation of the chemical shift, for a given nucleus in the solute, with the dielectric function $(\epsilon - 1)/(\epsilon + \beta)$ of the solvent ought to be linear, with a slope proportional to the product $k_E \mu \cos \theta$.

4.3. *Isolation of the Reaction Field from Other Factors*

To ascertain the chemical shift contribution from the reaction field effect, the following steps are necessary:

(a) *Choice of solvent.* It is impossible to satisfy strictly the criterion of a microscopically homogeneous, isotropic solvent. Nevertheless, such solvents which associate strongly with the solute are evidently to be eliminated: benzene and the other aromatics, which specifically solvate polar solutes by a dipole induced–dipole type interaction (see Section 13); water, alcohols or other proton donors like chloroform, having strong intermolecular hydrogen bonds (see Sections 6 and 8); solvents causing large van der Waals shifts, like CH_2I_2 (see Section 1). Their use would lead to extraneous shifts of the same order of magnitude, or even greater, than the reaction field effects being investigated.

A slowly varying range of dielectric constants is provided by changing the concentrations of the components in solvent mixtures.[17, 46, 79, 80] The dielectric constant has been accurately measured for a great number of mixed binary systems by Decroocq.[81] The bulk dielectric constant can then be calculated taking the concentration and the dielectric constant of the solute and reference into consideration (assuming additivity).[17, 46] Usually their concentrations are sufficiently low (solutions of 3–5% normally being adequate) to ignore this small contribution in the absence of quantitative knowledge on the dielectric constant of the pure solute and of the reference compound. For instance, the following sequence has been utilized to study the reaction field in solvents of dielectric constant between 2 and 40[80]

TABLE 17. DIELECTRIC CONSTANTS OF SOLVENT MIXTURES[81]

Composition of solvent (v/v)	ϵ
C_6H_{12}	2·02
CCl_4	2·24
$80\%C_6H_{12} + 20\%C_6H_{10}O$	4·17
$60\%C_6H_{12} + 40\%C_6H_{10}O$	6·78
$80\%CCl_4 + 20\%MeNO_2$	7·05
$60\%CCl_4 + 40\%Me_2CO$	12·47
$40\%C_6H_{12} + 60\%C_6H_{10}O$	12·67
$55\%CCl_4 + 45\%MeNO_2$	15·20
Me_2CO	20·7
$40\%CCl_4 + 60\%MeNO_2$	20·80
$20\%CCl_4 + 80\%MeNO_2$	29·00
$MeNO_2$	37·45

(see Table 17). These binary mixtures of polar and non-polar solvents can also be chosen to structurally resemble the solutes studied. For instance, the reaction field for α-bromocyclohexanones was studied in cyclohexane–cyclohexanone mixtures, since these molecules are rather similar to the solute.[79] One would hope, therefore, to minimize differences between self-association of solvent molecules and non-dipolar association with solute molecules, so that the solvent effects discussed here are due *only* to the variations of ϵ. The same holds for interactions between the solvent and the TMS standard. As we shall now see, the reference problem can be rather satisfactorily circumvented.

(b) *Choice of reference*. To avoid solvent effects upon the internal reference, Fontaine, Chenon and Lumbroso-Bader[17] have used a benzene external reference, and corrected the observed shifts by a precise determination of the magnetic susceptibilities. However, use of an internal reference, which may be tetramethylsilane[79, 80, 82] or cyclohexane,[46, 75, 77] is justified on account both of convenience and of the appreciable errors arising from bulk susceptibility corrections[83] when an external reference is used. Use of a *doubly* internal reference is at the same time convenient and satisfactory: the chemical shifts are measured from another resonance peak of the solute.[77, 80, 82] In this way, the methyl resonances C—8, C—10, and C—13 of epimanoyl oxide (I) have been internally correlated in a set of solvents.[82] Similarly, the differences in chemical shift for the methyl and methine

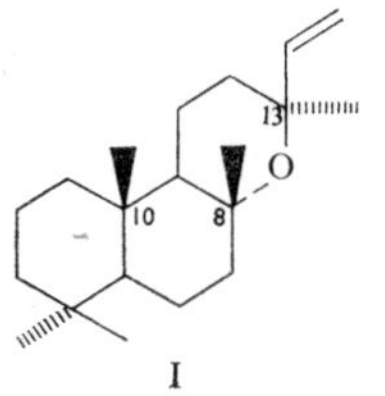

I

resonances of paraldehyde, in a series of solvents, have been used by Diehl and Freeman.[77] This procedure not only solves the problem of a choice between an external or an internal reference, it also suppresses the differences in the shape parameter β used by Diehl and Freeman[77] between solute and reference, and isolates the reaction field itself, to the exclusion of any other non-specific inter- or intramolcular interaction. These can be of the van der Waals type, or can occur with magnetically anisotropic solvent molecules. Then the Onsager cavity for the solute and for the reference is truly identical, the solute being referred to itself. An intermediate technique, which also proved to be satisfactory, is to use two stereoisomers, A and B, in equimolecular mixture in the medium, the resonances of A being measured relatively to those of B, and vice versa, the Onsager cavities being then qualitatively and quantitatively equivalent. The sample and its reference have there identical magnetic environments. 1,2-disubstituted ethylenes, as the configurational isomers IIA and IIB have been thus studied.[80]

$$
\begin{array}{cc}
\text{IIA} & \text{IIB}
\end{array}
$$

(c) *Choice of solute.* The ideal molecule for such a study would have a completely rigid, well-known structure, its general shape approximating to one of the ellipsoids defined by Diehl and Freeman.[77] Also it would have a large permanent dipole moment and possess various types of nuclei differently oriented with regard to this vector and having well-separated and identifiable resonances. Unfortunately, such constructions are not readily available. The shortcomings of the real situation (mainly arising from conformational mobility of the solute molecule) are deferred later until the section dealing with the interpretation of experimental results.

(d) *Elimination of the dispersion interaction or higher order terms.* The dipolar reaction field effects have been conveniently separated from dispersion interactions by the procedure of internal referencing to a non-polar molecule (as seen in the preceding paragraph) and/or by internally referring the solute to itself or to a very similar molecule dissolved in the same medium. In the cases of the 1,2-disubstituted ethylenes $XHC = CHX$, which we have studied, $X = CO\text{-}t\text{-}Bu$,[80] following Hruska, Bock and Schaefer:[46] $X = Cl$ or Br, the *cis* isomer is the only one to possess a permanent dipole moment, that is to experience a reaction field. The structurally similar *trans* form, devoid of a dipole, can be used as a model for higher order—e.g. quadrupolar —polar interactions, and dispersion shifts. Subtraction of the chemical shifts of the *cis* diastereomer from those of the *trans* diastereomer provides a good measurement of the effect of the reaction field upon the former.

In apparent contrast, the alternative approach of Fontaine, Chenon, and Lumbroso-Bader[17] leads to essentially the same results. These authors have used an external reference (see above) to measure the chemical shifts of their molecules in solution, and an internal reference to measure them for the pure solute in the gas phase. The solvent shift is defined as the difference of these values. It varies for acetone, for example, from -16.5 c/s in cyclohexane to -27.7 c/s (at 60 Mc/s) in carbon tetrachloride, intermediate values being observed with dioxane, tetrahydropyran, 1,2-dimethoxyethane, acetone and nitromethane. To obtain the desired reaction field shift the dispersion contribution is subtracted, after being calculated according to the procedure of Howard, Linder and Emerson[13] (cf. Section 1). Even so, for the acetone solute the dipolar shift varies between -6.7 c/s in cyclohexane and -17.3 c/s in carbon tetrachloride. In fact, the van der Waals shifts vary only slightly with the solvent, deviations from their mean value are not greater than 15–20%, and the variations of the solvent shifts reflect well the experimental reaction field.

4.4. *The Results*

The results of Diehl and Freeman[77] on paraldehyde merit detailed consideration (see Table 18).

TABLE 18. SOLVENT SHIFTS FOR THE METHYL AND METHINE PROTONS OF PARALDEHYDE I

Solvent	ϵ	$\delta_{(CH_3)}$[a]	$\delta_{(CH)}$[a]	Δ
C_6H_{12}	2·02	113·5	330·3	216·8
CCl_4	2·24	103·6	319·2	215·6
CS_2	2·64	97·5	314·1	216·6
Et_2O	4·34	108·7	330·4	221·7
$CHCl_3$	4·81	96·2	315·5	219·3
$C_6H_{10}O$	18·3	105·7	333·6	227·9
Me_2CO	20·7	102·5	331·1	228·6

[a] c/s from internal cyclohexane, at 60 Mc/s.

Let us first consider the isolated variations of either the methyl or methine protons with increasing dielectric constant of the medium. It is readily apparent that carbon tetrachloride, carbon disulphide and chloroform, are all in some way anomalous, and they cannot be considered with such confidence as other more regular solvents.

Looking at structure I, it is expected that:

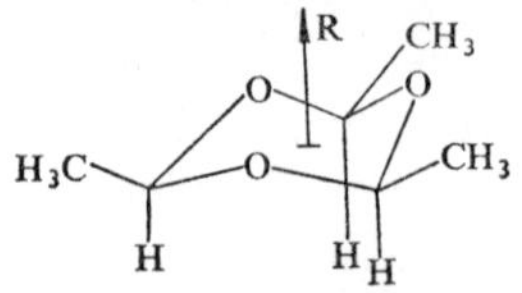

I

(a) The reaction field R being directed along the C_{3v} axis of symmetry, the CH methine protons will suffer it in full with $\cos \theta = +1$, and their chemical shift will tend to increase with the increasing dielectric function convenient for this disk-like molecule $(\epsilon - 1)/(\epsilon + 2\cdot 86)$.

(b) Conversely, the methyl protons should be shielded when the dielectric constant increases, because the projection of the reaction field along the C—H bond is now negative, with $\cos \theta = -1/9$: the magnitude of the variation ought to be one ninth of that for the methine protons.

These predictions are seen to be qualitatively, but not quantitatively, true. The individual slopes are $-12\cdot 3$ c/s per dielectric function unit (CH), and $-5\cdot 5$ c/s/dielectric function unit (CH$_3$), the scatter being considerable. Much better results are obtained when the *difference* $\varDelta$ is plotted against this same dielectric function, which isolates the reaction field from the other types of solvent interactions having similar orders of magnitude, e.g. van der Waals forces. This pioneer work is typical of the later studies in that an *internal* chemical shift—the difference of two chemical shifts for groups of nuclei in the same molecule—correlates approximately with the postulated reaction field, while the individual shifts do not. It is atypical in that the paraldehyde molecule I is seen *a posteriori* to have been a very good and lucky choice, since other studies did not find reaction field-induced shifts in opposite directions for the two possible signs of $\cos \theta$.

Murrell and Gil[84] have obtained the results given in Table 19 for 3,5-lutidine.

The reaction field picture obviously predicts a deshielding of the 4-proton relative to the 2-proton, which already resonates at a lower applied field, in qualitative agreement with the above results.

TABLE 19. SOLVENT SHIFTS FOR 3,5-LUTIDINE (II)

Solvent	ϵ	(H—2)–(H—4) (ppm)
C_6H_{12}	2·02	1·04
CCl$_4$	2·24	0·97
CDCl$_3$	4·81	0·95
Me$_2$CO	21·3	0·84
MeOH	33·62	0·70

II

Becconsall and Hampson[83] have simultaneously studied the ^1H and ^{13}C solvent shifts for methyl iodide and acetonitrile, the common *proton* internal reference being tetramethylsilane (see Table 20). The comparison between ^1H and ^{13}C chemical shifts shows a crude correlation, if one excludes the strongly anisotropic solvents, acetaldehyde and chlorobenzene, which give rise to anomalous shifts. In this correlation, the ^{13}C nuclei are affected much

TABLE 20. METHYL GROUP CHEMICAL SHIFTS AT INFINITE DILUTION

Solvent	ϵ^a	^1H (± 0.01 ppm)^b	^{13}C (± 0.08 ppm)^c
		Methyl iodide	
C_6H_{12}	2·02 (2·05)	2·011	156·18
TMS	1·89	1·991	156·8
CCl_4	2·24 (2·16)	2·150	153·05
$CHCl_3$	4·81 (2·07)	2·167	151·80
PhCl	5·71	1·733	153·05
Neat	7·00 (2·35)	2·181	149·10
CH_2Cl_2	9·08 (2·02)	2·161	152·12
MeCHO	21·8	2·181	153·22
		Acetonitrile	
CCl_4	2·24 (2·16)	1·973	126·73
CH_2Cl_2	9·08 (2·02)	1·979	126·76
MeCHO	21·8	2·033	128·46
Neat	38·8	1·963	127·51

^{a} Between parentheses, the effective dielectric constant (see text).
^{b} From TMS.
^{c} From benzene.

more than the protons, which eliminates any magnetic anisotropy origin of the effect, because it would be approximately the same at all points of the solute molecule. The contribution from van der Waals interactions may be appreciable for the ^{13}C shifts, which are referred to the ^1H tetramethylsilane internal reference, while for the protons, use of this same reference takes care of this factor: the existence of a correlation between both demonstrates that van der Waals forces cannot be the major factor. Specific effects, such as hydrogen bonding with the solvent, would affect the ^1H data more than the ^{13}C data, contrary to observation, which therefore strongly favours the electrostatic reaction field interpretation. When Becconsall and Hampson[83] attempt to plot the solvent shifts of methyl iodide and acetonitrile—referred to the pure liquid—against the dielectric function $(\epsilon - 1)/(2\epsilon + n_0^2)$ there is no straight line, or even smooth curve, through the experimental points. The interpretation they give is that the effective dielectric constant (local) in the immediate neighbourhood of the Onsager cavity is decreased, compared with the dielectric constant ϵ characteristic of the solvent. Two types of effects are operative: random thermal reorientation of the solute molecule is not long (an estimate[85] is 10^{-12}–10^{-11} sec) compared with the dielectric relaxation time of the medium (of the order of 10^{-10} sec), so that the effective dielectric constant is in fact intermediate between the static (low frequency) dielectric constant and the optical value n^2, where n is the refractive index of the medium. Also, the phenomenon of dielectric saturation caused by the intense electric fields in the vicinity of the polar solute results in a decrease of the dielectric constant proportional to the square of the electric field. The approach therefore used by Becconsall and Hampson[83] is to plot the solvent shifts against the new dielectric function $(n^2 - 1)/(2n^2 + n_0^2,)$ which does not lead either to smooth curves for methyl iodide and acetonitrile; a straight line is drawn that intersects each of the horizontal lines joining the "ϵ" and "n^2" points, defining for each solvent the values of the effective dielectric constant listed in Table 20.

The arguments they put forward[83] are well drawn, except for neglect of the important criticism of Coulson[86] regarding the "local" or "effective" dielectric constant approach; however, the precise and carefully calibrated experiments carried out were not very wisely chosen. Methyl iodide is one of the polarizable solutes most liable to show van der Waals shifts. The "infinite" dilution values for the ^{1}H and ^{13}C shifts quoted in Table 20 result from the extrapolation of "measurements made at four or five different concentrations, varying from the neat liquid to about 10 per cent molar"! The solvent shifts are referred to the pure liquid, which has much less significance than the comparison to the shift in an inert solvent, like tetramethylsilane or cyclohexane (but this evidently does not modify the qualitative result that the static dielectric constant does not account for the observed behaviour). Finally, the solvents themselves were a poor choice for such a study of methyl iodide, at least two were strongly anisotropic (MeCHO and PhCl), two hydrogen bonding ($CHCl_3$ and CH_2Cl_2), one would probably be anomalous (CCl_4), and the total dielectric constant range was accordingly too small. The recourse to a number of other solvents, in first place mixed solvents systems, could have been considered in this otherwise excellent study.

Hutton and Schaefer[87] have measured the following solvent shifts for para-nitroanisole (III) (see Table 21).

It is seen that the solvent shifts of the protons meta to the nitro group are

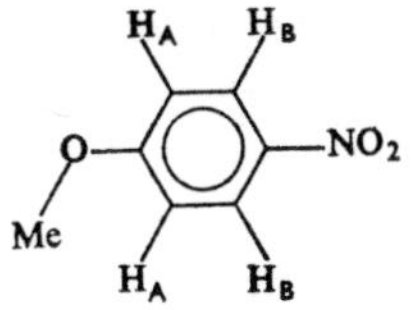

III

TABLE 21. SOLVENT SHIFTS FOR PARA-NITROANISOLE (III)

Solvent	ϵ	H_A	ΔH_A^a	H_B	ΔH_B^a
		(c/s from int. TMS at 60 Mc/s)			
C_6H_{12}	2·0	409·9	0·0	485·8	0·0
CCl_4	2·2	414·4	−4·5	487·1	−1·3
CS_2	2·6	412·8	−2·9	482·7	+3·1
Me_2CO	21	427·2	−17·3	492·4	−6·6
$MeOH$	33	420·6	−10·7	488·4	−2·6
$MeCN$	37	421·1	−11·2	489·3	−3·5

[a] Relative to shift in cyclohexane.

greater than those of the protons ortho to the nitro group. This is not even qualitatively in accord with the reaction field picture, since the permanent dipole moment is collinear with the nitro group, which determines its major part (the respective dipole moments of nitrobenzene and anisole are 4·01 and 1·30 D). One would have expected a shielding of H_B and a deshielding of H_A by equal amounts when the dielectric constant of the solvent is increased. Furthermore, the values of H_A and H_B obtained with acetone are seen to be rather exceptional, compared with acetonitrile or methanol, which have substantially higher dielectric constants. The explanation may come from the presence of a strong electron-attracting (NO_2) group and a strong electron-donating (OMe) group, so that the quinoid structure IV may contribute appreciably to the ground state. The amount by which IV is stabilized in polar solvents depends upon the nature and extent of the charge-transfer association which takes place.

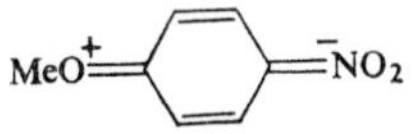

IV

Yamaguchi[88] has studied solvent effects upon the internal chemical shift $H_{ortho} - H_{meta}$ in para-substituted toluenes ($X = NO_2$, CN, Br, OH, OMe, NH_2, NMe_2), and has found that it decreases with increasing solvent dielectric constant; however, the reaction field fails to account qualitatively for the results.

Diehl[89, 90] has measured the solvent shifts of H—2 and H—5 in *meta*-dinitrobenzene. The solvents used were CCl_4 (reference), $nPrCO_2H$, Et_2O, AcOMe and Me_2CO. If the correlation obtained for the *internal* chemical shift between the 2- and the 5-proton is reasonably linear, Diehl[90] points out that its slope is too small by a factor of two, considering the magnitude of the molecular dipole and the shape factor for this solute. The inclusion of quadrupolar reaction fields would not be appropriate, since it would contradict the additivity of the effect observed in acetone solution: the carbon tetrachloride to acetone shifts observed for the mono-substituted nitrobenzene allow prediction of the analogous solvent shifts for the di-substituted case. Diehl[90] feels that a better quantitative treatment would be obtained using inhomogeneous reaction fields of dipoles excentrically placed in the Onsager cavity.[73] Schug and Deck[91] compared the ring proton chemical shifts and the pi-electronic distributions as established by Hückel-type calculations for hydroxybenzenes, by the following steps:

(a) they measured all spectra at equal molar concentration, so that the intermolecular ring current factor could be removed; they assumed that the intramolecular ring was current a constant for all molecules studied;

(b) they removed the reaction field contribution by extrapolation of the data to the unit dielectric constant vacuum, in order to get parameters pertaining to the isolated molecule (see Table 22);

TABLE 22. REMOVAL OF THE REACTION FIELD CONTRIBUTION TO RING PROTONS SHIFTS IN HYDROXYBENZENES

Molecule	Proton	MeOH ($\epsilon = 30\cdot6$)	Dioxane ($\epsilon = 2\cdot22$)	$\epsilon = 1$[a] ($\pm0\cdot013$ ppm)[b]
1-(OH)C₆H₅	H—2	0·525	0·600	0·648
	H—3	0·175	0·175	0·175
	H—4	0·500	0·475	0·458
1,4-(OH)₂C₆H₄	H—2	0·688	0·692	0·716
1,2-(OH)₂C₆H₄	H—3	0·539	0·630	0·688
	H—4	0·651	0·630	0·618
1,3-(OH)₂C₆H₄	H—2	1·025	1·150	1·232
	H—4	1·018	1·100	1·151
	H—5	0·350	0·395	0·395
1,2,3-(OH)₃C₆H₃	H—4	0·950	1·055	1·117
	H—5	0·793	0·843	0·874

[a] Extrapolation to 0 of $Z = (\epsilon - 1)/(2\epsilon + 2\cdot5)$
[b] Relative to C_6H_6 at the same concentration in the same solvent.

(c) they used the Buckingham equation:[10]

$$\Delta\sigma = -2 \cdot 10^{-12}E_z - 10^{-18}E^2$$

to account for the electric field at the protons due to the effective C—O (0·74 D) and O—H (1·51 D) dipoles; the doubly corrected (for inter- and intramolecular electric fields) final values were in fair agreement with the computed charge densities.

In this study, the specific interactions (hydrogen bonds) between the solutes and the oxygenated solvents, chosen presumably for solubility's sake, were completely neglected, and may have varied substantially between the different molecules. Paterson and Tipman[92] made an attempt, similar to that of Schug and Deck[91] (just discussed) to understand the ring proton shifts of para-substituted phenols; they differ in that they did not attempt to correct the experimental values for the changes in the reaction field from the solvent (carbon tetrachloride). It was assumed to remain constant, as the other solute–solvent interactions, when the para substituent changes in the series H, Cl, Br, I, OMe, CN, NO₂, etc. They also implicitly assumed that the pi-electron charge density would remain constant, and tentatively plotted the observed ring protons shift (H—2)−(H—3) against the calculated contribution to this shift from the electric field of substituent X (using the Buckingham expression of above). Actually, there is such a qualitative correlation, which is a rather good result considering the approximations involved. One notes a discrepancy between the calculated value for X = NO₂ in Fig. 4 of reference 92, and the value quoted in the text, 0·65 and 0·18 ppm, respectively. From this Fig. 4,[92] a value of 3·5 is suggested for k_E, rather than 2·0 as found by Buckingham.[10] The equatorial bromocyclohexanones V and VI were chosen by Laszlo and Musher[79] because of the relatively large values

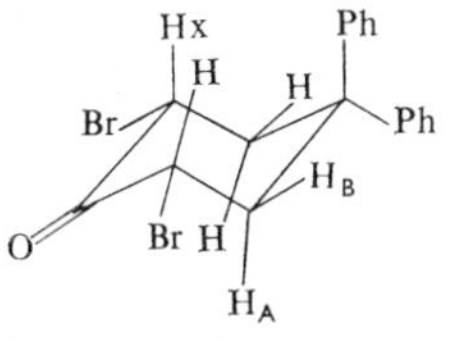

V 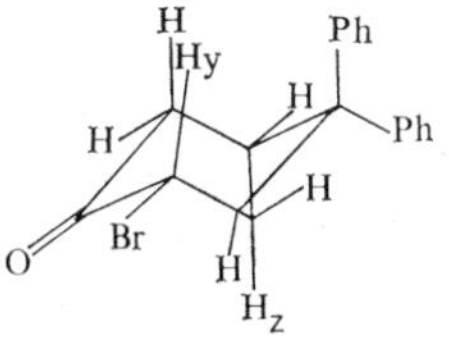   VI

of the permanent dipole moments (approximately 4·8 and 4·25 D respectively) and because the behaviour of the protons X, B, A, Y and Z, differently located with respect to the direction of the reaction field, can be simultaneously followed. For molecule V, the theoretical values of R_A, R_B and R_X are calculated for each ϵ from equation (27), and the corresponding chemical shifts variations for H_A, H_B, and H_X (σ_i's) are plotted versus these R_i in Fig. 4. The values of n^2 and α are taken as 2 and $25A^3$ respectively, by analogy with similar molecules. To quote from reference 79: "Were the reaction field adequately given by equation (27), these curves would all be straight lines with the slopes (corresponding to k_E) all being nearly equal. Instead of straight lines, however, smooth curves were obtained, of which only two out of three curve in the direction expected. Both the lack of linearity and the differing curvatures give dramatic indication of the inadequacy of the present concepts of the reaction field. If one calculated the proportionality constants

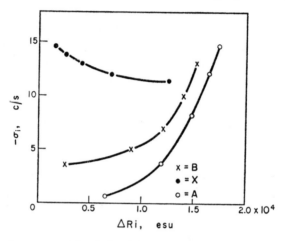

Fig. 4. Plot of $-\sigma_i$ versus ΔR_i calculated from equation (27). (Laszlo and Musher.[79])

k_E for pairs of points, one finds values varying from $1.0\text{–}42 \times 10^{-12}$ e.s.u., and therefore choosing of the best straight line fit would have little meaning. This range of values for k_E includes the -3.0 and -3.4×10^{-12} e.s.u. of Diehl and Freeman.[77] In addition the fact that the curve for σ_A bends in the opposite direction indicates that the reaction field *along* the C—A bond must have the same sign as that *along* the C—X bond. As these two bonds are antiparallel this means that the components of R, the electric field parallel to these bonds (supposedly constant over the entire molecule), *changes sign* from one position X to another position A. Simple calculations of orders of magnitude (using the procedure of Musher[93]) shows that only a negligible error is introduced in not considering the effect of the reaction field on the neighbouring bromine atom and phenyl groups which in turn would affect the long range shielding of these protons.

"In order to show that the inclusion of the spheroid shape corrections of Diehl and Freeman[77] does not correct the non-linearities of Fig. 4, σ_X is plotted in Fig. 5 versus $(\epsilon - 1)/(\epsilon + \beta)$ for $\beta = 1.0$, 2.86, and 0.5. The value of $\beta = 1.0$ is that used in Fig. 4 for a sphere with $n^2 = 2$ while the other two values are for the particular respectively oblate and prolate spheroidal examples studied by Diehl and Freeman.[77] It is seen that changing β over this rather wide range does not serve to make the curves more linear, although of course it does change the values of the tangents." We then summarize the arguments against substantial conformational changes in molecules V and VI, caused by the solvent.

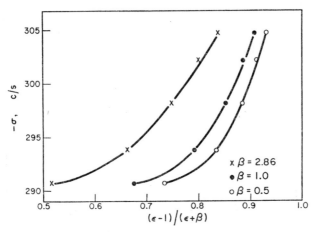

FIG. 5. Plot of $-\sigma_X$ versus $(\epsilon - 1)/(\epsilon + \beta)$ calculated for three values of β. (Laszlo and Musher.[79])

To quote again from reference 79: "Before inferring the necessary conclusions from these data it is desirable to give the experimental evidence available in confirmation of the assertion that the shielding σ is actually linear in some reaction field (the curves shown in Fig. 4 and the several significant deviations from the linear plots of reference 77 make an attempt to confirm this assertion as important and not trivial). Protons X (in V) and Y (in VI) are in positions of relatively similar local geometry and should therefore experience relatively similar reaction fields as a function of ϵ. In Fig. 6(a), σ_X is plotted versus σ_Y for the different values of ϵ and the reasonably straight line obtained indicates that $\sigma_X = c\sigma_Y$ which implies that both σ_X and σ_Y are proportional to the same $\rho(\epsilon)$. Since the dipole moments of V and VI are not equal and therefore the components μ_X and μ_Y are probably not equal, were there a contribution to the σ_i's quadratic in the μ_i's the plot of Fig. 6(a) would be curved and not a straight line. In Fig. 6(b) σ_A is plotted versus σ_Z as functions of ϵ for which the argument as given for σ_X and σ_Y holds despite the smaller magnitudes of the shifts involved. The slope of σ_X/σ_Y is 1·2 while that of σ_A/σ_Z is 0·9. The first of these is more significant than the second since the latter could be drastically changed within the experimental error. One would have expected both of these slopes to be roughly in the ratio of $\mu(V)/\mu(VI) = 1\cdot1$ as is observed. This should not be taken as a precise statement, since the former is directed along a different axis from the latter—although the angles to the respective axial bonds will be roughly the same. This does indicate, however, both the validity of reaction field arguments (that is, that the protons in both cases are shifted by the same interaction, presumably that with the reaction field, but the ϵ dependence of this reaction field is not that expected) for comparable local sites. In addition, there appears to be an absence of other significant contributions, e.g. electric field squared, London interaction, etc., to the σ_i's. In

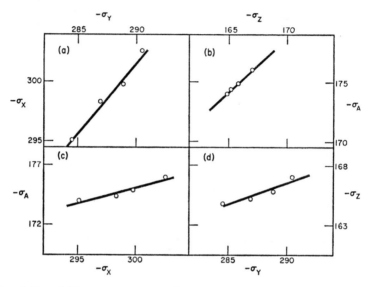

FIG. 6. Plot of (a) $-\sigma_X$ versus $-\sigma_Y$; (b) $-\sigma_A$ versus $-\sigma_Z$; (c) $-\sigma_A$ versus $-\sigma_X$; (d) $-\sigma_Z$ versus $-\sigma_Y$. All σ values are measured in c/s from tetramethylsilane. (Laszlo and Musher.[79])

Figs. 6(c) and (d) are plotted σ_A versus σ_X and σ_Z versus σ_Y, both of which also give relatively straight lines within experimental error and both of which have slopes of $\frac{1}{3}$. The linearity of these plots is a little less expected in that one might have thought the local (solute molecule) effects—A and X are in rather different environments—to play a non-multiplicative role. In any case, the experiments are not sufficiently precise to differentiate between the straight lines as drawn in these figures and the curves through the experimental points. The nearly identical slopes of $0\cdot3$ indicate that the reaction fields at A(Z) are roughly $\frac{1}{3}$ weaker and in the opposite direction than at X(Y). Since k_A probably differs from k_X by no more than $\sim10\%$, this difference as well as the sign is certainly significant."

Later, Fontaine, Chenon and Lumbroso-Bader[17] reached similar conclusions, and used analogous procedures, which only differ in their attempt (see Section 4.3) to separate the reaction field from the dispersion shifts. They have also used solvent mixtures to minimize the specific differences. The curves they obtain for the methyl groups (Fig. 7), the methine protons (Fig. 8), and their difference (Fig. 9), for the di-isopropylketone solute, versus the dielectric function $(\epsilon - 1)/(2\epsilon + n^2)$, in cyclohexane–acetone mixtures,

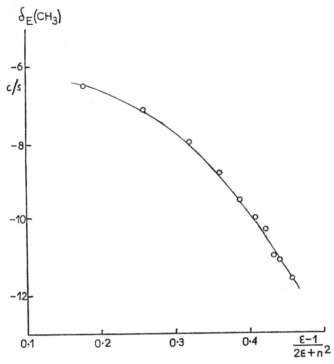

Fig. 7.   Plot of $\delta_E(CH_3)$ versus $(\epsilon - 1)/(2\epsilon + n^2)$ for di-isopropylk etone.
(Fontaine, Chenon and Lumbroso-Bader.[17])

are indeed far from being straight lines. The intermolecular comparison of the reaction field shifts obtained by Fontaine, Chenon and Lumbroso-Bader[17] for isopropyl-tert-butylketone and di-tert-butylketone (see Fig. 10 for tert-butyl resonances) also shows that the same polar factors are operative in both cases. Obviously, the variations in the relative populations of the rotamers VIa and VIb with solvent polarity must be considered in such conformationally mobile examples, which therefore are not clear-cut with respect to the reaction field situation. Nevertheless, it would have been expected that di-tert-butylketone would be more strongly influenced, due to

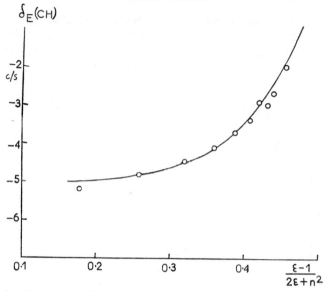

FIG. 8. Plot of $\delta_E(CH)$ versus $(\epsilon - 1)/(2\epsilon + n^2)$ for di-isopropylketone.
(Fontaine, Chenon and Lumbroso-Bader.[17])

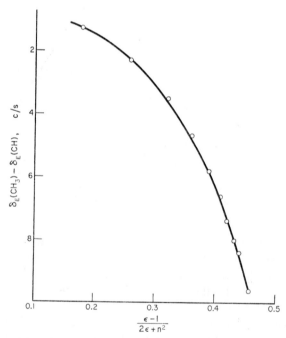

FIG. 9. Plot of $\delta_E(CH_3) - \delta_E(CH)$ versus $(\epsilon - 1)/(2\epsilon + n^2)$ for di-isopropylketone.
(Fontaine, Chenon and Lumbroso-Bader.[17])

an even greater predominance of conformation VIIa over conformation VIIb
when R = t-Bu than when R = i-Pr:

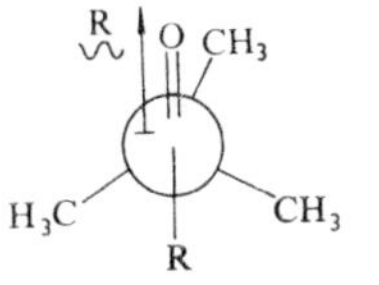

VIIa 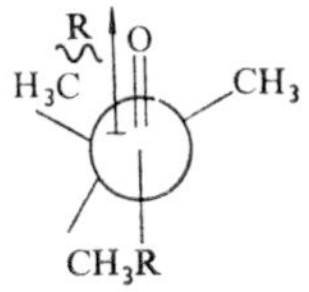   VIIb

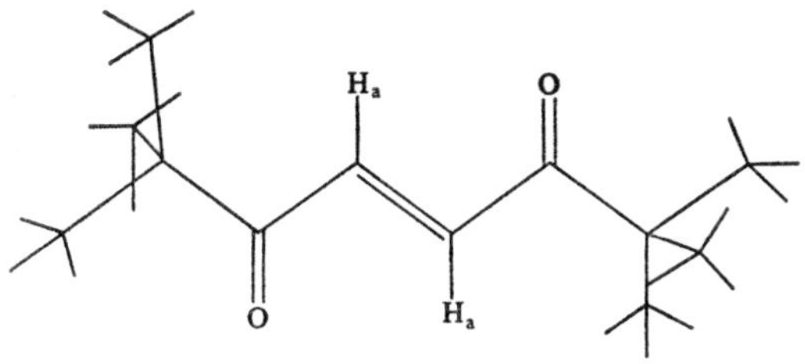

VIII

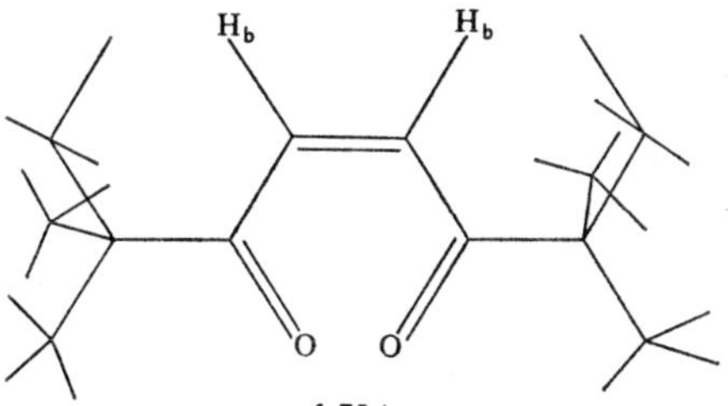

1.75 A

IX

The reverse is observed (see Fig. 10), as implied by the authors[17] at the end
of their paper. The present author[80] has also used as solute molecules
having bulky tert-butyl groups flanking the polar site and shielding it from
specific interactions on the part of the solvent molecules. With the configurational isomers VIII and IX, compared with the corresponding cis- and trans-
1,2-dihaloethylenes studied by Hruska, Bock and Schaefer,[46] it was possible
to demonstrate that the proportionality "constant" k_E actually varies with
the nature of the substituents, 0·6 (Cl) and 0·7 (Br), versus $1·95 \times 10^{-12}$
(CO-t-Bu). Molecule IX may not be planar, but somewhat twisted out of the
plane of the C=C double bond, which may be a solvent dependent phenomenon, but does not seem to have contributed appreciably to the experimental curve shown in Fig. 11 where the chemical shift difference H_a—H_b,

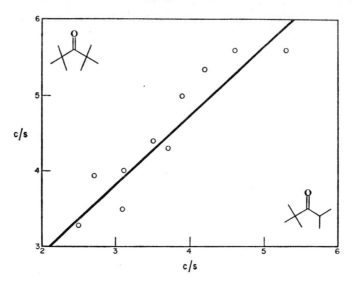

FIG. 10. Correlation between solvent shifts of ^{1}H resonances of t-butyl groups of isopropyl-tert-butylketone and di-tert-butylketone.

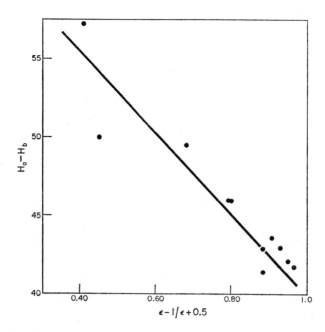

FIG. 11. Reaction field solvent shifts for molecule IX with molecule VIII used as a reference (Laszlo[80])

between VIII and IX, is plotted against the dielectric function appropriate for the prolate ellipsoids considered for VIII and IX.

Similar conclusions can be reached with polar aromatic solutes, whose effective dipole moments depend upon the solvent, as do the corresponding resonance forms. Taft *et al.*[42, 94] have provided a very detailed discussion of solvent effects for the fluorine resonance of meta and para substituted fluorobenzenes. Very small variations are due to the polarity of the solvent, as it is crudely measured by the dielectric constant and affects a given electronic distribution of the solute. The large observed effects are to be attributed unambiguously to the solvent favouring the contribution of *dipolar* resonance forms of the paraquinoid type, e.g. for the *para*-fluorobenzonitrile:

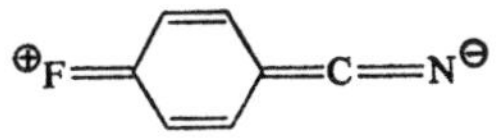

X

The necessary condition is that *both* atomic centres bearing the opposite formal charges be on the periphery of the molecule, a condition met only for $+R$ (electron-accepting) para substitutents; the derivatives with $-R$ para, or with meta, substituents, showing only negligible solvent shifts. These results are in good qualitative agreement with the Kirkwood model[95] which predicts the greater lowering of the energy of a dipolar resonance form $\frac{1}{2}(D_i - D)/(2D + D_i)(\mu^2/b^3)$, and therefore of the ^{19}F shielding, when the dielectric constant of the bulk solvent, D, increases relative to the internal dielectric constant D_i of the cavity of radius b, where the molecule with dipole moment μ is enclosed. Taft[42, 94] finds that the bulk dielectric constant D is an inadequate *quantitative* measure of the ability of the solvent to support charge separation in solute molecules.

4.5. Conclusions

All these experiments converge towards the idea that the reaction field indeed exists, and that it is apparent in NMR solvent shifts. However, it changes in magnitude and direction at different positions in the same molecule, i.e. it is much more specific an effect than expected from the simple theory; this observation is essentially in agreement with the conclusions of Lovell[74] on the essentially *macroscopic* character of the reaction field.

5. THE RELEVANCE OF CHARGE-TRANSFER TO SOLVENT EFFECTS

Charge-transfer complexes[96] have been little studied by NMR. One of the first observations to be reported was that of Cram and Singer,[97] who did not find any difference in the spectra of their paracyclophanes, when examined as either the free molecules or as part of donor-acceptor complexes (with these aromatic molecules being the electron donors). One may write generally the structure of these charge-transfer complexes as I, A the symbol representing the electron acceptor, for example molecular iodine, tetracyanoethylene, 1,3,5-trinitrobenzene, some cations, etc. The resultant chemical shifts,

I

for the acceptor A, as a crude first approximation, might be expected to be associated with the aromatic ring current of the underlying donor molecule (high field shift), and with the increase of the electronic charge, which would presumably also increase the shieldings in a general way. For the donor we might expect the reverse effect of a paramagnetic shift with the electron donation, combined with the effects from the anisotropy field of the A molecule. To a first approximation the reaction field, or more accurately the electric field along the charge-transfer axis, does not contribute at all to the chemical shifts of the protons situated in a plane perpendicular to that axis, such as those of the aromatic donor part, since the corresponding value of cos θ (see Section 4) is zero. The results of Cram and Singer[97] are then readily explained if this latter effect is dominant, the two charge-transfer molecules being separated by a large enough distance for the anisotropy effects to be small, and the actual degree of charge transfer also being small. This description is corroborated by Koch and Zollinger[98] who found only tiny differences (see Table 23) which do not correlate simply with the equilibrium constants as determined by the classical optical methods.[96] Even smaller shifts are found for the donor protons chemical shifts.

TABLE 23. CHARGE-TRANSFER SHIFTS FOR 1,3,5-TRINITROBENZENE (TNB) COMPLEXES

Complex composition	Solvent + ref.	Shifts[a]	K_c[b]
TNB alone (0·417 M)	CDCl$_3$/TMS	—	—
TNB-pyridine (1 : 1)	CDCl$_3$/TMS	0	>10·0
TNB-mesitylene (1 : 1)	CDCl$_3$/TMS	6	3·2
TNB-C$_6$Me$_6$ (1 : 1)	CDCl$_3$/TMS	12·5	5·7
TNB-naphthalene (1 : 1)	CDCl$_3$/TMS	23	21

[a] c/s at 60 Mc/s, relative to the aromatic resonance at 9·39 ppm for the 0·417 M TNB molecule.
[b] Equilibrium constant.

Biellmann[99] has found relatively large shifts, of the order of 0·30–0·50 ppm, for pyridinium salts of the type (II), but these may well be in fact true addition compounds.

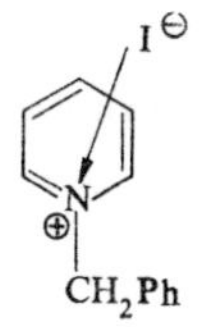

II

Chemical shifts of a similar order of magnitude are reported by Katritzky, Swinbourne and Ternai[100] for the α-, β- and γ-protons of pyridine and of methyl-substituted pyridines, when the solvent is changed from cyclohexane or carbon tetrachloride (inactive) to the Lewis acids $AsCl_3$, PCl_3 or SO_2. Since the solvent shifts are similar with trifluoroacetic acid, in which the existence of the pyridinium cation has been proven, and since α-methyl groups have an unfavourable steric effect on the solvent interaction, it is also concluded that the π-electronic system does not participate: instead the nitrogen lone pair is involved in the formation of an addition compound with arsenic trichloride as in III. The covalent bonding, and the corresponding

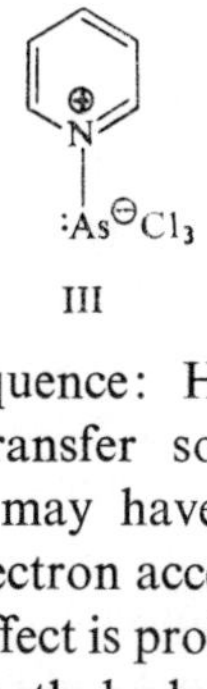

III

charge transfer, follow the sequence: $H^+ > AsCl_3 > SO_2 > PCl_3$. It is therefore a feature of charge-transfer solvent shifts to exhibit a strong geometrical dependence, which may have diagnostic importance, because they are substantial only if the electron acceptor is in or near the plane of the electron donor. An electric field effect is probably responsible. In the following Section we shall consider at length hydrogen bonding solvents, hydrogen bonding being but a case of charge transfer, but where the hydrogen bond shifts are necessarily much greater for the proton directly engaged in such a bond.

6. HYDROGEN BONDING PHENOMENA

We shall delve neither in the theory of hydrogen bonding nor into the origin of the so-called hydrogen bond shift in observed nuclear magnetic resonance. We shall limit ourselves to the techniques available to reach this hydrogen bond shift and to its interpretation.

6.1. *Computational Techniques for the* NMR *Study of Hydrogen Bonding Phenomena in Solution*

The method of Saunders and Hyne[101] for studying hydrogen bonded complexes by NMR, or other associative processes, is based upon the Gutowsky and Saika[102] theory of fast exchange between the bound and the free chemical species. There results a unique resonance peak whose position v is the weighted average of all contributors:

$$v = \frac{v_1M_1 + 2v_2K_2M_1^2 + 3v_3K_3M_1^3 + \cdots}{M_1 + 2K_2M_1^2 + 3K_3M_1^3 + \cdots} \tag{32}$$

where M_1 is the concentration of monomer, and C the total concentration and K_i the equilibrium constants. An approximation that can frequently be made is that all K_i's but one are negligible, and that the system studied is adequately described by the monomer n-mer equilibrium, in which case equation (32) reduces to:

$$v = \frac{v_1M_1 + nv_nK_nM_1^n}{C} \tag{33}$$

with $C = M_1 + nK_nM_1^n$.

The solutions must be ideal in the individual species for the K's to be thermodynamic equilibrium constants. As long as ratios of activity coefficients do not vary in the concentration range studied, this treatment will yield "observed" constants. The technical procedure followed is to arbitrarily assume a set of v_1, v_n, and K_n, compute for each given M_1 the values of C and v, and by trial-and-error fit the experimental curve, v versus C. Mavel,[103] using the Gutowsky–Saika chemical exchange laws,[102] and following closely the previous work of Redlich and Kister,[104] has explicitly derived analytical expressions and curves for complex formation and disruption phenomena. He finds that for a binary mixture it is possible, to a first approximation, to account for separate elementary processes using their independent equilibria. He also considers the equilibria monomer + dimer (acetic acid in 1,1-dichloroethane), monomer + n-mer (from which it appears that water is essentially a dimer, with $K = 8$) and monomer + all polymers, for the self-association cases. The complexes studied are of the $1 : 1$, $1 : n$, and $n : 1$ types, and theoretical curves are drawn for equilibrium constants K between 0 and 100.

Berkeley and Hanna[105] have further refined a treatment of hydrogen bonding effects given by Huggins, Pimentel and Shoolery.[106] Assuming a $1 : 1$ complex AD between the hydrogen bond donor D (initial concentration X_D^0) and the acceptor A (initial concentration X_A^0), the observed dilution shift $\Delta = \delta_0 - \delta_D$, where δ_0 is the chemical shift of the donated proton in the infinitely dilute donor. The dilution shift is related to the hydrogen bond shift $\Delta_{AD} = \delta_{AD} - \delta_D$ through a function of the equilibrium constant K (with the usual assumption of an ideal mixture):

$$\Delta = \delta_0 - \delta_D = \frac{K}{1+K}\Delta_{AD}. \tag{34}$$

The calculation of the limiting slope at infinite dilution of donor S_0 gives the relationship (35),

$$S_0 \equiv \left(\frac{d\delta}{dX_D^0}\right)_{X_D^0 \to 0} = -\Delta_{AD}\frac{K}{(1+K)^2} \tag{35}$$

which together with (34) is sufficient to determine Δ_{AD} and K uniquely from the infinite dilution shift δ_0 and the limiting slope S_0. To correct for the non-specific part of the solvent effect, Berkeley and Hanna[105] have adopted the empirical equation of Bothner-By,[12] which eliminates the influence of differential solvent effects on donor and reference. Also, it automatically includes the weak chloroform self-association, when the hydrogen bonded complexes of chloroform are studied, for example. The determination of the association constant K is rendered easier by adopting a suggestion made by Foster and Fyfe;[107] an alternative expression for (34) is, according to Hanna and Ashbaugh:[108]

$$\Delta = \frac{X_D^0}{1+KX_D^0} \cdot \Delta_{AD} \tag{36}$$

which can also be written as:

$$\frac{1}{\Delta} = \frac{1}{K\Delta_{AD}} \cdot \frac{1}{X_D^0} + \frac{1}{\Delta_{AD}} \tag{37}$$

or:

$$\frac{\Delta}{X_D^0} + K \cdot \Delta = \Delta_{AD} \cdot K. \tag{38}$$

From this last expression, it is seen that the plot of Δ/X_D^0 against Δ should be a straight line which gives K directly as the negative gradient.[107] Rather than the usual methods which evaluate the equilibrium constants for complex formation from the study of the variation of the chemical shift in a binary mixture as a function of the mole fraction of the solute, Lussan[109] has developed a more rigorous method. This is based on the utilization of ternary mixtures: the solute and the complexing species are titrated in a dilute solvent; the constant K is deduced from the values of the infinite dilution chemical shift of the solute in different mixtures of the inert solvent with the active solvent. This procedure is applicable also to aromatic solvents, since the effect of the intermolecular anisotropy is eliminated by this means. For instance, the literal treatment of the equilibria between a monomeric species

AH, the open dimer AH ... AH, and the bonded species AH ... B to the complexing solvent B, in an inert solvent S, is the following:

$$2AH \rightleftharpoons AH \ldots AH \qquad (K_A, \delta_A) \qquad (39)$$

$$x_A^0 - 2n_2 - c_1 \quad n_2$$

$$AH + B \rightleftharpoons AH \ldots B \qquad (K_B, \delta_B) \qquad (40)$$

$$x_A^0 - 2n_2 - c_1 \quad x_B^0 - c_1 \quad c_1$$

$$x_A^0 + x_B^0 + x_S^0 = 1.$$

If $u = \dfrac{n_2}{x_A^0}$ and $v = \dfrac{c_1}{x_A^0}$; $S = 1 - x_S^0$

$$K_A = \frac{u[1 - x_A^0(u + v)]}{x_A^0[1 - (2u + v)]^2} \qquad (41)$$

$$K_B = \frac{v[1 - x_A^0(u + v)]}{[1 - (2u + v)][1 - x_S^0 - x_A^0(1 + v)]} \qquad (42)$$

and the chemical shift measured relative to the monomer is:

$$\delta = u\delta_A + v\delta_B \qquad (43)$$

δ_B being the hydrogen bond shift.

Extrapolation to infinite dilution gives:

$$\delta(x_A^0 \to 0) = \frac{K_B S}{1 + K_B S} \delta_B \qquad (44)$$

which does not depend upon the value of K_A. The experimental procedure is to plot, for a series of values of x_S^0 (the total concentration in the inert solvent), the dilution curves for the solute's chemical shift, and the infinite dilution shifts are entered into equation (44). For instance, the following

TABLE 24. INFINITE DILUTION SHIFTS (ppm) FOR N-METHYL ANILINE

x_S^0 (mole fraction)	$\delta(x_A^0 \to 0)$	
	CCl$_4$	Freon-112
1·00	0	0
0·77	−0·67	−0·68^5
0·70	−0·79	—
0·50	−1·03	−1·07
0·30	−1·18	−1·22
0·00	−1·37^5	−1·40

values were obtained by Chenon and Lumbroso-Bader[110] for N-methyl aniline hydrogen bonded to acetone, in either carbon tetrachloride or Freon-112 solution (see Table 24). From this table average values of K_2 and δ_B are readily computed. Similar treatments have been applied to a cyclic dimer or trimer.[109] The formalism of LaPlanche, Thompson and Rogers[111] applies to chain association equilibria. The assumptions are clearly defined, and not as restrictive as they may appear at first sight:

(a) The equilibrium constant of the solute with its dimer, K_{12},

$$\overset{K_{12}}{A + A \rightleftharpoons A \ldots A}$$

differs from $\bar{K}$, for the equilibrium of the monomeric solute A with *any* of its n-mer ($\bar{K}$ is the same for all n's >1):

$$\overset{\bar{K}}{A + A_n \rightleftharpoons A \ldots A_n.}$$

(b) The same holds with K'_{12} and $\bar{K}'$, when in the above equilibria the left-hand A is replaced with the solvent S.
(c) The solute A associates by forming linear chain-like polymers A_n.
(d) Each n-mer molecule can associate with only one solvent molecule, by hydrogen bonding at the end of the chain.
(e) No *intra*molecular associations are present.
(f) There is no self-association of the solvent.
(g) Only two distinct cases are considered: either the solvent does not form hydrogen bonds with the solute (*inert solvent*), or it forms bonds at only *one* site (*strongly hydrogen bonding solvent*).
(h) the solute contains only two hydrogen bonding sites, one donor and one acceptor.

The notation is now summarized:

$\bar{Y} = \sum_n Y_n$ $\qquad\qquad$ $Y_n =$ mole fraction of unsolvated n-mer.

$Y = \sum_n n \cdot Y_n$

$\bar{Z} = \sum_n Z_n$ $\qquad\qquad$ $Z_n =$ mole fraction of solvated n-mer.

$Z = \sum_n n \cdot Z_n$

$\qquad\qquad\qquad\qquad$ $S =$ mole fraction of solvent.

$\bar{Y} = \bar{Z} + \bar{S} = 1$

$L_{12} = Z_2 \cdot S/Z_1^2$ $\qquad\qquad$ $K_{12} = Y_2/Y_1^2.$

$\bar{L} = Z_{n+1} \cdot S/Z_1 \cdot Z_n$ $\qquad$ $\bar{K} = Y_{n+1}/Y \cdot Y_n$ $\qquad n > 1.$

K

These K and L values are the equilibrium constants for self-association, and for complexation, respectively.

If, ν_M and ν_D are the chemical shift of the *monomeric* and *dimeric* protons liable to hydrogen bonding, respectively, and ν_C is the chemical shift in the hydrogen bonded *complex* with the solvent; and ν_F is the chemical shift for the *unbonded* protons in the solvent–solute complex, the solvent attaching itself to the solute molecule at another site than the proton–donor site, then the final equations are, for the observed chemical shift δ_0,

Inert solvent:

$$\delta_0 = \bar{Y}/Y(\nu_M - \nu_D) + \nu_D. \tag{45}$$

Strongly hydrogen bonding solvent:

$$\delta_0 = \bar{Z}/Z(\nu_F - \nu_D) + \nu_D. \tag{46}$$

A computer technique is necessary to solve these equations for the unknown chemical shifts and equilibrium constants.

The concentration dependence of the proton chemical shifts in hydrogen bonded systems is interpreted by Satake et al.[112] in terms of statistical mechanics *without the assumption of any particular hydrogen bonded complex in the solution*. The quasi-chemical approximation of Guggenheim[113] is applied to the ethanol–carbon tetrachloride (A), water–acetone (B), and water–dioxane (C) systems.

The theoretical treatment considers the binary mixture of molecules A and B, for which the protons of type i in the A molecules are in Barker contact[114] with points of type j in the B molecules, the number of such pairs being denoted N_{ij}^{AB}, such that:

$$2N_{ii}^{AA} + \sum_{B, j} N_{ij}^{AB} = Q_i^A N_A \tag{47}$$

N_A being the number of molecules A, and Q_i^A denoting the number of contact points in the ith class of the A molecules. The quasi-chemical equation is written:

$$(N_{ij}^{AB})^2 = 4N_{ii}^{AA}N_{jj}^{BB} \exp(-U_{ij}^{AB}/kT) \tag{48}$$

where U_{ij}^{AB} is the energy of interaction for each possible contact. By introducing the parameters:[114]

$$\eta_{ij}^{AB} = \eta_{ji}^{BA} = \exp(-U_{ij}^{AB}/kT)$$

$$X_i^A = (N_{ii}^{AA}/N)^{1/2}; \quad \text{and} \quad X_j^B = (N_{jj}^{BB}/N)^{1/2},$$

N being the total number of molecules in the system:

$$N_{ij}^{AB} = 2X_i^A X_j^B \eta_{ij} N \tag{49}$$

X_i^A and X_j^B being given by the reciprocal expressions:

$$X_i^A \cdot \sum_{B,j} \eta_{ij} X_j^B = Q_i^A x_A/2, \qquad (50)$$

$$X_j^B \sum_{A,i} \eta_{ij} X_i^A = Q_j^B x_B/2 \qquad (51)$$

where x_A and x_B are the mole fractions.

The chemical shift is, as usual,[102] the weighted average of the different contributions:

$$\delta_i = \sum_j N_{ij} \delta_{ij} / \sum_j N_{ij} \qquad (52)$$

and the theoretical dilution curve can be constructed from equation (52), the values of N_{ij}^{AB} being estimated from equations (49), (50) and (51), provided that the values of the η_{ij}'s and the Q_i^A's be known. Conversely, the comparison of the calculated and the experimental curves will yield the interchange energies. In the examples treated, assumptions are first made regarding the lattice model (EtOH as a trimer; CCl_4 a pentamer, in A; H_2O, a monomer; Me_2CO, a trimer, in B; H_2O, monomer; dioxane, pentamer, in C), from which the number of contact points is evaluated. The theoretical dilution curves can be derived as functions of the interchange energies and the proton fraction. The values of the interchange energies giving the best fit with the experimental curves are $-3 \cdot 09$ kcal mole^{-1} bonds (A), $-4 \cdot 20$ and $-3 \cdot 68$ kcal mole^{-1} bonds (B), and $-4 \cdot 40$ and $-3 \cdot 20$ kcal mole^{-1} bonds (C). The strength of hydrogen bonding is found to decrease in the order

$$H_2O—H_2O > H_2O—Me_2CO > H_2O—dioxane > EtOH—EtOH.$$

The degree of confidence to place upon these different methods descriptive of hydrogen bonded associations can be empirically estimated by the amount of divergence between the results for the more common systems. Chloroform has been extensively used as proton donor, and from its mixtures with acetone, ether, triethylamine and dioxane, the respective range of values are deduced:

For the equilibrium constant K: $1 \cdot 8$–2; $3 \cdot 75$; $2 \cdot 76$–$4 \cdot 70$; —.

For the hydrogen bond shift (ppm): $0 \cdot 93$–$1 \cdot 42$; $0 \cdot 54$–$2 \cdot 68$; $1 \cdot 42$–$1 \cdot 88$; $0 \cdot 85$–$1 \cdot 20$.

For the enthalpy of formation: $1 \cdot 2$–$2 \cdot 7$; $6 \cdot 6$; $4 \cdot 0$–$5 \cdot 5$;—(kcal mole^{-1}).[109, 115–17]

6.2. Amount of Solvent Self-association (or Solute Self-association)

Springer and Meek[118] have studied the concentration dependence of the diethylamine NH proton resonance at 40°C in acetonitrile and in cyclohexane. The purified acetonitrile used contained less than 10^{-5} mole fraction of water, an amount quite sufficient to cause rapid exchange of the NH proton and wash out the quadrupole coupling with ^{14}N, but insufficient to cause a detectable shift in the concentration range for the study. Acetonitrile is a more basic molecule than diethylamine, which forms stronger hydrogen bonds with it than with itself. While the methyl triplet of the amine resonates at 1.01 ± 0.03 ppm, the NH singlet is displaced towards lower fields when the concentration of Et_2NH is lowered, from 0.71 ppm in the pure liquid to 1.17 ppm for a 0.0166 mole fraction in MeCN. The dilution curve of diethylamine in cyclohexane, where the NH resonance varies from 0.71 ppm (1.000 mole fraction) to 0.40 ppm (below 0.1 mole fraction), is interpreted by means of the Saunders–Hyne,[101] Gutowsky–Saika[102] treatment. The best fit corresponds to the joint existence of the monomer (0.40 ppm) and a tetramer, presumably cyclic[119] (its chemical shift is 1.18 ppm), with an equilibrium constant $K = 1.75 \times 10^{-3}$ M^{-3}. The agreement with the earlier investigation of Feeney and Sutcliffe[119] is reasonable since these authors had proposed previously a tetramer and had found $K = 2.5 \times 10^{-4}$ M^{-3} for another solvent (CCl_4) and for another temperature ($25°C$). Feeney and Sutcliffe[119] have calculated an enthalpy of formation $\Delta H_f = -6.8$ kcal mole^{-1} for the tetramer, since K increases to 5.0×10^{-3} M^{-3} at $-37°C$; these authors have also proposed a monomer–tetramer equilibrium for ethylamine, with $K = 2.5 \times 10^{-3}$ M^{-3}, at $25°C$. However, if the self-association shift of 0.78 ppm (~ 1 ppm according to Feeney and Sutcliffe[119]) determined by Springer and Meek[118] is consistent with the value of 0.66 ppm of Lemanceau et al.,[109, 115, 120] the equilibrium constants cannot be compared since the model chosen for the self-association in cyclohexane is not identical, namely a tetramer[118] versus a polymer chain.[120] Springer and Meek[118] rightly indicate that a dilution shift of 0.3 ppm, as they obtained for Et_2NH, can equally be observed with non-hydrogen bonded protons. Therefore, some caution must be attached to these parameters for self-association, which are very difficult to obtain accurately. The self-association of a number of protic molecules, like chloroform and the alcohols, act as perturbations when the parameters for the liquid phase association with electron donors are to be determined. One can separately evaluate the amount of self-association for the pure AH solute in an inert solvent, and then correct for it when it is mixed with a proton acceptor B. This is, for instance, what Jumper, Emerson and Howard[121–2] did for chloroform, for which they proposed that $K_A = 0.13 \pm 0.03$ M^{-1} and $\delta_A = -1.8 \pm 0.3$ ppm. As pointed out by Souty and Lemanceau,[115] this procedure is unreliable, since the dilution curve for the chemical shift of chloroform in cyclohexane would equally well be fitted by any value of K_A between 0.01 and 0.16 M^{-1}. A better procedure is afforded by the method of Lussan[109] (see Section 6.1), who examined the ternary mixture AH + B + S, S being an inert solvent. The method allows the independent determination of the constant K_B for the equilibrium:

$$AH + B \rightleftharpoons AH \ldots B$$

without any complication from the self-association equilibrium. Evaluation of the thermodynamic parameters for the latter requires a preliminary hypothesis upon the real nature of the self-association, that is, whether the dimeric, trimeric, etc., polymeric structures predominate, and if they are of the open or closed form. An independent study of the self-association of chloroform is due to Kaiser.[117] This author subtracted from the total experimental chemical shift the reaction field contribution, in a solvent of dielectric constant ϵ:

$$\delta = -0.23 \left(0.60 - \frac{\epsilon - 1}{\epsilon + 1.53}\right). \tag{53}$$

His results are shown in Table 25.

TABLE 25. EVALUATION OF CHLOROFORM SELF-ASSOCIATION IN VARIOUS SOLVENTS

Solvent	ϵ	$\delta \times 10^{-2}$	Total dilution shift	Δ (ppm)
n-Hexane	1·89	−7·8	−0·29	−0·21
C_5H_{10}	1·97	−7·4	−0·29	−0·22
C_6H_{12}	2·02	−7·1	−0·29	−0·22
CCl_4	2·24	−6·2	−0·29	−0·23
Dioxane	2·21	−6·4	−0·28	−0·22
NEt_3	2·42	−5·5	−0·28	−0·22
Et_2O	4·34	−0·7	−0·23	−0·22
Me_2CO	20·7	+6·6	−0·15	−0·22

This total self-association shift Δ of chloroform, between the pure liquid and an infinitely dilute solution in an inert solvent, is therefore −0·22 ppm, in very good agreement with the careful determination of Souty and Lemanceau,[115] based upon the work of Lussan[109] which gave −0·20 ppm. It should be quite clear that this *experimental* datum is not identical with the variation of the shielding between the monomer and the n-mer (probably a dimer), of the order of −1·8 ppm,[122] for only a minority of molecules are bound in the pure liquid. The following parameters describe the self-association of a series of alcohols, amines, plus acetic acid, according to Lemanceau *et al.*;[109, 115, 120] the apparent enthalpies of dimerization, between 20 and 60°C, are taken from Davis, Pitzer and Rao[121] (Table 26). The self-association of amines in carbon tetrachloride has also been studied by Bystrov and Lezina.[123] Saunders and Hyne,[101] by application of their theoretical treatment to the dilution curves down to 10^{-2} M of methanol, ethanol, tert-butanol and phenol, in carbon tetrachloride solution, could show that the MeOH–CCl$_4$ system is best described by a monomer–tetramer equilibrium, both t-BuOH–CCl$_4$ and PhOH–CCl$_4$ by monomer–trimer equilibria. However a conjunction of trimers and tetramers, together with the monomeric species, is necessary to fit the ethanol dilution curve (see Table 27).

TABLE 26. SELF-ASSOCIATION OF OH AND NH PROTONS IN INERT SOLVENTS

Solute	Solvent	$K_A(\text{M}^{-1})$	δ_A (ppm)	Apparent $-\Delta H$ (kcal mole^{-1})
MeOH	CCl$_4$	70[a]	$-4\cdot88$	$9\cdot4 \pm 2$
EtOH	CCl$_4$	51[a]	$-4\cdot9$	$7\cdot6 \pm 2$[d]
t-BuOH	CCl$_4$	32[a]	$-4\cdot16$	$4\cdot4 \pm 2$
Et$_3$COH	CCl$_4$	8[a]	$-3\cdot5$	
(t-Bu)(i-Pr)CHOH	CCl$_4$	4[b, c]	-3	
(t-Bu)$_2$CHOH	CCl$_4$	$1\cdot2$	$-5\cdot75$[b]	
			$-2\cdot87$[c]	
piperidine	C$_6$H$_{12}$	$2\cdot37$[a]	$-1\cdot15$	
Et$_2$NH	C$_6$H$_{12}$	$2\cdot75$[a]	$-0\cdot66$	
AcOH	CCl$_4$	1000[c]	-10	
		16[a]	$-6\cdot60$	

[a] Polymer chain.
[b] Open dimer hypothesis.
[c] Closed dimer hypothesis.
[d] $5\cdot1 \pm 1$ kcal mole^{-1} for EtOH in C$_6$H$_6$.

TABLE 27. CORRECTED EQUILIBRIUM CONSTANTS AND FREQUENCIES[a] FOR VARIOUS MONOMER n-MER THEORETICAL CURVES, FIT TO THE EXPERIMENTAL AT 21°C

System	MeOH–CCl$_4$	EtOH–CCl$_4$	t-BuOH–CCl$_4$	PhOH–CCl$_4$
$k_3(1^2/\text{M}^2)$	—	$5\cdot19$	$5\cdot60$	$4\cdot78$
$k_4(1^3/\text{M}^3)$	$28\cdot4$	$44\cdot9$	—	—
ν_1 (c/s)	$+169\cdot5$	$+174\cdot0$	$+164\cdot0$	$+19\cdot5$
ν_3 (c/s)	—	$-36\cdot5$	$-2\cdot7$	$-123\cdot8$
ν_4 (c/s)	$-18\cdot3$	$+4\cdot6$	—	—

[a] Relative to an external H$_2$O reference.

6.3. Hydrogen Bonding in Binary Mixtures

Chenon and Lumbroso-Bader[110] have applied the method of Lemanceau and Lussan[109, 115, 124] (see Section 6.1), which uses a ternary mixture of an inert solvent, a self-associating substrate, and a strong complexant, to deduce the hydrogen bond shift of N-methyl aniline (PhNHMe) dissolved in Me$_2$CO, without any perturbation from its self-association. Whether carbon tetrachloride or Freon-112 were used as inert solvents, coherent results were obtained, yielding average values of $K_B = 2\cdot2_7 \pm 0\cdot13$ (at 36°C) for the equilibrium:

$$\text{PhNHMe} + \text{Me}_2\text{CO} \rightleftharpoons \text{Complex}$$

and the hydrogen bond shift $\delta_B = -1.98 \pm 0.05$ ppm. These data suggest that N-methyl aniline is intermediate as a proton donor between tert-butanol ($K_B(Me_2CO) = 4.3$; $\delta_B = -2.76$ ppm[124]) and chloroform ($K_B(Me_2CO) = 2$; $\delta_B = -0.93$ ppm), all determined at 25°C.[115] The magnitude of the disputed[125-7] self-association of N-methyl aniline could not be determined in this study, the curvature of the dilution plot in an inert solvent was too weak to give reliable results. Infrared studies,[126] could not be used either but Pannetier and Abello[128] were recently able to measure it by a thermodynamic study using the Prigogine theory of ideal associated solutions: the constant for the dimerization equilibrium is $K_A = 0.18$ at 25°C, while a value of $K_n = 0.32$ characterizes the higher order associations at the same temperature.

Mateos, Cetina and Chao[129] have studied the monomer–dimer equilibrium in amino NH bonds, using the NH resonance dilution curve, analysed by the Huggins, Pimental and Shoolery[106] procedure, for concentrations 0.5–0.005 mole fraction in carbon tetrachloride or in pyridine, cyclohexane being the external reference (see Table 28). There is no simple relation between

TABLE 28. HYDROGEN BOND COMPLEX FORMATION OF AMINES

Compound	δ (NH monomer)	δ (NH . . . NC$_5$H$_5$)	K	pK_a
Ethylenediamine	52·7	73·6	6·65	9·98
Dimethylamine	28·6	51·9	4·23	10·64
Piperidine	61·8	99·8	3·88	11·13
Pyrrolidine	76·2	124·2	1·04	11·37
m-Chloroaniline	216·7	317·7	0·52	3·32

the equilibrium constant K and the amine basicity as measured by the pK_a values. The authors assert that the chemical shifts for the dimer ought to equal those for complexation in pyridine, but this may only be a quite crude first approximation. LaPlanche, Thompson and Rogers[111] have applied their chain-association computational technique to the precise determination of the amount of self-association and of complexation of the solute N-isopropyl acetamide (see Table 29). This amide is then highly self-associated, as shown by the values of K_{12} and $\bar{K}$ in carbon tetrachloride and in cyclohexane; $\bar{K}$ is larger than K_{12}, as was predicted by Savolea-Mathot.[130] It is also noted that the values of ν_D experience similar variations to those of ν_ρ, temperature dependent. From this study, it appears that the hydrogen bond strengths probably follow the sequence:

NH . . . OSMe$_2$ > NH . . . OCEt$_2$; NH . . . dioxane;

NH . . . Cl$_3$CH > NH . . . NH.

TABLE 29. VALUES OF EQUILIBRIUM CONSTANTS K AND L (MOLE FRACTION) AND CHEMICAL SHIFTS FOR MeCONH-i-Pr

Solvent	K_{12} or L_{12}[a]	K or L[a]	ν_C [a, b]	ν_D [a, b]	ν_P [a, b, c]
		Inert solvents			
CCl$_4$	$16 \cdot 0 \pm 2 \cdot 0$	150 ± 10	284	484	484·0
C$_6$H$_{12}$	$10 \cdot 5 \pm 2 \cdot 0$	450 ± 50	278	501	490·5
		Associative solvents			
Dioxane	$3 \cdot 5 \pm 0 \cdot 5$	$9 \cdot 0 \pm 0 \cdot 5$	371	490	485·0
Et$_2$CO	$4 \cdot 5 \pm 1 \cdot 5$	$6 \cdot 5 \pm 0 \cdot 5$	410	491	485·0
Me$_2$SO	$0 \cdot 6 \pm 0 \cdot 1$	$0 \cdot 9 \pm 0 \cdot 1$	451	480	480·0
CDCl$_3$	$4 \cdot 5$	$8 \cdot 0$	319	510	490·0

[a] The notation has been summarized in Section 6.1.
[b] c/s from TMS at 60 Mc/s.
[c] Chemical shift of the NH proton in the pure amide, varies because of the small difference in the spectrometer probe temperature from day to day.

Hruska, Kotowycz, and Schaefer[131] have successfully correlated the 100% hydrogen bonded proton shifts of chloroform (516 c/s), 1,1,2,2-tetrachloroethane (416 c/s), and 1,2,2-trichloroethane (H − 1 = 262 c/s; H − 2 = 409 c/s), in the complex with dimethyl sulphoxide and referred to internal tetramethylsilane,[132-3] with the occupation number n[134] of the $1s$ hydrogen orbital. These shifts in the pure complex were arrived at by McClellan and Nicksic,[132] using the method of Huggins, Pimental and Shoolery;[106] the values obtained by this method depend on a laborious curve fitting procedure to determine the association constant K. The linear correlation obtained by the Canadian authors[131] is depicted in Fig. 12.

This linear correlation predicts a zero hydrogen-bonded shift for methane—or, more exactly, its extrapolation leads to a − 10 c/s shift for CH$_4$ relative to internal TMS—, which suggests[131, 135] that correlations of this sort will be useful in NMR hydrogen bond studies, by reducing the labour necessary to measure the hydrogen bond shift by the classical methods.[106, 132-3] The slope for this linear correlation (Fig. 12) is − 67 ppm per unit charge, much greater than the usual proportionality constant of ∼ 10 ppm per unit charge, appropriate when charge enters the $p\pi$ orbital of a trigonal carbon atom. In this approach of Hruska, Kotowycz and Schaefer[131] the contributions of the solvent effect (for example, the intermolecular hydrogen bond with dimethyl sulphoxide) have to be separated from the substituent effect. In fact, for compounds of the type CHXYZ the proton chemical shift appears to correlate equally well with the single occupation number n, whatever the solvent chosen (cyclohexane, carbon tetrachloride or dimethyl sulphoxide). The good straight lines obtained for cyclohexane and carbon tetrachloride are shown in Figs. 13 and 14.

Comparison of the slopes shows that the hydrogen bonding phenomenon is accounted for by a somewhat larger (by 20%) slope for the dimethyl sulphoxide solution than with the inert solvents. Berkeley and Hanna[105, 136-7] have determined the actual shifts Δ_{AD}, upon complex formation between chloroform and a series of nitrogen bases, the corresponding equilibrium constants, and enthalpies of formation (see Table 30). The ΔH's are based

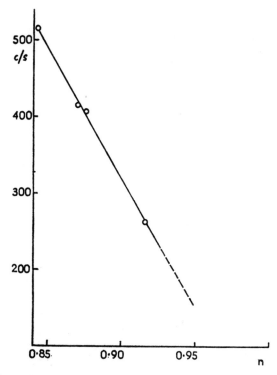

Fig. 12. Correlation of hydrogen bond shifts in dimethyl sulphoxide with the occupation number n. (Hruska, Kotowycz and Schaefer.[131])

upon temperature variations of the order of 10–20°. Loewenstein and Margalit[138] have compared the strengths of the hydrogen bonded complexes between the isomeric acetonitrile MeCN and methyl isocyanide MeNC, on the one hand, and methanol MeOH, on the other. Qualitatively, it is observed that the ^{14}N resonance in MeNC and the ^{13}C resonance in MeCN are concentration independent, while the ^{13}C resonance in the former and the ^{14}N resonance in the latter show high field shifts typical of an hydrogen bonding process. The atom accepting the proton is the nitrogen for MeCN, and the terminal carbon for the isocyanide. From the dilution shift of methanol in either nitrogen base, and having grossly corrected for its self-association, the authors[138] found the data listed in Table 31. It is seen that the enthalpy of formation is reversed for the cyanide–isocyanide pair when the methyl-substituted plus methanol[138] or the cyclohexyl-substituted plus chloroform systems[136] are compared. Berkeley and Hanna[137] have tried to derive the hydrogen bond *length* from the actual hydrogen bond *shift*, in the examples

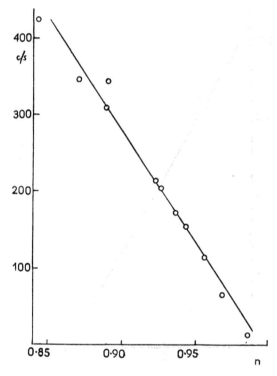

FIG. 13. Correlation between ^1H shifts and the occupation number n for substituted methanes CHXYZ in C_6H_{12}. (Hruska, Kotowycz and Schaefer.[131])

of N-methyl pyrrolidine, MeHC=N-i-Pr, and cyclohexyl cyanide. The $\varDelta_{AD}$ shift is split into two components; the Buckingham electric field effect is:

$$E = E_z = \frac{4 \cdot 80 \times 10^{-10}}{(0 \cdot 529)^2 \times 10^{-16}} (E' - 2/R^2) \qquad (54)$$

R being the interatomic nitrogen–hydrogen distance, it being assumed that the C—H bond lies along the axis of symmetry of the nitrogen lone pair, and E' is the electric field due to the nitrogen lone pair calculated with Slater atomic orbitals. The latter has values of $0 \cdot 9 \times 10^{-6}$ e.s.u. and $0 \cdot 3 \times 10^{-6}$

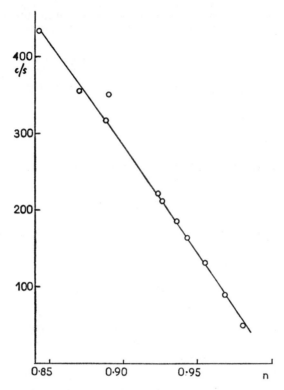

FIG. 14. Correlation between ^1H shifts and occupation number n for substituted methanes CHXYZ in CCl_4. (Hruska, Kotowycz and Schaefer.[131])

TABLE 30. CHARACTERISTICS OF THE 1:1 COMPLEXES BETWEEN $CHCl_3$ AND NITROGEN BASES, AT 37·5°C

Proton acceptor	K	Δ_{AD} [a]	$-\Delta H_f$ (kcal mole^{-1})
Pyridine	0·69	−234	2·1
N-methyl pyrrolidine	2·2	−123	3·9
MeHC=N-i-Pr	5·2	−119	—
MeCN	3·2	−37·6	—
Cyclohexyl cyanide	1·9	−46·6	2·6
Cyclohexyl isocyanide	1·8	−45·7	2·1

[a] c/s at 60 Mc/s.

TABLE 31. CHARACTERISTICS OF THE HYDROGEN BONDED COMPLEXES BETWEEN METHANOL
AND EITHER ACETONITRILE OR METHYL ISOCYANIDE

Temp. °C	K (MeCN)	K (MeNC)	$-\Delta H_f$ (MeCN) (kcal mole^{-1})	$-\Delta H_f$ (MeNC) (kcal mole^{-1})
30	2·6	3·0		
8	2·4	2·1	0·9	2·0
−18·5	2·0	1·6		

e.s.u., respectively, when the hydrogen bond length varies from 2·2 to 3·2 Å; the corresponding chemical shift is then:

$$\Delta_E = -3·0 \times 10^{-12} E_z - 0·738 \times 10^{-18} E^2 \qquad (55)$$

the variation of k_E from 2·6 to 3·4 bringing an error of $\pm 0·08$ Å into R for a given Δ_E. The neighbour anisotropy effect is estimated by the dipolar approximation of Pople[139] and McConnell:[140]

$$\Delta_N = \frac{\Delta\chi}{3 R_N^3} (1 - 3 \cos^2 \gamma) \qquad (56)$$

where R_N is the radius vector from the proton to the centre of anisotropy, $\Delta\chi$ is the difference in magnetic susceptibility of the anisotropic group parallel and perpendicular to the bond axis, and γ is the angle between the latter axis and the radius vector. The results of the calculations are shown in Table 32. It should be noted that these calculations are not very sensitive to the

TABLE 32. CONTRIBUTIONS TO THE HYDROGEN BOND SHIFT (c/s) AND ESTIMATION OF THE HYDROGEN BOND LENGTH

Compound	Δ_{AD} (exptl)	Δ_N (calc)	Δ_E (calc)	R (Å)
N-methyl pyrrolidine	−123	0	−123	2·42
MeHC=N-i-Pr	−119	−8·2	−111	2·53
Cyclohexyl cyanide	−46·6	+35·9	−82·5	2·72

effect of hybridization and of the type of atomic orbital chosen, the error arising from this choice is within $\pm 0·05$ Å. Berkeley and Hanna[136] have further provided a semi-classical model for these weak hydrogen bonds, with a classical electrostatic energy of attraction and a quantum mechanical energy of repulsion due to overlap. The latter is proportional to the product of an average ionization potential and the square of the overlap integral. This simple approach is successful in explaining the 0·3 Å longer hydrogen bond for cyclohexyl cyanide compared with N-methyl pyrrolidine, the ionization potential for the amine being significantly lower than for the nitrile. This difference in bond lengths accounts for the *ca.* 1·3 kcal mole^{-1} difference in electrostatic energy, hence in the enthalpy of formation, observed between both classes of compounds.

Creswell and Allred[141] have thermostated their samples to $\pm 0.3°$ in order to measure the thermodynamic constants for hydrogen bonding of the haloforms to tetrahydrofuran in cyclohexane solution. If A, B and S are the initial numbers of moles of proton donor (haloform), base (tetrahydrofuran) and inert solvent (cyclohexane), C the number of moles of complex (1 : 1) at equilibrium, the constant K for the equilibrium:

$$A + B \rightleftharpoons C \qquad (K)$$

is given by:

$$K = \frac{C\,(A + B + S - C)}{(A - C)(B - C)} \qquad (57)$$

and the chemical shift δ_{obsd} by:

$$\delta_{obsd} = C/A\varDelta + \delta_{free} \qquad (58)$$

where δ_{free} is the proton chemical shift of unassociated haloform, and $\varDelta = (\delta_{complex} - \delta_{free})$ is the hydrogen bond shift. A trial and error procedure yields the best value of K giving a linear variation with calculated values of C/A. The concentration of A was kept below 0·02 mole fraction to eliminate its self-association, and that of the proton acceptor below 0·25 mole fraction in the hope of minimizing solvent effects. From least-squares plots of log K versus $1/T$, the parameters listed in Table 33 were obtained. The hydrogen

TABLE 33. THERMODYNAMIC CONSTANTS FOR HYDROGEN BOND FORMATION BETWEEN THE HALOFORMS AND TETRAHYDROFURAN IN CYCLOHEXANE, AND RELATED CHEMICAL SHIFT DATA

	CHF$_3$	CHCl$_3$	CHBr$_3$	CHI$_3$
$-\varDelta H$ (kcal mole^{-1})	2·4 $\pm$ 0·2[a]	3·6 $\pm$ 0·4	2·6 $\pm$ 0·2	1·6 $\pm$ 0·4
$-\varDelta G_{25°}$ (kcal mole^{-1})	0·90	1·04	0·93	0·91
$-\varDelta S_{25°}$ (e.u.)	5·2 $\pm$ 0·6	8·6 $\pm$ 0·4	5·6 $\pm$ 0·4	2·3 $\pm$ 0·6
Average $-\varDelta$ (ppm)	0·69	0·85	0·83	0·43

[a] $\varDelta H = -2.593 \pm 0.024$ from the more accurate ^{19}F determination.[142]

bond shifts $\varDelta$ parallel the $\varDelta H$ values, although they are not linearly related. The order defined here for the hydrogen bonding propensities of the haloforms, Cl $>$ Br $>$ F $>$ I, does not coincide with the relative hydrogen bond strengths as defined by Allerhand and Schleyer [66] using the $\varDelta\nu$ infrared shifts in Me$_2$SO or C$_5$D$_5$N (Br $>$ I $>$ Cl; F(?)). The former order has been found independently by Paterson and Cameron[144] using various solvents. With chloroform and bromoform in dimethyl sulphoxide solution, the inter-molecular hydrogen bonds are somewhat stronger, as indicated by the hydrogen bond shifts obtained by McClellan, Nicksic, and Guffy,[132, 145] (1·20 and 0·97 ppm, respectively) using the method of Huggins, Pimental and Shoolery.[106] The dilution shifts of chloroform measured relative to an internal TMS signal by Kaiser[117] as a function of concentration in its binary mixtures with dioxane, acetone, triethylamine and ether, were corrected for the reaction field and for the self-association of chloroform, to yield the results shown in Table 34.

TABLE 34. THE HYDROGEN BOND OF CHLOROFORM WITH VARIOUS ACCEPTORS

Acceptor	H bond shift (ppm)	$-\Delta H_f$ (kcal mole^{-1})	K_1 (mole fraction) at 28°C
Me$_2$CO (1 : 1 complex)	$-1\cdot29$	$2\cdot7 \pm 0\cdot5$	$1\cdot8 \pm 0\cdot6$
NEt$_3$ (1 : 1 complex)	$-1\cdot88$	$4\cdot0 \pm 1\cdot9$[a]	$3\cdot0 \pm 1$[a]
Et$_2$O (1 : 1 complex)	$-2\cdot68 \pm 0\cdot22$	$6\cdot6 \pm 0\cdot6$	$0\cdot36 \pm 0\cdot04$
Dioxane (1 : 1 complex)	$-1\cdot20$		
Dioxane (2 : 1 complex)	$-0\cdot85$		

[a] Reference 106.

The accuracy of the data is not sufficient to demonstrate conclusively the existence of a linear correlation between the NMR line shift and the energy of the hydrogen bonds which are believed to cause the formation of complexes in these systems, although their variations are parallel, as was independently noted by Creswell and Allred.[141] The parameters relative to hydrogen bonded chloroform and tert-butanol, in various solvents, as obtained by Lemanceau, Lussan and Souty[109, 115] are summarized in Table 35.

Fratiello and Douglass[20] have studied, by means of dilution curves, the complexes between water and dioxane, and between water and pyridine. For the water-dioxane system, at the low water concentrations, the data have been corrected subsequently by Frost and Hall[22] for an error in the concentrations. In this case, the (uncorrected) hydrogen bond shifts in the complex are $-2\cdot20$ and $+0\cdot194$ ppm respectively, for the water and the dioxane protons, with an equilibrium constant $K = 0\cdot15 \pm 0\cdot03$ for water concentrations in the range 1·6–40% and an unknown n-meric complex. In this study, the hydroxylic proton dilution curve has a vertical tangent at

TABLE 35. EQUILIBRIUM CONSTANTS AND HYDROGEN BOND SHIFTS FOR VARIOUS SYSTEMS

Solvent	CHCl$_3$			t-BuOH	
	K_B [a]	$-\delta_B$ [b]	$-\varDelta H$ [c]	K_B	$-\delta_B$
Me$_2$CO	2	0·93	1·2	4·2	2·76
Et$_2$O	3·75	0·543	—	—	—
Et$_3$N	2·76	1·42	5·5	—	—
Bu$_3$N	1·14	1·02	—	—	—
Dioxane	0·91	0·88	—	2·6	2·9

[a] M^{-1}, $\pm 0·2$.
[b] $\pm 0·02$ ppm.
[c] kcal mole^{-1}.

infinite dilution, with either dioxane or pyridine, so that the precision in the determination of the hydrogen bond shift is limited. In addition, an estimate was made of the magnitude of the pyridine ring current perturbation, and the extent of the associations between these bases and water was confirmed by the measurement of self-diffusion coefficients for the non-aqueous component, by spin echo techniques. The ^{14}N chemical shift of pyridine in a methanolic solution suffers a paramagnetic shift due to hydrogen bonding with the solvent: from 306 ± 1 ppm for neat pyridine to 297 ± 2 ppm for the 0·5 mole fraction; the OH proton shift of methanol shows a maximum shift of 0·5 ppm to low field at the same concentration.[146] Saito et al.[146] have attributed the 9 ± 3 ppm ^{14}N shift to the contribution of the valence bond charge-transfer structure:

$$\text{MeO}^- \text{ H—Pyr}^+ \qquad \text{Pyr} = \text{pyridine}$$

with a $7 \pm 3\%$ weight, and the total shift for the pyridinium ion, as measured by Mathias and Gil[147], being 123 ± 11 ppm. Coppens, Nasielski and Sprecher[148] have studied the hydrogen bonding between methanol and monoaza-aromatic bases in carbon tetrachloride solution. There is no relation between the equilibrium constants they determined and the pK's of the bases. Furthermore, the hydrogen bond equilibria considered are quasi-identical for all the derivatives. This suggests an identical "intrinsic" basicity of the non-bonding lone pair on nitrogen in the series, which would ultimately depend only upon the nitrogen atom hybridization. Paterson and Tipman[92] report for para-substituted phenols (Table 36) that the OH chemical shift at infinite dilution in a particular solvent is almost independent of the nature of the ring substituent. Granächer, in a careful and detailed study,[149] has provided a theoretical model for the hydrogen bond of phenols. Mavel[150] had established the correlation of the infra-red absorption band with the

TABLE 36. CHEMICAL SHIFT OF PHENOLIC HYDROGEN IN PARA-
SUBSTITUTED PHENOLS

p-substituent X	Solvent	OH inf. diln.[a]
H	CCl_4	2.84 ± 0.03
Cl	CCl_4	2.90 ± 0.03
Br	CCl_4	2.91 ± 0.03
I	CCl_4	2.92 ± 0.03
H	C_6H_6	2.32 ± 0.02
Cl	C_6H_6	2.22 ± 0.02
Br	C_6H_6	2.21 ± 0.02
I	C_6H_6	2.24 ± 0.02
OMe	C_6H_6	2.14 ± 0.02

[a] ppm from internal cyclohexane.

chemical shift of water in a series of proton acceptor solvents. A very similar proportionality relationship is obtained when the chemical shift of the phenolic proton (relative to the gas phase) is plotted against the relative shift $\Delta k / k_0 \times 10^3$ of the infra-red OH band at $k_0 = 3654$ cm^{-1} in the gas phase, with a slope of approximately -4.8 ppm for 365 cm^{-1} (or -1.3 c/s per cm^{-1}, at 100 Mc/s). The potential function of Lippincott and Schröder[151] is then used to derive the hydrogen bond shift of phenol $\Delta \sigma = \sigma - \sigma_0$ as a function of the lengthening Δr of the OH bond:

$$\sigma_{\text{ppm}} = \sigma_0 \exp(- 6.8 \, \Delta r) \, \text{Å} \qquad (59)$$

where σ_0, the shielding of PhOH in the gas phase, has the value 27 ppm.

$$\begin{array}{c} \leftarrow r \rightarrow \\ O - H \ldots O \\ \leftarrow R \rightarrow \end{array}$$

In Table 37, the values of $R, r, \Delta r$ are summarized together with the corresponding values of ΔK, and $\Delta \sigma$ (equation (59)) and of the hydrogen bond binding energy E. Granächer has supplemented these empirical calculations

TABLE 37. NMR AND IR HYDROGEN BOND SHIFTS OF PHENOLIC OH AS A FUNCTION OF THE BOND LENGTHENING

R (Å)	r (Å)	Δr (Å)	Δk (cm^{-1})	$\Delta \sigma$ (ppm)	E (kcal mole^{-1})
3.03	0.980	0.010	140	-1.8	2.2
2.85	0.990	0.020	300	-3.5	4.6
2.79	1.00	0.030	400	-5.1	
2.75	1.01	0.040	500	-6.5	
2.70	1.02	0.050	600	-7.9	8.6
2.68	1.03	0.055	680	-8.5	

of the NMR hydrogen bond shift, using the theoretical evaluations of the intermolecular magnetic shift (Hameka[152]), of the electrostatic field (Buckingham[10, 153]), and of the dispersion shift (Stephen[154], Marshall and Pople[155]), together with the hydrogen bonding model of Paoloni.[156] Granächer[149] used electric dipoles to calculate electric field magnitudes, a procedure criticized by Berkeley and Hanna[137] because the uncertainty in the location of the dipole moment may change the electric field E_z along the C—H bond of acetonitrile MeCN from $(\mu/36 \cdot 6) \times 10^{24}$ (if the dipole is taken to be at the nitrogen nucleus) to a value lower by a factor of two, $(\mu/16 \cdot 4) \times 10^{24}$ (if the dipole is moved by 0·6 Å to the midpoint of the triple bond). This uncertainty, which originates with the r^3 dependence of the electric field upon the dipole moment, is obviously greatest whenever the uncertainty in the location of μ is not very small compared with the distance of the proton from the dipole. To calculate this polarization shift, Berkeley and Hanna[137] recommended a procedure based upon the appropriate (simple A.O.'s) orbitals to describe the lone pair electron distribution, which they have used in their theoretical estimate for hydrogen bonding of chloroform to nitrogen bases. Granächer[149] has also studied the ring protons shifts in various solvents and at various concentrations for *ortho*-chlorophenol[158], *ortho*-nitrophenol, *ortho*-cresol and the para isomeric molecules[149], and discussed the spreading of the perturbation by the OH group throughout the molecule. The paper by Shapet'ko *et al.*[157] also establishes and discusses the correlation between NMR and IR hydrogen bond shifts, for which similar correlations as discussed here had been established for the *intra*molecular cases by Forsén and Åkermark.[159]

6.4. *Steric Effects on Hydrogen Bonding*

Ouellette[160] has confirmed and applied the anterior remark of Becker, Liddel, and Shoolery[161] that the chemical shift of the hydroxylic proton of ethanol is linear with the concentration below 0·015 mole fraction, so that extrapolation to the infinite dilution chemical shift of the pure monomer appears feasible. This same linearity is indeed again observed with cyclohexanol, *cis*- and *trans*-4-t-butylcyclohexanol; the extrapolated chemical shifts of the axial and equatorial hydroxyl protons, in carbon tetrachloride solution, are 25·7 and 46·4 c/s respectively (TMS reference, at 60 Mc/s). From the position of the OH proton in cyclohexanol itself (41·6 c/s) the equilibrium constant calculated by the Eliel method is $K = 3 \cdot 3 \pm 0 \cdot 2$ at 40°C, corresponding to a free energy change of 0·75 kcal mole^{-1}. Both pyridine and dimethyl sulphoxide shift the OH resonance to low field, due to hydrogen bonding. Repetition of the same procedure yields the equilibrium constant for the axial-equatorial interconversion $K = 3 \cdot 7 \pm 0 \cdot 2$ at 40°C in both solvents, not significantly different from that in the inert solvent. Ouellette[160] also made the very interesting observation that not only the chemical shift of the OH proton is a function of stereochemistry, but also the *slope* of the dilution line, 2510 and 3160 c/s $(N_{\text{ROH}})^{-1}$, respectively, for *cis*- and *trans*-4-t-butylcyclohexanol. The explanation probably involves some steric hindrance to hydrogen bonding in the axial position, whereas the equatorial hydroxyl group points away from the rest of the molecule. Similarly, the slopes for 1-ethynyl-*cis*-4-t-butyl-cyclohexanol and 1-ethynyl-*trans*-4-t-butylcyclohexanol are 690 and 1060 c/s $(N_{\text{ROH}})^{-1}$, respectively. The smaller values for the ethynyl compounds compared with the parent cyclohexanols reflect the decreased susceptibility of tertiary alcohols to hydrogen bonding. Ouellette, Booth and Liptak[162] have measured the limiting slopes for fifteen bicyclic alcohols with the bicyclo-(2.2.1)-heptane skeleton, which vary between 700 c/s $(N_{\text{ROH}})^{-1}$ for isonorborneol (I), hindered considerably by the methyl group in the 1-position and by the bridge methyl groups at the 7-position, to 1860 c/s $(N_{\text{ROH}})^{-1}$ for the unsubstituted norborneol (II). It is shown that these steric factors may be expressed in terms of an additive, free energy increment. For example, *exo*- and *endo*-3-methyl groups correspond respectively to slope ratios of 1·12 and 1·57, so that the slope for camphenilol (III) with two geminal methyl groups at position 3, is predicted to be that for norborneol (II) (1860) divided by the product of the above

I   II   III

factors, hence a value of 1055 is calculated, in good agreement with the observed value of 1050. These additive parameters are a function of the dihedral angle between the methyl and the hydroxyl groups, when they are β to each other. Yamaguchi[163-7] has evolved a technique similar to that employed by Ouellette[160] for the cyclohexanols, to study hydrogen bonding of phenols, and his results nicely complement the more physical study by Granächer.[149] For *para*-cresol (IV), the OH proton is shifted to high field from *ca.* $-2\cdot5$ ppm to *ca.* $+1$ ppm (reference: external H_2O), in carbon tetrachloride solution, when the mole fraction of the solute is varied from one to zero. The intermolecular hydrogen bonds between the phenolic molecules are less probable with increasing dilution, a polymer species being dissociated above 0·3 mole fraction, and a dimeric species below 0·05 mole fraction. A similar curve is obtained for the benzene solution, where OH . . . π interaction and the aromatic ring anisotropy also contribute to the observed shift. In dioxane and acetone, lines of small slope are obtained, which suggests that the strength of the hydrogen bond towards these molecules is comparable with that of the self-association. Conversely, for pyridine which forms a strong hydrogen bonded complex, the shift is towards low field, approximately − 6 ppm (against an external water reference) at infinite dilution. The methyl and ring protons—an AA′BB′ spin system for the dilute solutions, coalescing

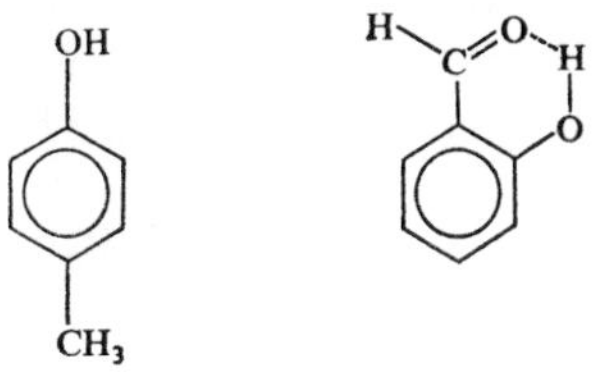

IV   V

to a single line when more concentrated—are less affected, as one would expect.[166] Also expected is the concentration independence of the phenolic resonance in salicylaldehyde V, since the OH proton takes part in a strong intramolecular hydrogen bond ($\sim$ − 6 ppm downfield from external water), in cyclohexane, carbon tetrachloride, chloroform, methanol, benzene and acetone. The basicity of pyridine, again, is much higher, and the observed shifts are consistent with a stronger intermolecular hydrogen bond of salicyl-aldehyde with pyridine than its intramolecular hydrogen bond. The solvent shifts of the ring protons served to assign the resonances of the four types of hydrogens.[163]

The hindering effect of ortho methyl groups upon the hydrogen bond shift from self-association of methyl- and dimethylphenols was also studied by Yamaguchi[165] using the dilution curve of the OH proton in carbon tetra-chloride solution, as summarized in the structures VI, VII and VIII:

| OH free: | + 0·3 ppm | + 0·3 ppm | + 0·3 ppm |
| OH bonded: | + 0·1 ppm | − 1·3 ppm | − 2·5 ppm |

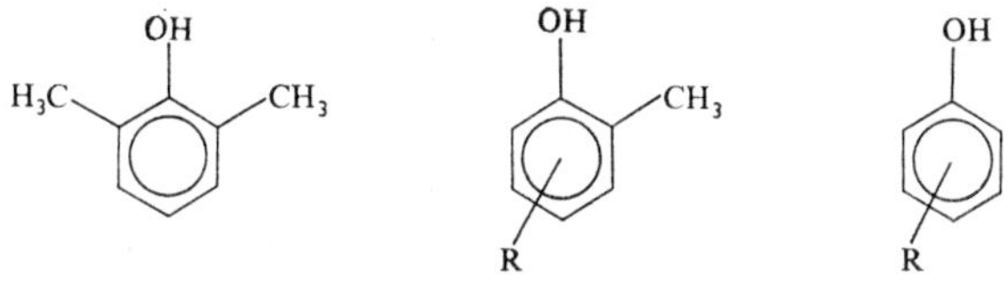

VI   VII   VIII

At infinite dilution, all of the curves approach about $+ 0.3$ ppm, to high field of the external water resonance, while for the pure liquids the values indicated are characteristic of the number of adjacent methyl groups, as shown (cf. the analogous results obtained by Friedrich on the $CHCl_3-NR_3$ system[169] and also discussed in the present section). The concentration dependence found for the acetone solution is more complex and suggests that this solvent is competing with the self-association of the solute. The shifts for the methyl and ring protons have also been interpreted.[163] The more detailed and quantitative study of hindered phenols by Somers and Gutowsky[168] established that both dimerization constants K and association constants for phenol–dioxane complexes K_c decrease with increasing steric bulk of the ortho substituents (Table 38). The temperature dependence

TABLE 38. EQUILIBRIUM CONSTANTS FOR HYDROGEN BONDING OF HINDERED PHENOLS

Phenol	K (dimerization)	K_c (dioxane)
Unsubstituted	13 ± 7	—
2-Isopropyl	1.7 ± 0.5	14
2-t-Butyl-4-methyl	1.37	—
2,6-di-Isopropyl	1.3 ± 0.5	7.1
2-t-Butyl	1.0 ± 0.5	6.7
2,4-di-t-Butyl	0.96	—
2-Methyl-6-t-butyl	$\leqslant 0.05$	5.6
2,6-di-t-Butyl	$\leqslant 0.05$	$\leqslant 0.7$

of the OH shifts parallels these K and K_c equilibrium constants, and is largest for the 2-isopropyl- and smallest, by a sixfold factor for 2,6-ditert-butyl- in the ortho substituted series. The K and K_c values also follow approximately the observed dilution shifts. The K_c's are in the same sequence as the K's but about tenfold larger, consistent with the greater ease with which the smaller dioxane molecule approaches the donated phenolic proton. In this leading reference, Somers and Gutowsky[168] have also examined the hydrogen bonding between hindered phenols and ethanol.

Friedrich,[169] using a series of amines substituted by alkyl groups of varied steric volume, obtained the following chemical shifts for the chloroform proton in the equilibrium (60) (see Table 39).

$$\begin{matrix} R_1 \\ \diagdown \\ R_2-N + H-CCl_3 \rightleftharpoons R_2-N \ldots H-CCl_3. \\ \diagup \\ R_3 \end{matrix} \qquad (60)$$

TABLE 39. CORRELATION BETWEEN STERIC HINDRANCE OF THE AMINE $R_1R_2R_3N$ AND CHLOROFORM CHEMICAL SHIFT

R_1	R_2	R_3	Taft steric parameters[170] ΣE_S	δ (c/s)[a]
H	H	H	+3·9	520
Et	Et	H	+1·1	519
t-Bu	H	H	+0·9	517
n-Pr	n-Pr	H	+0·5	505
n-Bu	n-Bu	H	+0·5	503
i-Pr	i-Pr	H	+0·3	506
Me	Me	Me	0	504
Et	Et	Et	−0·2	495
Cyclohexyl	Cyclohexyl	H	−0·3	494
i-Bu	i-Bu	H	−0·6	480
Cyclohexyl	Et	Et	−0·9	480
n-Pr	n-Pr	n-Pr	−1·1	466
n-Bu	n-Bu	n-Bu	−1·2	468
Cyclohexyl	n-Bu	n-Bu	−1·6	454
n-Bu	i-Bu	i-Bu	−2·3	435
$CHCl_3$ + indifferent solvent = n-hexane				430

[a] From TMS at 60 Mc/s; chloroform was present as 0·5% in the pure amine.

It is seen that the predominant effect appears to be the change in the position of the equilibrium (60) towards uncomplexed chloroform when the solvent bulk is increased. Even though the hydrogen bond shift itself is unknown, in each case, and the equilibrium constants have not been established, the existing linear correlation with the sum ΣE_S of the Taft steric parameters[170] suggests this interpretation. The electronic influence of the different alkyl substituents appears to be minor.

To conclude this section, solvent shifts[171] (see Table 40) have been interpreted for the aprotic solvents (Me_2CO) as due to the reaction field, or to an association between benzene and the azine ring in which benzene avoids the negatively charged nitrogen atoms (C_6H_6) (see Section 13). In the cases of methanol and trifluoroacetic acid, the changes are consistent with an increase in the molecular dihedral angle, concomitant with protonation of the nitrogen(s) lone pair(s) leading to an increased CH ... $^+$HN steric repulsion compared with the CH ... N interaction.

6.5. The Hydrogen Bond from the Proton Acceptor Point of View: Resonances Other than ^{1}H

The varying tendencies of protic solvents to form hydrogen bonds to the negative end of the carbonyl group $^{13}C=O$ of ketones have been examined by Maciel and Natterstad[39] using carbon-13 chemical shift measurements.

TABLE 40. SOLVENT SHIFTS OF *ortho*-PROTONS IN AZABIPHENYLS

Molecule	Solvent				
	CCl_4	Me_2CO	MeOH	CF_3COOH	C_6H_6
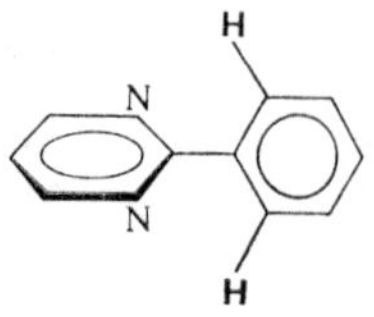	−1·18	−1·12	−1·03	−0·99	−1·68
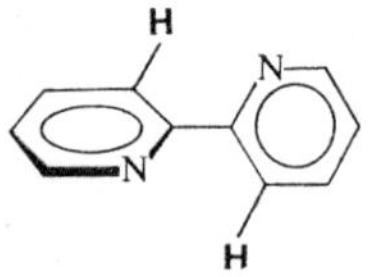	−1·25	−1·09	−0·89	(−0·48)[a]	−1·90

[a] This value refers to 4,4′-dimethyl-2,2′-bipyridyl.

The carbonyl-containing solutes were in 1 : 5 mole ratio solutions in a variety of solvents. The largest effects occur for acetone,[39, 47] the variations (relative to the pure liquid) range from + 2·4 ppm in cyclohexane to − 39·2 ppm in sulphuric acid. Intermediate values of − 2·3, − 8·7, and − 14·1 ppm, respectively are observed with chloroform, phenol, and trifluoroacetic acid as solvents. The aprotic solvents cause only small variations. The relative sensitivities of the carbonyl compounds to solvent effects are: acetone, 1·05; pinacolone (methyl-tert-butylketone), 1·00; fenchone, 0·90; di-tert-butyl-ketone, 0·81; ethyl acetate, 0·51; methyl formate, 0·37; dimethyl carbonate, 0·15; hexachloroacetone, 0·03. This behaviour is that expected in terms of the inductive donation or withdrawal by the substituents X and Y. The basicity of the carbonyl oxygen was related by Cook[172] to the ionization potential, for a series of ketones. Hexachloroacetone is indeed weakly basic, according to these correlations. Similarly, the esters are predicted to be less sensitive to hydrogen bonding by the solvent than acetone. It is to be remembered, however, that the strength of the hydrogen bond, as suggested from the ^{13}C measurements, or another spectral shift, e.g. in the infra-red, does not follow the heat of formation ΔH of the hydrogen bonded complex: this thermodynamic parameter is of the same order of magnitude for acetone and for ethyl acetate.[173] The association constants between n-butanol and acetone, pinacolone and di-tert-butylketone are respectively 1·64, 2·13 and 1·42 as determined by infra-red techniques.[174] Gramstad[175] has determined the association constants of phenol with the same ketones as 12·31, 9·12, and 9·20 respectively. In their study, Maciel and Natterstad[39] conclusively dismiss enolization, or a formal protonation—except in the special cases of extremely acidic solvents, sulphuric acid being the only instance—as contributing to the observed shifts. These are very satisfactorily correlated with the data of Christ and Diehl[48] for the solvent shifts of the ^{17}O resonance of acetone, with the ^{31}P chemical shifts of triphenylphosphine measured by Maciel and James[176] in the same solvents, with the pK_a's of the more acidic solvents. There is a monotonic relationship with the carbonyl stretching infra-red frequencies reported by Bellamy and Williams[49] for ketones in various solvents. However, if for these last measurements the parameter $\Delta v/v$ for a given ketone plotted against the corresponding value for acetophenone is a straight line, thus implying similar local association effects, and the respective sensitivities: acetyl chloride, 1·55; benzophenone, 1·12; acetone, 1·08; acetophenone, 1·00; di-isopropylketone, 0·96; methyl acetate, 0·81; cyclohexanone, 0·74; dimethyl formamide, 0·59; and acetaldehyde, 0·55, there is no simple relation with the proton accepting power of the C=O group as measured, e.g. by the N—H shift of pyrrole, by the carbonyl frequency or the ionization potential. A tentative correlation was also found by Maciel and Natterstad[39] between the ^{13}C solvent shifts and the $n \rightarrow \pi^*$ solvent effects for acetone, and by Stothers[177] who also measured the solvent shifts for acetophenone. Stothers[177] has also noted a correlation with the solvents' Y-values. Savitsky, Namikawa, and Zweifel[178] have found a correlation between the ^{13}C carbonyl chemical shifts for cyclic ($n = 4$–10, 12) and bicyclic ketones, and the $n \rightarrow \pi^*$ transition energies: these are approximately proportional to the average excitation energy ΔE which enters into the expression of the paramagnetic contribution to the chemical shift. Incidentally, the rate constants for borohydride reduction of cyclanones as measured by Brown and Ichikawa,[179] when plotted against the ring size, show a marked resemblance with the analogous ring size graph for the ^{13}C chemical shifts of their carbonyl groups, and this could be a supplementary indication that "steric approach control"[180] rather than "product development control"[181] determines the stereochemistry of the reduction of cyclic ketones.

Parallel to their investigation of the solvent shifts induced by protic solvents upon the ^{13}C resonance of the carbonyl group,[39, 47] Maciel and James[176] have studied the analogous phenomenon with the more sensitive and precise ^{31}P resonance of triphenylphosphine oxide Ph_3PO. The 1 : 20 mole ratio of solute to solvent was referred to an external reference of 85% phosphoric acid in a concentric spherical sample container to eliminate bulk susceptibility effects (see, however, Section 2). The observed shifts are presented in Table 41.

The pertinent commentary of Maciel and James[176] stresses the marked

TABLE 41. ^{31}P SOLVENT SHIFTS FOR $(C_6H_5)_3$ ^{31}PO

Solvent	Chemical shift, ± 0.2 ppm[a]
Dioxane	24·8
CCl$_4$	24·9
C$_6$H$_6$	26·1
Me$_2$CHOH	29·8
MeOH	32·6
AcOH	33·3
m-Cresol	36·4
HCOOH	37·3
Cl$_2$HC—COOH	41·7
CF$_3$COOH	48·1
96% H$_2$SO$_4$	59·8

[a] All negative with respect to the zero.

similarity with the ^{13}C (acetone) shifts obtained in the same solvents; the correlation is linear to a good approximation, with the exception of dioxane. The interpretation involves a displacement of the equilibrium:

$$Ph_3PO + AH \rightleftharpoons Ph_3PO \ldots HA \rightleftharpoons Ph_3PO^+H + A^-$$

in the protic solvents AH. The predominant effect is the increase in the number and strength of the hydrogen bonded complexes, since Hadzi[182] has previously shown by infra-red measurements on crystalline adducts that, even with trichloroacetic acid, a formal proton transfer does not take place: it only occurs with the much stronger acid hydrogen bromide. To end this paragraph, we note that Maciel[183] reports solvent effects for the ^{7}Li resonance in LiBr (0·092 M, LiClO$_4$ reference) of the same order of magnitude as chemical shifts!

6.6. Solvent Scales

No empirical solvent polarity scale based on NMR has yet been proposed, which is somewhat surprising on account of its amazing intrinsic sensitivity, compared with other spectroscopic methods. The nuclei which would best qualify are ^{19}F and ^{13}C, for which the chemical shifts are large and measurable with high accuracy. To illustrate this possibility, we have plotted in Fig. 15 the ^{13}C chemical shift of acetone in a variety of solvents (taken from the data of Maciel and Natterstad[39]) against the E_T values of Dimroth, Reichardt, Siepmann and Bohlmann.[184] The latter are the transition energies corresponding to the solvatochromic band of pyridinium N-phenol beta-ines, and which measure at the same time the polarity and the hydrogen bond donor ability of the solvent. The correlation obtained (Fig. 15) is actually quite good, the scatter may be due to the variation of the conformer populations for the acetone solute in the different solvents; if the carbonyl ^{13}C shift for the rigid fenchone molecule I is chosen instead of that of acetone,

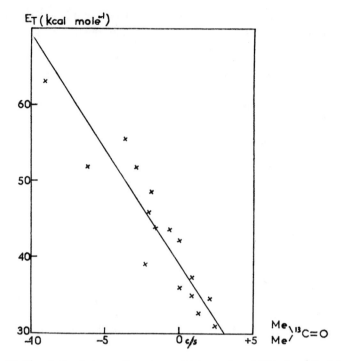

FIG. 15. Correlation between E_T values and the carbonyl ^{13}C chemical shifts for acetone in various solvents.

the correlation appears to be worse, which may be attributed to the high hindrance of the carbonyl group, flanked by three methyl groups, in this molecule, prohibiting the close approach of bulky solvents. Where carbonyl ^{13}C shifts were used for the determination of a solvent scale, the obvious choice would accordingly be a rigid ketone, with an unhindered carbonyl group, such as the 7-bicyclo-(2.2.1)-heptanone II:

Delpuech[185] has used the solvent shift of the chloroform resonance, without attempting to derive the true hydrogen bond shift for the 100% bound species, to measure the polarity of a few solvents; the correlation with the E_T values[184] is indicated in Table 42. One also ought to recall at this point the correlation established by Stothers[177] for the ^{13}C shifts with the Winstein Y-values. It is not necessary to stress the crude character of such scales.

TABLE 42. SOLVENT POLARITY AND THE NMR SPECTRAL SHIFT
OF CHLOROFORM

Solvent	E_T	$\Delta\delta$ (ppm)[a]
MeCN	46·0	0·31
Ethylene glycol		
carbonate	(46·6)	0·41
Sulpholane	44	0·48
Me$_2$NCHO	43·8	0·98
Me$_2$SO	45	1·05
Hexamethylphosphotriamide	—	1·89

[a] Relative to pure chloroform.

6.7. Spectral Assignment Through Protonation Shifts

Ottinger et al.[186] have used the example of caffein (I) to demonstrate the possibility of attributing spectral lines to the individual proton groups by the semi-empirical usage of solvent shifts. The particular problem they had to

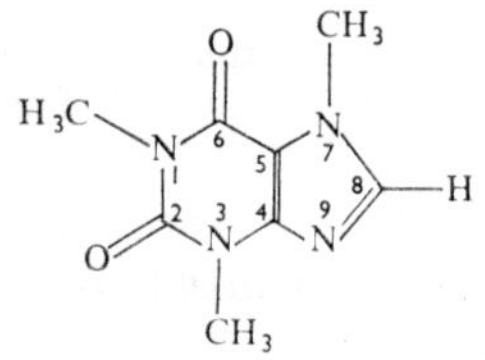

I

solve was to assign each of the three singlet N-methyl peaks, temporarily noted A, B and C from high to low field, which resonate between 3 and 4 ppm in deuterochloroform, to the 1-, 3- and 7-position of I. The solution of this problem was made possible by the observation that the magnitude of the shifts, in protonating medium, or in an aromatic solvent, follow the sequence: $\Delta\delta_A < \Delta\delta_B < \Delta\delta_C$. First, the separations between peaks A and B (Fig. 16B), and between peaks A and C (Fig. 16C) are plotted as a function of the pH in mixtures of sulphuric acid and deuterium oxide, and compared with the corresponding change in the ultraviolet absorption (Fig. 16A). Examination of Fig. 16 shows that the modifications of the chemical shifts for the N-methyl groups are correlated with the ultraviolet spectral shift, and therefore with the protonation of the imidazole moiety in caffein; this

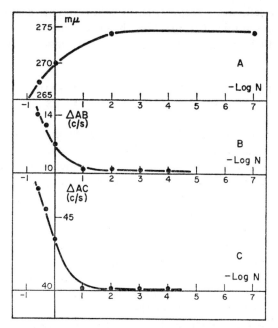

FIG. 16. The pH dependence of the ¹H resonances of the N—Me group of caffein.
(Ottinger, Boulvin, Reisse and Chiurdoglu.[186])

protonation is expected to affect preferentially the positions of the resonances
for H—8 and the Me—7 group which, for this reason is identified with
peak C. In Fig. 16 the A peak (which is practically invariant or, rather, is
the least affected of the three N-methyl resonances when the pH varies)
serves as reference, since it is impossible to use TMS as a standard (internal)
in the D_2O—H_2SO_4 mixtures. It is apparent that the change is more
pronounced for this C peak than for the B peak. To rationalize the smaller
perturbation induced by the protonation of the imidazole ring upon A
compared with B, the authors[186] stress the greater "distance"—whether
this is a field or an inductive effect—to the imidazole ring in the former case,
which it is thus preferable to attribute to the N-methyl group at position 1.
This suggested assignment of the N-methyl peaks (identical with that of
Alexander and Maienthal,[187] who compared the spectra of caffein and its
ethyl homologues) appears to be confirmed by the ASIS (cf. Section 13)
measured between deuterochloroform and benzene: $+4$ c/s (A); $+12$ c/s
(B); and $+51.5$ c/s (C). Thus it appears that the imidazole ring acts as the
electron-acceptor in the postulated complex with benzene, and that the same
"distance" criterion as above is operative. The effects of other solvents, of
more complex interpretation, have also been studied, using pyridine, acetic

acid, trifluoroacetic acid, acetonitrile, cyclohexylamine, acetone and nitro-benzene.[186] Peterson[188] has measured the proton chemical shifts of alicyclic ketones and ethyl derivatives in CCl_4, CF_3COOH and $CF_3COOH—H_2SO_4$.

7. TEMPERATURE DEPENDENCE OF CHEMICAL SHIFTS AND COUPLING CONSTANTS

Apart from the modifications of equilibrium constants between, for example, conformational or rotational isomers, discussed briefly in Section 3, and for which the review by Reeves[54] is adequate, temperature can separately affect the absolute values of NMR parameters themselves for a given molecular system in solution. In the formalism of the distinction between the intramolecular perturbations and the intermolecular phenomena, the former consist in the excitation of low lying torsional or vibrational states and the latter are association phenomena. Evidently, were these perfectly understood, there would be no need for the present section, since the inter-molecular phenomena result also from temperature dependence of equilibrium constants, and the intramolecular ones do not belong to a study of solvent effects. Unfortunately, the state of the art is not yet so accomplished, and neither is it possible to distinguish unambiguously between both factors for a definite temperature dependence of molecular parameters as those specific to nuclear magnetic resonance spectroscopy. For this reason, the results available will be presented here as uninterpreted raw data, but with reference (see Section 8) to the impertinence (in the etymologic sense) of some hasty conclusions based on variable temperature evidence.

7.1. *Temperature Variation of Coupling Constants*

The order of magnitude for the temperature variation of coupling constants in a given solvent is comparable with the variation in coupling constants caused by a change of solvent. For example, Ng, Tang and Sederholm[189] have provided a number of measurements on bromotrifluoroethylene (see Table 43). The arguments of these authors, who measured these coupling constants to within ± 0.10 c/s in order to separate the influence of vibrational excitation from the association phenomena, are the following:

(a) The temperature dependence ought to be the largest for those solvents capable only of *weak* molecular interaction with the solute; the *strong* molecular interactions are supposed to affect every molecule of the solute, so that they cannot be further increased by a lowering of the temperature. This is verified by the behaviour of J_{13}, which is barely modified by the 125° temperature change in the strongly associative

TABLE 43. TEMPERATURE VARIATION OF COUPLING PARAMETERS
FOR BROMOTRIFLUOROETHYLENE

$$F_1 \quad F_2$$
$$\diagdown \quad \diagup$$
$$C = C$$
$$\diagup \quad \diagdown$$
$$F_3 \quad Br$$

Solvent	J_{13}		J_{12}		J_{23}	
	20°C	−105°C	20°C	−105°C	20°C	−105°C
CS$_2$	71·7	71·8	56·4	55·4	123·1	123·4
Neat	73·7	72·8	56·6	55·5	122·8	123·4
CFCl$_3$	73·3	72·1	56·8	55·3	123·4	123·2
Et$_2$O	73·8	73·8	56·0	53·0	123·1	123·1
EtOH	73·9	73·9	55·9	53·9	122·9	123·1
MeCHO	74·2	74·1	55·1	53·5	122·6	123·2

solvents, and therefore is not modified by thermal excitation of vibrational or torsional modes; there are two exceptions, the neat sample and CFCl$_3$, for which weak molecular interactions may be operative.

(b) *Trans* J_{23} also shows little dependence on temperature in agreement with the postulated mechanism through the carbon–carbon double bond, which would be little affected by population of the torsional mode.

(c) "A good deal of the vicinal *cis* J_{12} must come about through non-bonded interactions ('through-space' coupling),[190-1]) and this would depend strongly upon the excitation of the torsional mode, as observed."[189]

The major criticism that can be levelled against this approach is that it seeks to bring an indirect proof of the "through-space" coupling mechanism,[190] and that it conveniently omits for this purpose the adverse possibility of a zero temperature effect by internal compensation between opposing intra- and intermolecular factors; there is not enough support for the affirmation that J_{12} is modified by torsional excitation, instead of the solvent interaction. Evidently, the relaying role of the solvent transmitting nuclear spin information between F_1 and F_2, but not between F_1 and F_3, nor F_2 and F_3, by an unknown mechanism, is also favourable to the "through-space" interaction[190-1] of F_1 with F_2. Ng, Tang and Sederholm[189] have provided a very interesting interpretation of the large number of experiments they have conducted, one may only regret in conclusion of this brief summary of their work, also described in Section 12, that the concentration of the solute

(50% per volume) ought to have been varied, and anyhow, was too great. The work of Ramey and Brey[192] also deals with $^{19}F-^{19}F$ spin–spin interactions, in the trifluorovinyl molecules $F_aF_bC=CF_xY$ ($Y = CF_3$, CF_2Cl, CF_2Br and COF). The changes with temperature of the coupling constants involving the nuclei of the Y group are discussed in terms of the possible conformational isomers.[192] Assuming that the coupling constants internal to the trifluorovinyl group, that is: $^2J_{ab}$, $^3J_{ax}$, and $^3J_{bx}$, do not depend upon rotational isomerism about the C—Y bond, the vibrational interpretation adequately explains the observations: their values decrease with increased bulk of the substituent and with decreased temperature. The data for $Y = CF_2Cl$, at several temperatures and in several solvents, are given in Table 44. It is seen that the variations

TABLE 44. TEMPERATURE VARIATION OF THE TRIFLUOROVINYL GROUP COUPLING CONSTANTS OF $F_2C=CFCF_2Cl$[a]

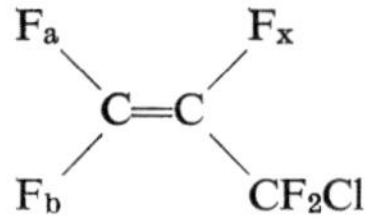

Coupling const (c/s)		Pure liquid	50% CFCl$_3$	50% MeCCl$_3$
J_{ab}	58°	56·9 ± 0·3	57·2 ± 0·3	56·7 ± 0·3
	30°	56·8 ± 0·4	56·5 ± 0·3	56·3 ± 0·4
	−50°	55·8 ± 0·3	55·7 ± 0·2	55·9 ± 0·3
	−90°	54·9 ± 0·4	55·0 ± 0·3	54·9 ± 0·3
J_{ax}	58°	39·2 ± 0·2	39·2 ± 0·2	38·9 ± 0·3
	30°	39·0 ± 0·3	38·6 ± 0·4	38·7 ± 0·4
	−50°	37·5 ± 0·3	37·4 ± 0·3	37·6 ± 0·3
	−90°	37·0 ± 0·3	36·9 ± 0·3	37·1 ± 0·4

[a] Due to inferior precision, the values of J_{bx} were not given in reference 192, but appear from Fig. 5 of that paper to vary between 116 c/s at 85°C and 115·3 c/s at −100°C.

caused by temperature are greater than the differences between solvents; Ng, Tang and Sederholm[189] have pointed out, however, that the solvents studied by Ramey and Brey[192] differ little in polar character from the pure liquid, so that this result is not completely unexpected. The same authors[192] using the value of 358 cm^{-1} assigned to the CF_2 in-plane deformation of the $CF_2=CFY$ olefins,[193] and the Boltzmann distribution, have calculated the rather considerable increase of the fraction of molecules in the excited state for this vibrational mode (from 5% at −100° to 26% at +100°). They, furthermore, remark that in the ground state the system spends the larger amount of time near the equilibrium configuration, while for the excited states it spends most of the time near the extremes of the vibration. The

observed variation is the result of opposite tendencies, the marked anhar-
monicity which would characterize the F—C—F symmetric in-plane vibration
tending to pull apart the fluorine atoms, and the separate "through-bond"
and "through-space" coupling mechanisms predicting enhanced coupling
when these coupled nuclei are in van der Waals contact. The latter factor
has been shown by experiment to be more important. Finally, the point is
made that *for both s-cis* and *s-trans* conformers of $F_2C=CFCOF$, J_{ab} and
J_{ax} increase with temperature,[192] which argues for a vibrational excitation
origin of the effect. Thus the original assumption is supported that this is
not a phenomenon arising from rotational isomerism for the CF_3-, CF_2Br-
and CF_2Cl-substituted molecules.

The present author[194] has investigated the sensitivity to temperature of
the one bond ^1J coupling parameters for various molecules as pure liquids
in order to ascertain the precision of the correlation proposed with the
endocyclic interatomic angles in alicyclic molecules (Fig. 17). A striking
feature of Fig. 17 is the great similarity of the curves for the different compounds studied, with the exception of dimethyl sulphoxide. The total *decrease*
in ^1J for an *increase* in temperature of *ca.*100° is in the range 1–5 c/s. The

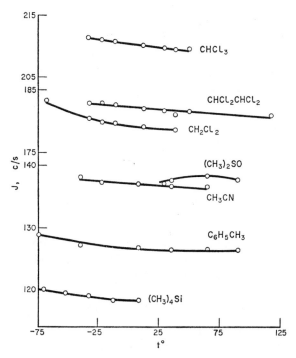

FIG. 17. Temperature variation of ^{13}C—^1H coupling constants (Laszlo[194]).

only temperature coefficient previously determined for these ^1J's was reported as $-(0.97 \pm 0.35) \times 10^{-3}$ c/s $(°C)^{-1}$ for acetonitrile by Frei and Bernstein,[195] and their measurement is consistent with ours. These results can be visualized as a combination of intramolecular and intermolecular effects. The former could be due to excitation of different vibrational states in the C—H bond, considered as the anharmonic oscillator, while the latter would arise from dispersion forces or self-association in the liquid phase. This association alone does not account for the observations. Table 45 presents a comparison of

TABLE 45. CONCENTRATION DEPENDENCE OF $^1J_{CH}$

Molecule	ΔJ^{Me_2SO} [a]	$\Delta \nu^{Me_2SO(143)}$	ΔJ^{diln} [a]
$CHCl_3$	$+8.2$	29	0.0
$CHCl_2CHCl_2$	$+4.5$	(25)	-0.6
CH_2Cl_2	$+3.3$	(1–4)	-1.4
$(CH_3)_2SO$	—	—	$+0.2$
CH_3CN	0.0	0	-0.4

[a] At 29°C.

the hydrogen bonding capabilities of some of the molecules studied. If ΔJ^{Me_2SO}, the change in going from the pure liquid to a 17% v/v solution in Me_2SO, roughly parallels $\Delta \nu^{Me_2SO}$ from the infrared measurements of Allerhand and Schleyer,[66] then ΔJ^{diln}, the modification incurred by ^1J from the pure liquid to a 10–15% v/v solution in carbon tetrachloride, is:

(a) of the same order of magnitude as the experimental errors (their limit is ± 0.4 c/s),
(b) irregular, and
(c) apparently independent of ΔJ^{Me_2SO}.

If there were any significant degree of self-association due to hydrogen bonding, it should be less for CH_2Cl_2 and $CHCl_2CHCl_2$ than it is for $CHCl_3$, with its three strongly activating groups. It should be pointed out that the disputed weak self-association of chloroform (see Section 6.2) has recently been ruled out by an infra-red study.[196]

In Table 46 we compare the sign of the temperature dependence (positive sign for an increase of the absolute magnitude of the coupling constant with increasing temperature) with the actual sign of the coupling.

7.2. Effect of Temperature on Chemical Shifts

Apart from the consequences of variations in conformational equilibria or solvation equilibria, there may be more specific variations of the chemical shifts with temperature, which have to be separated from the former. The

TABLE 46. THE SIGN OF COUPLING CONSTANTS AND THE SIGN OF
THEIR TEMPERATURE DEPENDENCE

Coupling type	Temperature dependence	Sign of coupling
$^{13}C-^{1}H$	−	+
$H-C-H$	+	−
$^{29}Si-^{19}F^a$	+	−[b]
$^{19}F-C-^{19}F$	+	+[c]
$^{19}F \quad\quad ^{19}F$ $\diagdown \quad \diagup$ $C-C$	+	+[c]
^{19}F $\diagdown$ $C-C$ $\diagdown$ ^{19}F	−	−[c]

[a] SiF_4 in CF_3Cl.
[b] Reduced coupling constant (see reference 197).
[c] Reference 198.

existing theory of Buckingham[8] accounts for the observed temperature coefficients of 10^{-10} ppm per degree by separately considering the excitation of vibrational modes, that of rotational states (centrifugal stretching), and intermolecular forces, in gases.

For protons involved in hydrogen bonding, a temperature variation of the hydrogen bond chemical shift is also to be expected since several excited vibrational levels of the hydrogen bond (stretching mode) should be highly populated, even at temperatures as low as $200°K$, and the potential well is anharmonic. Furthermore, the hydrogen bond shift itself is not symmetrical with respect to a small change in the position of the proton either side of its equilibrium position, both electric field and anisotropy terms being different. Muller and Reiter[199] were able to show such an effect experimentally, similar to the calculated averages over the vibrational levels. Berkeley and Hanna[105, 136-7] have demonstrated conclusively the existence of a fairly large temperature coefficient for the hydrogen bond shifts, Δ_{AD}, which they have measured for the pure 1 : 1 complexes between chloroform and nitrogen bases (see Table 47). The infinite dilution shift of fluoroform CHF_3 in cyclohexane apparently undergoes a slight temperature variation, not elaborated upon by Creswell and Allred:[191] from 4·800 ppm at 38·3° to 4·815 ppm at −8·3°. No change was detected for the resonances of the other haloforms between 55·6° and 2·9°C, but the same authors have also reported a similar temperature variation for the chloroform–triethylamine system.

The changes represented in Fig. 18 may also be due to vibrational factors, and/or a weak association with the carbon disulphide solution, they cannot

L

be attributed to conformational mobility, since camphor was chosen precisely on account of its rigidity for this study. These data are relevant to the discussion presented in the following section.

TABLE 47

Acceptor molecule	Temp., ±2°C	Δ_{AD}, ±0·1 c/s
N-methyl pyrrolidine	−10·0	118·9
,, ,,	7·5	115·7
,, ,,	23·5	127·9
Cyclohexyl cyanide	8·8	46·7
,, ,,	14·5	48·2
,, ,,	20·0	48·2
Pyridine	7·5	210
,,	17·5	220
,,	29·5	224
Cyclohexyl isocyanide	0·5	43·1
,, ,,	9·8	43·5
,, ,,	21·7	45·2

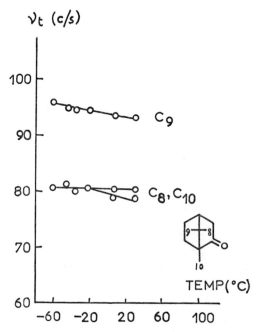

FIG. 18. Temperature variation of the ¹H chemical shifts at 60 Mc/s of the methyl groups of camphor in CS_2 solution (Fétizon, Golfier and Laszlo[200]).

8. CASES OF UNDUE CONFIDENCE IN A SINGLE SOLVENT, THE SOLVENT SHIFTS BEING MISTAKEN FOR STRUCTURAL EFFECTS

In most sections of this study of NMR solvent effects, these are considered clearly as a topic important to study *per se*; therefore we will discuss some of their principal causes and manifestations. The complementary viewpoint, where solvent effects are but parasitical perturbations of the intramolecular phenomenon studied, has not and should not be forgotten. We have pointed out in Sections 3 and 7, the extent to which studies of ethane rotamers are limited in precision by the antagonistic effects of solvent and temperature on the chemical shifts and coupling constants. We have shown how Schug and Deck[91] proceeded to remove the shifts due to the reaction field (Section 4) in order to compare the charge distributions for hydroxybenzenes from the theory with the experimental chemical shifts. We have reported the arguments of Berkeley and Hanna[105] to remove the non-specific solvent effects from the hydrogen bond shifts (Section 6), etc. We have placed the present section in the mid-part of this work so that the reader, who might by this point have started to share our enthusiasm, be warned again that solvent effects can be frequently a nuisance!

This negative importance is, for each solvent, proportional to the degree of ill effects it presents multiplied by its frequency of use. For this reason, we have written at length elsewhere on the disadvantages of chloroform,[201] stressing the general necessity that for structural studies to be valid results must be obtained for several solvents. A typical example discussed there is that of the anisotropy of the oxirane ring. Tori *et al.*[202] have discussed the proton chemical shifts for rigid polycyclic molecules containing an epoxidic ring, *in chloroform solution*. They concluded that there is a diminished shielding for the protons in the plane of the anisotropic three-membered oxygenated ring, and an increased shielding for protons above this plane, *except if* these are near enough to the heteroatom, in which case the screening constant is also decreased. The validity of this approach can be appreciated by the comparison presented in Table 48, for the norbornene oxide molecule (I). It is then evident that the value for H_a—H_c, in deuterochloroform solution, is attributable only in part to epoxide anisotropy; neither is it appropriate to use that solvent, as was done, to compare the chemical shifts of epoxidic (e.g. structure I) and non-epoxidic molecules.[202] These authors took account of this criticism, and have subsequently published new data relative to the anisotropy of the oxirane and aziridine ring in carbon tetrachloride solution[203] (see Table 49).

It may well be, as the Japanese authors[203] point out, that slight geometrical

TABLE 48. CHEMICAL SHIFTS (ppm) FOR NORBORNENE OXIDE
IN VARIOUS SOLVENTS

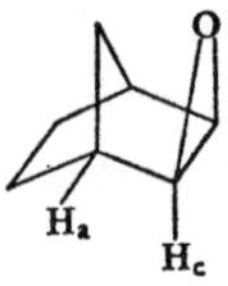

I

Solvent	Conc. (w/v)	H_a	H_c	$|H_a - H_c|$
CCl_4	5%	2·36	2·84	0·48
C_6H_{12}	5%	2·30	2·81	0·51
MeCN	5%	2·36	2·99	0·63
$CHCl_3$	5%	2·41	3·02	0·61
$CDCl_3$	10% (202)	2·44	3·16	0·74

distortions are responsible for the large variations observed on introducing an olefinic or aromatic double bond inside the bicyclo-(2.2.1)-heptane skeleton; this assumes that the anisotropy and the associated "ring current"[202] be localized in the three-membered ring, and that they do not interact with those of the other parts, unsaturated, of the molecule. We prefer to consider that the results reported in Table 49 demonstrate such an interaction, and these complex molecules as a whole, specially since there exist arguments for an intrinsic anisotropy of the hydrocarbon skeleton itself, associated with a substantial pi-electronic delocalization.[204]

Another interpretation affected by solvent effects is relative to conformational mobility of ring A, between the chair II and the boat III, in Δ^5-3-keto-4,4-dimethyl steroids:

II   III

In a number of examples, the 19-methyl group resonates at high fields, which has been attributed[205-6] to the predominance of the boat conformation III, in which the 19-methyl group is in the shielding part of the carbonyl "anisotropy cone"; we have indicated elsewhere[200] the conceptual and physical weaknesses of this argument. However, the authors of this study[205] have also conducted a variable temperature experiment, *in deuterochloroform*

TABLE 49. CHEMICAL SHIFTS CONSECUTIVE TO THE INTRODUCTION
OF A THREE-MEMBERED RING

Molecule	Y	H—1	H—7 anti	H—7 syn (ppm).
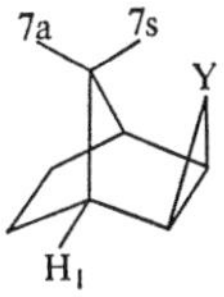	NH	−0·09	+0·68	∼+0·04
	O	−0·21	+0·63	−0·06[a]
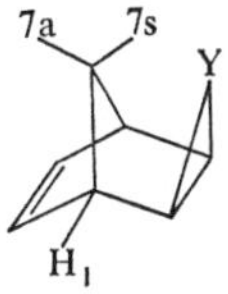	NH	+0·08	+0·37	−0·49
	O	−0·06	+0·12	−0·49
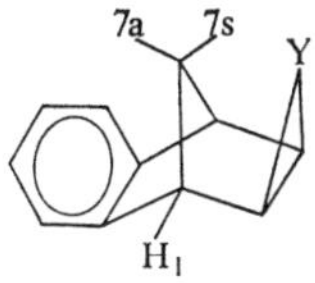	NH	+0·12	+0·46	−0·45
	O	−0·01	+0·26	−0·45[b]

[a] ∼+0·50 in $CDCl_3$[202].
[b] −0·01 in $CDCl_3$.

solution, and, by plotting the variation of the 19-methyl resonance against $1/T$, have determined the thermodynamic parameters relative to the equilibrium of II and III. To stress the danger, and the large errors, resulting from such an ambiguous experiment, where the association of chloroform with, for example, the 3-carbonyl group may be considerable, we have made the test[200] of following the methyl resonances of the rigid camphor molecule, as a function of temperature: these vary, somewhat more than in carbon disulphide solution and in the same direction (Fig. 18), and their order of magnitude, if of the opposite sign, is comparable with that found in the steroidal examples.[205]

When some signals vary with temperature, other signals remaining invariant, their shift as well as their immobility may attest an alteration of the environment of the corresponding protons. One should not therefore decide about the existence of a conformational equilibrium from variations of the chemical shift with temperature of a magnitude comparable to that of the solvent effects; for example, 0·10 ppm for a temperature range of 100–150°.

9. RECOURSE TO LIQUID CRYSTALLINE SOLVENTS FOR SPECTRAL AND STRUCTURAL ANALYSIS

This is an extremely important subject which is touched upon briefly here; a full account is given in Chapter 2 of Volume 2 of this Progress series.

9.1. *The Nematic State of Matter*

There exists a state of matter intermediate between the crystalline solids and the isotropic liquids, and substances in this state are alluded to as "liquid crystals"; this may occur for rod-like molecules, rather polar, which orient themselves by their mutual polarization with parallel longitudinal axis. With respect to NMR studies, the important feature of these nematic phases is their alignment in the magnetic field of the spectrometer so that the axis of minimum magnetic susceptibility is parallel to the applied field, thus constituting an anisotropic micro-environment for the solute molecules. It is possible to dissolve in a liquid crystal other rod-like, polar molecules, which generally tend to orient themselves parallel to the solvent. The nematic solvent itself shows its peculiar properties in a relatively narrow temperature range, above the solid crystalline phase and below the normal isotropic liquid phase. For certain compounds, there is another transition, from the crystal to a smectic two-dimensional phase; there is yet another class of oriented molecules, the cholesteric; we shall confine our discussion here to a number of preliminary experiments where nematic mesophases serve as an ordering matrix for the solute.

The characteristics of a number of such solvents are summarized in Table 50, which is taken from a paper by Rowell *et al.*[207] Introduction of a second component lowers the transition temperature for the nematic phase to liquid phase transformation, and it therefore reduces the temperature range for which the binary mixture is anisotropic to between 70 and 170°C (Table 50). This effect obviously depends upon the resemblance, in shape and bulk, between the solute and the solvent molecules.

9.2. *Nuclear Magnetic Resonance in Liquid Crystals*

Molecular diffusion is still present in these media so that the *inter*molecular spin–spin interactions are averaged to zero and disappear as in an ordinary liquid. The unidimensional ordering, however, leads to large (of the order of a thousand c/s) *intra*molecular dipole–dipole interactions between nuclei.

Table 50. Structures, Symbols, Nematic Temperature Ranges, and Reorientation Axes for Nematic Mesophases[207]

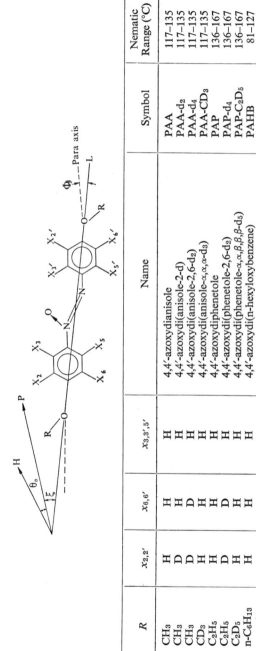

R	$x_{2,2'}$	$x_{6,6'}$	$x_{3,3',5'}$	Name	Symbol	Nematic Range (°C)
CH_3	H	H	H	4,4'-azoxydianisole	PAA	117–135
CH_3	D	H	H	4,4'-azoxydi(anisole-2-d)	PAA-d$_2$	117–135
CH_3	D	D	H	4,4'-azoxydi(anisole-2,6-d$_2$)	PAA-d$_4$	117–135
CD_3	H	H	H	4,4'-azoxydi(anisole-α,α,α-d$_3$)	PAA-CD$_3$	117–135
C_2H_5	D	D	H	4,4'-azoxydiphenetole	PAP	136–167
C_2H_5	H	H	H	4,4'-azoxydi(phenetole-2,6-d$_2$)	PAP-d$_4$	136–167
C_2D_5	H	H	H	4,4'-azoxydi(phenetole-α,α,β,β,β-d$_5$)	PAP-C$_2$D$_5$	136–167
n-C$_6$H$_{13}$	H	H	H	4,4'-azoxydi(n-hexyloxybenzene)	PAHB	81–127

R	$x_{2,6}$	$x_{3,5}$	Y	Name	Symbol	Nematic Range
n-C$_3$H$_7$	H	H	D	4-(n-propyloxy)benzoic acid-d	PPBA-d	145–154
n-C$_3$H$_7$	H	D	H	4-(n-propyloxy)benzoic acid-3,5-d$_2$	PPBA-d$_2$	145–154
n-C$_3$D$_7$	H	H	H	4-(n-propyloxy)benzoic acid-α,α,β,β,γ,γ,γ-d$_7$	PPBA-C$_3$D$_7$	145–154
n-C$_6$H$_{13}$	H	H	H	4-(n-hexyloxy)benzoic acid	PHBA	106–154
n-C$_6$H$_{13}$	H	H	D	4-(n-hexyloxy)benzoic acid-d	PHBA-d	106–154
n-C$_6$H$_{13}$	H	D	H	4-(n-hexyloxy)benzoic acid-3,5-d$_2$	PHBA-d$_2$	106–154

This direct spin–spin coupling is averaged to zero in ordinary liquids, where only the indirect (through the valence electrons of the molecular frame) coupling is apparent. The NMR spectrum of a solute in a nematic solvent, compared with the spectrum in the liquid above the transition temperature, will possess relatively sharp lines because of the fast intermolecular motions and of the lattice expansion, but will be far more complicated on account of the additional spin–spin interactions, through the direct dipolar mechanism. The spectral lines are symmetrical with respect to the centre of the spectrum, the line width being minimum at some point of the spectrum, near its centre for protons but which may be more removed for other nuclei, and increasing to both lower and higher fields, due to temperature gradients in the sample.[208] Typical figures for the half-widths are 2 and 50 c/s for the centre and the wings respectively.

The spectrum for the solvent is extremely broad, over a range of a few gauss, the spectral lines being very broad because the *inter*molecular dipole–dipole interactions are not averaged to zero in the rather rigid arrangement of the nematic solvent molecules. The nematic compound then displays a very broad signal, underneath the sharp solute resonances, which is not detected under these conditions. The multiplicity of the interactions can be reduced, for either solute or solvent, by selective deuteration.

The NMR spectrum of a solute may be described by the Hamiltonian:

$$\mathcal{H} = \mathcal{H}_0 + \mathcal{H}_1 \tag{61}$$

$\mathcal{H}_0$ being the normal "chemical shift" Hamiltonian, and

$$\mathcal{H}_1 = \sum_{i>j} (J_{ij} + D_{ij})I_z(i)I_z(j)$$
$$+ \sum_{i>j} (\tfrac{1}{2}J_{ij} - \tfrac{1}{4}D_{ij})I_+(i)I_-(j) + I_-(i)I_+(j), \tag{62}$$

I_+ and I_- are the usual displacement operators:

$$I_{\pm} = I_x \pm iI_y \tag{63}$$

J_{ij} is the usual indirect spin–spin coupling constant between nuclei i and j, whereas

$$D_{ij} = \gamma_i \cdot \gamma_j \cdot \frac{\hbar}{2\pi} \cdot r_{ij}^{-3}\langle 1 - 3\cos^2\theta_{ij}\rangle \tag{64}$$

is the direct dipole–dipole nuclear spin coupling constant, with r_{ij} the internuclear distance, θ_{ij} the angle between the r_{ij} and the magnetic field axis. This $\langle 1 - 3\cos^2\theta_{ij}\rangle$ mean value term can be expressed as equal to[209]:

$$\tfrac{1}{2}(1 - 3\cos^2\gamma)(1 - 3\cos^2\theta')S \tag{65}$$

where γ is the angle between r_{ij} and the axis for molecular rotation (the *para* axis in, for example, 4,4'-azoxydianisole), and θ' the angle between this secondary axis and the major molecular axis for molecular reorientation. These terms are molecular constants for a given pair of nuclei i, j. The symbol S represents the degree of micro-order; it varies between zero in an isotropic system, and unity for a perfectly ordered system, and is equal to $(\frac{3}{2} \cos^2 \xi - \frac{1}{2})$, ξ being the angle between the long molecular axis and the direction of preferred orientation (the optical axis, with which the axis of greatest polarizability will tend to be collinear). The direct coupling constants D_{ij} are then proportional to S/r_{ij}^3, the degree of micro-order S (usually in the range 0·4–0·7, it also depends upon the temperature) is the same for all pairs of nuclei in the molecule, so that the analysis of the NMR spectrum is possible. Apart from the absolute value of S, *relative* values may be obtained for the r_{ij} internuclear distances (relative, because some standard length is needed to measure the internuclear distances, for instance the carbon–hydrogen bond length will serve as a unit). Since the value of S is determined primarily by the dispersion forces between molecules, this may constitute a privileged means for studying this type of intermolecular force. Spiesecke[210] provides the very illustrative example of oriented $H^{13}CN$ in *p*-capronyloxy-*p'*-ethoxy-azobenzene. The doublet splitting at 63°C with a concentration of approximately 10 mole% $H^{13}CN$ is 1112 c/s. With J_{CH} (indirect) = 270 c/s, and positive, the degree of micro-order S for the longitudinal axis can be calculated as:

$$S = D_{12} \cdot r^3/60 = (1112 - 270) \cdot (1·0646)^3/60 \times 10^6 = 0·017. \quad (66)$$

Surprisingly enough an experiment with *p*-capronyloxy-*p'*-ethoxy-azoxy-benzene yielded a splitting of 116 c/s, with S therefore equal to $-0·003$ only. This negative sign for the direct coupling constant D_{12} implies a *perpendicular* orientation of the rod-like HCN molecule with respect to the longitudinal axes of the nematic host substance. It is not to be explained by chemical exchange, since HCN does not exchange with other acids in solution nor with itself within the limits of the J_{CH} time scale, 270 c/s. In contrast to the results reported by Pauling in *The Nature of the Chemical Bond*, HCN must be monomolecular in nematic phases since the spectra are typical two-spin spectra. This puzzling phenomenon remains unexplained; dipolar interaction with the azoxy group may be the cause.

Determination of internuclear distances, for nuclei with non-zero spins, is evidently of enormous value to define accurately molecular structure, complementary to the more classical methods, such as microwave, X-ray or electron diffraction methods. For instance, the structures of cyclopropane and cyclobutane have been derived from the analysis of the liquid crystal spectrum in *p,p'*-di-n-hexyloxy-azoxybenzene, as shown in (Fig. 19).

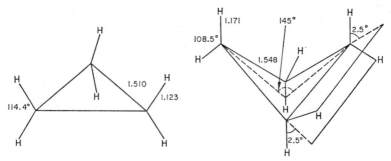

FIG. 19. Nematic phase determinations of the structures of cyclopropane and cyclobutane. (Snyder and Meiboom.[211])

Snyder and collaborators[211–15] have elaborated a computer programme to produce the entire nematic phase spectrum; it is compared with the experimental spectrum, and the spectral parameters are modified until convergence is achieved within the band widths.

The signs of nuclear spin–spin coupling constants have also been determined by this method, for instance Snyder and Anderson[213] evaluated the coupling constants for benzene, fluorobenzene, and hexafluorobenzene[215] listed in Tables 51 and 52.

TABLE 51. DIRECT DIPOLAR COUPLING CONSTANTS (c/s)

D_{ij} (i, j)	C_6H_6	C_6H_5F	C_6F_6
1,2	−639·45	−491·94	−1484·26
1,3	−123·06	−201·15	−285·64
1,4	−79·93	−165·42	−185·53
2,3		−1612·45	
2,4		−239·10	
2,5		−62·77	
2,6		−25·40	
3,4		−502·16	

Experimental values of the anisotropy of the chemical shifts are afforded by the comparison between the resonances of the nematic phase (partial orientation, the degree of molecular ordering being defined by the S-matrix) and of the isotropic phase (random orientation of the solute). Englert and Saupe,[217–20] who were the instigators of all these liquid crystal studies, determined the anisotropy of the chemical shift of benzene and 1,3,5-trichlorobenzene as −2·88 and −4·91 × 10⁻⁶, respectively. Englert[221] has also studied the less favourable cases of 2,3,5,6-tetrachlorotoluene, 2,3,4,6-

TABLE 52. INDIRECT COUPLING CONSTANTS (c/s)

J_{ij} (i,j)	C_6H_6	C_6H_5F	C_6F_6
1,2	+6[a]	+9·4	−2·2
1,3	+2	+5·8	−4
1,4	+1	0	+6
2,3		+8·9	
2,4		+2·2	
2,5		0	
2,6		+2·2	
3,4		+8·9	

[a] This value appears to be abnormally small, and does not agree with the independent determination of Reavill and Bernstein[216] of 7·7 c/s, using benzene-^{13}C-1,2-d_4-3,4,5,6.

tetrachloroanisole, and 2,3,5,6-tetrachloroanisole, for which the aromatic protons also suffer high field shifts in the nematic phase (40–60 c/s, at 60 Mc/s), typical of the preferred ordering of the solute parallel to the magnetic field, as the solvent molecules; the methyl group of the substituted toluene also shows such an effect, because of its average location in the molecular plane (16 c/s), whereas the methyl groups of the anisoles, twisted out of the molecular plane, show negligible or even down-field shifts. From his data for fluoro-benzene and hexafluorobenzene Snyder[215] has derived the anisotropy of the diamagnetic shielding constant of ^{19}F in the C—F bond:

$$\sigma_{yy} - \sigma_{xx} = +2·59 \times 10^{-4}$$

$$\sigma_{zz} - \sigma_{yy} = +0·30 \times 10^{-4}$$

The fluorine is diamagnetically shielded when the applied field is perpendicular to the molecular plane, and least shielded when the applied field is perpendicular to the carbon–fluorine bond and in the molecular plane of the ring. The experimental shifts for the fluorine resonance between the nematic and the isotropic phase, the latter occurring at higher fields, are +1650 and +205 c/s (at 56·4 Mc/s, presumably), respectively, for the hexafluoro- and the monofluoro derivatives. The complementary investigation of Nehring and Saupe[222] gave the following results:

(a) The degree of order in a nematic solution decreases in the sequence hexafluorobenzene ($S = -0·20$), 1,3,5-trifluorobenzene ($S = -0·15$), and benzene ($S = -0·10$).

(b) The scalar coupling constants J_{HH}^{meta}, J_{FF}^{meta} and J_{HF}^{ortho} are positive, J_{HF}^{para} is negative, the values and the relative signs agree with the independent determination of Jones, Hirst and Bernstein.[223] J_{FF}^{meta} has the opposite sign in hexafluorobenzene, but a very similar magnitude![214]

(c) The shift anisotropy for 1,3,5-trifluorobenzene $\sigma_F = +99 \times 10^{-6}$ differs strongly from the value for hexafluorobenzene ($+154 \times 10^{-6}$).

Combining it with the shielding anisotropy of monofluoro-benzene given by Snyder,[215] the shielding tensor components are:

$$\sigma_{yy} - \sigma_{xx} = +1 \cdot 16 \times 10^{-4}$$

$$\sigma_{zz} - \sigma_{yy} = +0 \cdot 41 \times 10^{-4}$$

in fair agreement with the findings of Snyder.[215]

Other important applications[217–20] include determination of deuterium quadrupole coupling constants[207] e^2qQ/μ_D, proportional to $(1 - 3\cos^2\theta)$, θ being the angle between the electric field gradient experienced by the deuterium quadrupole moment Q, and a molecular axis for internal rotation (for a methyl group, θ is the tetrahedral angle). These deuterium quadrupole splittings will provide information about the valence angles, and also about organization in the nematic mesophase.

10. SOLVENT PERTURBATION OF CHEMICAL SHIFTS, WITH LITTLE ALTERATION OF COUPLING CONSTANTS: AN AID TO ANALYSIS BY INSPECTION AND TO ANALYSIS SENSU STRICTO

Sometimes NMR spectra provide examples of coalesced resonances for two (or more) types of nuclei in the same molecule which are not isochronous by symmetry, but accidentally isochronous. There is a wealth of such observations, which in most instances are not understood and for which a classification appears difficult. One might be able to separate such super-imposed resonances by recourse to a much higher spectrometer frequency, something seldom feasible economically, so that NMR spectroscopists have gradually come to realize the utility of solvent effects for this end. Also, solvent effects may serve to change markedly the appearances of a complex spin interaction system, thus permitting either:

(a) a check on the more rigorous "computer" analysis; the usual postulate, sometimes verified, is the independence of coupling constants to change of the solvent;

(b) a simplification of the analysis, which may even become possible through mere spectral inspection.

For instance, an ABC system may be transformed by this means into an ABX system, and anyway the precision available is seldom sufficient to derive coupling constants to better than 0·1 c/s, and the chemical shifts within 0·005 ppm, so that this procedure can be quite useful, at least to start with. Figs. 20, 21, 22 and 23 serve to illustrate this point. Figs. 22 and 23 were spectra obtained by Grant, Hirst and Gutowsky[225] for solutions of veratrole in benzene. Farges and Dreiding[226] have noted the isochrony of the two different types of hydrogens in the *spiro* molecule (I) in carbon tetrachloride, deuterochloroform, or trifluoroacetic acid solution, whereas they become nonequivalent in acetone or benzene (see Table 53). Trager, Nist and Huitric[227] report upon the apparent first order spectrum obtained for 4-hydroxycyclo-

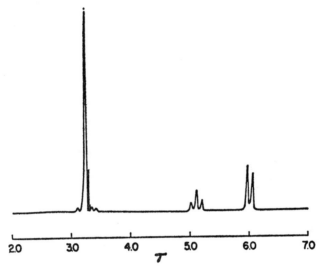

Fig. 20. The ¹H NMR spectrum of neat benzyl alcohol ($J = 5·6$ c/s). (McGreer and Mocek.[224])

hexanone-2,2,6,6-d_4 in D_2O, pyridine and deuterated chloroform solutions leading to deceptively simple spectra; related observations have been made for *cis*-1,4-cyclohexanediol-3,3,4,5,5-d_5,[228] and for the 1,4-diphenyl- and 1,4-dimethyl-1,4-cyclohexanediols.[229] In Fig. 24, the spectra obtained by Courtot, Kinastowski and Lumbroso[229] in deuterochloroform and in pyridine are reproduced. Hutton and Schaefer[230] have changed the solvent in order to convert the cyclopropylamine spectrum from an AA′BB′X to an AA′A″A‴X type, and to change the spectral parameters of 1-phenyl-cyclopropyl-carboxylic acid[231], an AA′BB′ spin system.

Several studies have further considered solvent effects on the non-equivalence of diastereotopic[232] nuclei, and because of the complexity of their task—in the absence of any good theory of the chemical shift—they had to remain very qualitative and descriptive. Singer and Mislow[233] have

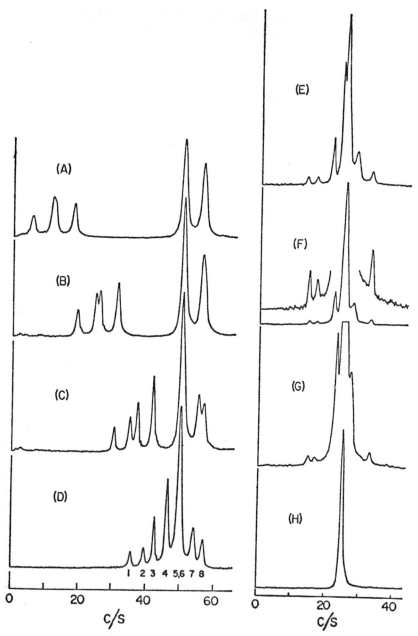

Fig. 21. Effect on the ^{1}H spectrum of benzyl alcohol of the addition of acetone (A) 22% w/w acetone ($J/\Delta = 0\cdot14$). (B) 33% w/w acetone ($J/\Delta = 0\cdot21$), (C) 45% w/w acetone ($J/\Delta = 0\cdot37$), (D) 49% w/w acetone ($J/\Delta = 0\cdot60$), (E) 55% w/w acetone ($J/\Delta = 1\cdot0$), (F) 57% w/w acetone ($J/\Delta = 1\cdot3$), (G) 58% w/w acetone ($J/\Delta = 1\cdot9$), (H) 60% w/w acetone. (McGreer and Mocek.[224])

found that only benzene and thiophene are efficient in separating the diastereomeric resonances of the methylene groups for the diketone II, among thirty-five different solvents (the majority of which were aromatics). For the biphenylic sulphide (III), many such solvents produce the expected non-equivalence, but there is no simple relationship between the nature of the solvent and the amount of non-equivalence, as measured by the internal chemical shift $H_A - H_B$ of these methylenic protons. In another work,

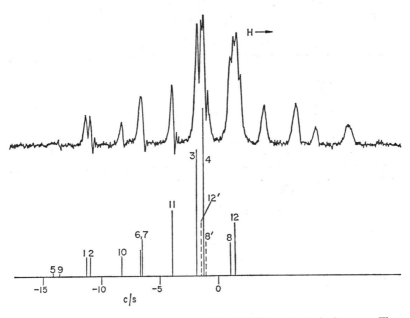

FIG. 22. ^1H resonance spectrum at 60 Mc/s of 16% veratrole in benzene. The chemical shift $\nu_0\delta = 9.1$ c/s while $K = 7.9$ c/s, $L = 6.5$ c/s, $M = 7.1$ c/s and $N = 9.4$ c/s. As a result of the strong coupling, the multiplets have interlaced as shown by lines 8 and 12. (Grant, Hirst and Gutowsky.[225])

Mislow and collaborators[234] had reported the respective effects of carbon disulphide, carbon tetrachloride, deuterochloroform, benzene, pyridine and nitrobenzene, on the methylene, C-methyl and N-methyl resonances of the bridged biphenyls (IV–VII). In some cases there is some kind of a correlation with the solvent dielectric constant, benzene being quite abnormal, and pyridine producing solvent shifts markedly different from those of benzene.

There have been attempts to discuss these non-equivalences in conformational terms, the populations of the various conformers being modified by the solvent (see Section 3). For example, Roberts and collaborators[235] have measured the solvent shifts given in Table 54 for the H_A and H_B resonances

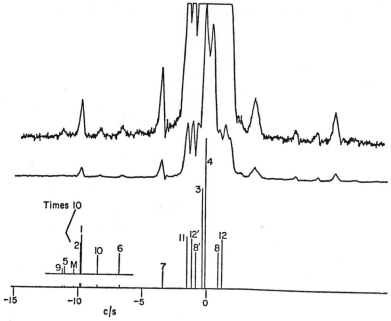

FIG. 23. ^{1}H resonance spectrum at 60 Mc/s of 70% veratrole in benzene. The chemical shift $\nu_0\delta = 4\cdot0$ c/s and the coupling parameters are the same as those for Fig. 22. Because of the very strong coupling, the spectrum has mainly coalesced to the centre. The small outer lines, including a mixed transition labelled M, form the basis for the analysis. (Grant, Hirst and Gutowsky.[225])

TABLE 53. SOLVENT SHIFTS FOR THE PROTONS OF MOLECULE

I

Solvent system	H_A	H_B (ppm)
CCl_4	6·42	6·42
$CDCl_3$	6·48	6·48
CF_3COOH	6·80	6·80
C_6H_6	5·49	6·07
Me_2CO	6·44	6·71
$Me_2CO/CHCl_3(1:1)$	6·37	6·61
$Me_2CO/CHCl_3(1:2)$	6·38	6·60
$Me_2CO/CHCl_3(1:4)$	6·39	6·58

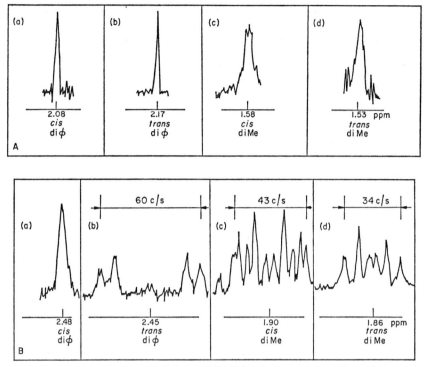

FIG. 24. ¹H NMR spectra of diols at 25°C (A) in CDCl₃ (B) in pyridine. (Courtot, Kinastowski and Lumbroso.[229])

of 1-phenylethyl benzyl ether (VIII). With the exception of the hydrogen bonding solvents, there is a qualitative relationship between $H_A - H_B$ and the solvent dielectric constant; it is also rather remarkable that the replace-

TABLE 54. SOLVENT VARIATION OF $H_A - H_B$ FOR 1-PHENYLMETHYL BENZYL ETHER

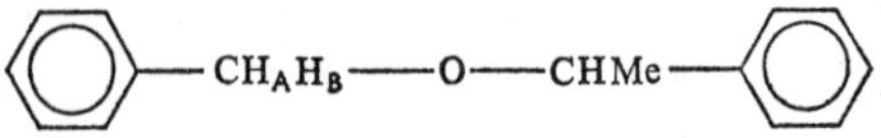

$$\text{\textcircled{O}}—CH_AH_B——O——CHMe—\text{\textcircled{O}}$$

VIII

Solvent	ϵ	$H_A - H_B$ (c/s, at 60 Mc/s)
n-C_5H_{12}	1·84	10·1
C_6H_{12}	2·05	10·1
Dioxane	2·21	6·9
CCl_4	2·24	10·2
C_6H_6	2·28	10·9
Et_2O	4·33	9·0
$CHCl_3$	5·05	8·6
PhCl	5·95	8·2
AcOH	6·3	10·3
MeI	7·0	6·0
o-$C_6H_4Cl_2$	7·47	7·2
$PhNH_2$	7·25	7·8
t-BuOH	10·9	5·7
C_5H_5N	12·5	6·1
Me_2CO	21·4	2·5
EtOH	24·3	7·4
$PhNO_2$	36·1	3·8
Me_2NCHO	36·7	2·5
$MeNO_2$	37·5	3·1
Me_2SO	48·9	2·0
HCOOH	58·3	9·9

ment of one of the phenyl rings by an alkyl group considerably decreases this solvent dependence of $H_A - H_B$. Substitution of a halogen para or meta to the CH_AH_B group has almost no effect on $H_A - H_B$ in the solvents studied, whereas when one ortho hydrogen atom is substituted this difference becomes, on the contrary, so small as to escape detection, and when *both* are substituted the original non-equivalence of the unsubstituted derivative is regained. Simple arguments of conformational analysis are used to rationalize this phenomenon, indicating that the non-equivalence of H_A and H_B results from diastereomeric relationships with the phenyl ring adjacent to their non-asymmetric carbon atom. In the preferred conformation of the unsubstituted molecule, this phenyl ring is tilted into the bisectional plane for the H_ACH_B angle when an ortho substituent is introduced. This conformational dependence, however, is not a general rule. Snyder[236] has drawn attention to the rather crude parallel between the solvent shifts upon the conformationally

mobile CH₂BrC(Me)(COOH)Br and the rigid ethylenic molecule
$H_2C=C(Me)(COOH)$. In another investigation, Snyder[61] does not find
any correlation between the chemical shift difference of the methylene
hydrogens in the alicyclic series and that of the olefins from which they
formally derive. For twelve ethanes of the type $CHXYCH_2Z$, Snyder[61]
reports relatively small solvent variations of the three bond ^3J vicinal coupling
constants, concomitant with large solvent variations of the chemical shifts.
From these observations, he concludes that the latter should not be considered
suitable for deriving information about conformer distribution, a statement
which cannot be accepted as logical without further hypotheses about the
relative sensitivities of chemical shifts and coupling constants towards
conformational changes. Our conclusion to this section will be a quotation
from the recent important communication of Raban:[237] "Although the
conformation population approach leads to an understanding of the usual
decrease in magnetic non-equivalence with temperature, we should not lose
sight of the fact that the decrease in the conformer population contribution
is superimposed over an intrinsic diastereomerism contribution which can
be quite large in some cases."

11. SOLVENT MODIFICATION OF THE RESIDENCE TIME
FOR EXCHANGING PROTONS

The hydroxylic proton of methanol shows up as a quartet provided that
the residence time in a CH_3OH molecule is long compared with the reciprocal
of the coupling constant to the vicinal methyl group time scale; this slowing
down of intermolecular exchange occurs when the pure liquid or its solutions
in inert solvents are freed from the traces of acids that catalyse proton
exchange. Strongly acidic or basic molecules thus catalyse exchange; thus
apart from the carboxylic acids and the phenols, many amines, such as
cyclohexylamine, triethylamine, morpholine and piperidine,[238] also effect
the spin decoupling of an exchanging OH or NH proton. The use of a strongly
hydrogen bonding solvent, in order conversely to observe and measure the
hydroxylic proton splitting, seems to have been first recommended by
Holmes, Kivelson and Drinkard,[239] who used acetone solutions; subsequently, Chapman and King[240] demonstrated such an effect for dimethyl
sulphoxide solutions. In these, the chemical shift of the OH protons and their
coupling constants with the adjacent hydrogens are sufficiently characteristic
that Rader[241] could use the former, Uebel and Goodwin[242] the latter, with
success for the conformational analysis of cyclohexanols. The coupling
constant of the equatorial hydroxyl (compounds I and III) is larger than for

I
$^3J_{CHOH}$ 4.53

II
3.22

III
4.64

IV
3.60

V
4.40 c/s

the axial hydroxyl in the isomeric cyclohexanol (compounds II and IV),[242] which is to be attributed to the loss in the axial position of a degree of liberty corresponding to the rotamer with a 180° angle between the hydroxylic and methine protons. In cyclohexanol itself, with the intermediate value of 4·40 c/s, Uebel and Goodwin[242] deduce an energy difference of 0·6–1·0 kcal mole^{-1} for the hydroxyl group, in good agreement with the chemical shift determinations. Perlin[243] has also used hydroxylic signals in dimethyl sulphoxide solution to determine the ring size of a variety of cyclic sugars, and to differentiate between the cyclic and the acyclic forms, and to characterize partially acetylated sugars. Page and Bresler[244] have used pyridine solutions, which like aniline[240] also diminish the exchange rate, to study primary alcohols. Casu et al.[245] have also studied the hydroxyl proton resonances of sugars in dimethyl sulphoxide. Anderson and Silverstein[246] have shown the inadequacy of this same solvent for studying amines, but show that the chemical shifts and coupling constants of amine salts yield useful information; the amine salts are prepared by addition of trifluoroacetic acid to the carbon tetrachloride or deuterochloroform solution of the amine. This study spans the borderline between solvent effects, and in situ chemical reactions (e.g. reference 247) run in the NMR tube. Feeney and Heinrich[248] have very astutely utilized the trace amounts of impurity water in commercial supplies of deuterochloroform, at 1·5 ppm, to identify absorption bands from groups which undergo slow reversible chemical exchange with water. This exchange phenomenon is used to detect and characterize phenolic OH groups at 5–6 ppm by irradiation of the water signal in a double irradiation frequency swept experiment which causes collapse of this phenolic peak.

12. SOLVENT EFFECTS ON COUPLING CONSTANTS

12.1. *One Bond Coupling Constants* (^1J)

Evans[249] has measured the increase in the $^1J_{CH}$ coupling constant of chloroform in cyclohexane where it is uncomplexed ($^1J = 208 \cdot 1$ c/s) and in ether where the hydrogen bond is strong ($^1J = 213 \cdot 7$ c/s), and in dimethyl sulphoxide, a still more basic solvent ($^1J = 217 \cdot 7$ c/s). In Tables 55 and 56, the data relative to acyclic bases,[249] are supplemented with the values obtained for some cyclic saturated and unsaturated solvents.[194]

Parenthetically, these experimental results are direct evidence against the theoretically derived proportionality of $^1J_{CH}$ with the square p^2 of the bond order for the C—H bond,[250] and they suggest that there is no lengthening

TABLE 55. CORRELATION BETWEEN SOLVENT BASICITY
AND $^1J_{CH}$ FOR $CHCl_3$ [a]

Solvent	$^1J_{CH}$ (± 1 c/s)
C_6H_{12}	208·1
CCl_4	208·4
Neat	209·5
C_6H_6	210·6
AcCl	211·8
$MeNO_2$	213·6
Et_2O	213·7
Et_3N	214·2
MeOH	214·3
MeCN	214·6
Me_2CO	215·2
Me_2NCHO	217·4
Me_2SO	217·7

[a] Mole fraction of chloroform kept below 0·15.

TABLE 56. CORRELATION BETWEEN SOLVENT BASICITY
AND $^1J_{CH}$ FOR $CHCl_3$ [a]

Solvent	$^1J_{CH}$ (± 1 c/s)	Ionization potential I (eV)
C_6H_{12}	208·1	(9·88)
Nitrogen heterocycles		
Pyridine	215·0	9·76
Pyridazine	215·0	9·86
s-triazine	211·0	10·07
Pyrazine	210·5	10·01
Oxygen heterocycles		
Tetrahydrofuran	214·0	9·54
Dioxane	213·0	9·13
2,6-dimethyl-1,4-pyrone	212·0	?
Paraldehyde	212·0	?
Dihydropyran	210·5	8·34

[a] Mole fraction of chloroform kept below 0·15.

of this bond in the intermolecular hydrogen bonded complex.[249] There is no absolute measure for the basicity of the heterocycles of Table 56; the ionization potential values I are unfortunately rather imprecise, as can be appreciated by the spread of the existing data for pyridine.[251-2] Hoffmann[253] has also pointed out that the highest occupied σ orbital is not necessarily the nitrogen lone pair in aza-aromatics. The values of the pK_a show a similar variation (pyridine = 5·23; pyridazine = 2·0; pyrazine = 0·3[254]). In the expectation that other good, or better proton donors, would display analogous behaviour, Evans[249] employed phenylacetylene which failed to show an enhanced C—H bond acidity, since the ^1J value changes from 250·8 ± 0·3 in carbon tetrachloride to 252·0 ± 0·3 c/s in dimethyl sulphoxide.

Becker[255] has shown conclusively that ^{15}N-aniline, acting as hydrogen bond donor towards a number of acceptors (see Table 57), has the $^1J_{15NH}$ coupling parameter increased by a substantial amount.

TABLE 57. CORRELATION OF $^1J_{NH}$ FOR ^{15}N-ANILINE[a]
WITH SOLVENT BASICITY

Solvent	$^1J_{NH}$ (c/s)
$CDCl_3$	78·0
C_6D_{12}	78·0
CCl_4	78·0
C_5H_5N	81·4
Me_2CO	82·1
Me_2NCHO	82·3
Me_2SO	82·3

[a] 0·2 M.

The equilibrium constants are unknown for some of these systems, Becker[255] quotes M. W. Hanna to the effect that in pyridine at room temperature nearly 90% of the aniline is bound. Reuben and Samuel[256] also found that the ^{17}O—^{1}H coupling constant of methanol varies with the solvent by up to 5%. Martin, Castro and Martin[257] could reproduce qualitatively the observations made by Evans[249] with a solute activated by only two electron-withdrawing substituents (see Table 58). Yet more astonishing is the observation by De Jeu, Angad Gaur and Smidt[258] that the methyl groups of acetone and dimethyl sulphoxide are themselves capable of similar variations (see Table 59). The interpretation provided[258] is based on the linear correlation with the chemical shift of ^{17}O in the carbonyl group as measured by Christ and Diehl,[48] for several mole fractions of acetone in water, with a slope of 4×10^{-2} c/s (ppm)$^{-1}$: hydrogen bonding with the protic solvents SH increases the relative importance of polar resonance structures for the carbonyl group:

$$\text{C=O} + \text{H—S} \leftrightarrow \text{C}^+\text{—O}^- \ldots \text{H—S} \leftrightarrow \text{C}^+\text{—O} \ldots \text{H—S}^-.$$

TABLE 58

Solvent	$^1J_{CH}$ (c/s)	
	$ClCH_2OCHMe_2$	$ClCH_2OCH_2CH_2Cl$
C_6H_{12}	173·0	173·5
Neat	175·0	176·0
Tetrahydrofuran	175·5	176·5
Me_2CO	176·0	177·5
Me_2NCHO	177·0	178·0
Hexamethylphosphotriamide	177·0	178·0

TABLE 59. SOLVENT VARIATION OF THE $^1J_{CH}$ COUPLING
PARAMETERS FOR METHYL GROUPS

Solvent	Mole fraction of solute	$^1J_{CH}$ ($\pm 0·2$ c/s)	
		Me_2CO	Me_2SO
CCl_4	0·35	126·7	137·6
CS_2	0·20	126·6	—
Me_2CO	—	126·6	—
$PhNO_2$	0·35	127·0	—
$MeCN$	0·20	127·1	—
$PhNH_2$	0·35	127·3	—
$CHCl_3$	0·35	127·4	—
H_2O	0·35	127·4	139·2
$PhOH$	0·35	128·0	138·0
$HCOOH$	0·20	128·1	—
H_2O	0·20	128·1	—

A positive charge is placed on carbon, and therefore $^1J_{CH}$ is increased by a mechanism similar to the operation of electronegative groups. The dipolar aprotic solvents are thought to favour also the above polarization of the carbonyl group.[258] However, it is also found that the long range $^4J_{HH}$ coupling parameter for acetone varies with the solvent by up to 30%, from 0·51 c/s in pure acetone to 0·66 c/s at infinite dilution in water, in a way analogous to that for $^1J_{CH}$. There is apparently no good reason to prefer the above rationalization[258] to one invoking the intervention in variable amounts of the eclipsed and staggered rotational isomers for acetone, since in pure acetone itself the difference in free energy between the rotamer Ia, where carbonyl and hydrogen are eclipsed, and the less stable rotamer where they are staggered Ib, is but 0·8 kcal mole^{-1} (see reference 67 and references cited therein). Takahashi[67] has indeed shown that conformational factors

Ia   Ib

are responsible for the concentration effect of the long range spin–spin coupling constants in ketones (see Section 3). These same conformational factors are probably responsible of the $\sim$0·15 c/s change in the ^3J coupling of acetaldehyde over a temperature range of $\sim$150°.[259] Watts, Loemker and Goldstein[260] have also noted small, but experimentally significant, variations of $^1J_{CH}$ in some dihaloethylenes dissolved in cyclohexane (a solvent less polar than the solute) and in dimethylformamide (a solvent more polar than the solute). They have studied too the concentration dependence of these effects which are linear with the concentration of the solute in the range 50–100 mole % for the cis- and trans-1,2-dichloroethylene and trichloro-ethylene (see Table 60). Similar variations, but smaller, are obtained with

TABLE 60. SOLVENT EFFECTS ON $^1J_{CH}$ FOR THE DIHALOETHYLENES

Solvent	Concentration of solute (mole %)	$^1J_{CH}$ ($\pm$0·2 c/s)
	cis-$C_2H_2Cl_2$	
Neat	100	197·89
C_6H_{12}	91·9	197·72
,,	81·9	197·50
,,	77·1	197·45
,,	68·9	197·35
,,	53·5	197·17
,,	49·0	197·12
Me$_2$NCHO	79·3	198·64
,,	71·3	199·27
,,	59·1	199·86
,,	43·8	200·70
,,	41·8	200·62
	Trichloroethylene	
Neat	100	200·92
C_6H_{12}	85·6	200·74
,,	70·3	200·68
,,	67·1	200·73
,,	49·2	200·50
Me$_2$NCHO	81·2	201·88
,,	74·7	202·74
,,	68·8	202·98
,,	49·7	204·20
,,	37·2	204·48

cis- and *trans*-1,2-dibromoethylene, *trans*-1,2-dichloroethylene and 1,1-dichloroethylene. It is seen that the variations with the concentration in cyclohexane are very small; they may be due to some degree of self-association between the haloethylene molecules. The vicinal coupling constants $^3J_{HH}$, in the same conditions, are practically solvent independent. The chemical shifts suffer variations which also correlate linearly with the concentration, and consequently with the $^1J_{CH}$ values. The origin of the effect may be localized in the carbon–halogen bond, as suggested by comparison of the NMR with the infra-red evidence (see Table 61). Both the C—H and C—X stretching

TABLE 61. COMPARISON OF NMR[a] AND IR[b] SOLVENT DEPENDENCE[c]

Solute	ΔJ_{CH} (c/s)	$\Delta \nu_{CH}$ (cm^{-1})	$\Delta \nu_{CX}$ (cm^{-1})
cis-$C_2H_2Cl_2$	2·97	9·6	9·7
trans-$C_2H_2Cl_2$	2·40	16·8	10·5
cis-$C_2H_2Br_2$	2·70	19·9	6·2
trans-$C_2H_2Br_2$	2·00	19·6	8·5
1,1-$C_2H_2Cl_2$	0·73	14·2	3·5

[a] 50 mole % concn.
[b] 4–9 mole % concn.
[c] From C_6H_{12} to Me_2NCHO.

frequencies are smaller in dimethyl formamide than in cyclohexane and the bands broadened; the C—X stretching mode variations, being somewhat more sensitive, parallel the $^1J_{CH}$ variations. Solvent effects of similar magnitudes had been measured for C—X bonds of halocyclohexanes by Reisse and Chiurdoglu,[261] and have been related to the polarity of the carbon–halogen bond: contrary to the present observations of Watts, Loemker and Goldstein,[260] they found a greater solvent sensitivity for the bromo- than for the chloro-derivative. However, since the *cis* and *trans* configurational isomers are affected by comparable amounts, the simple reaction field explanation does not hold, and is dispelled by the observed concentration dependence which is not more pronounced for the *cis* isomers in cyclohexane than in dimethyl formamide, as this would imply dependence upon the dielectric constant of the medium.[260] Coyle *et al.*[262] have obtained extremely interesting results by studying the behaviour of a molecule with a zero dipole moment, SiF$_4$ (see Tables 62 and 63). Addition of *any* solvent increases the ^1J value relative to that for the pure SiF$_4$: 169·00 ± 0·08 (gas, 30 atm); 168·84 ± 0·03 (gas, 110 atm); and 169·97 (liquid, −52°C). There is no correlation with the magnitude of the solvent dielectric constant, contrary to other observations of solvent dependent J's. However, for series of related solvents of formula SiX_nF_{4-n} or CX_nF_{4-n}, ^1J increases monotonically with increasing n, which

TABLE 62. SOLVENT DEPENDENCE OF $^1J_{^{29}Si-^{19}F}$ IN SILICON TETRAFLUORIDE

Solvent	^1J (c/s)	Solvent	^1J (c/s)
Si_2OF_6	170·51	SiF_2Br_2	174·51
CF_3CN	170·66	Me_4Si	174·68
$CClF_3$, gas	170·78	CCl_3F	175·03
$MeSiF_3$	171·12	CH_2Cl_2	175·23
SiF_3Br	171·51	$CHCl_3$	176·12
$EtSiF_3$	172·01	$Si_2Cl_5F + Si_2Cl_6$ (1 : 3)	176·14
$H_2C=CHSiF_3$	172·05	$SiFBr_3$	176·45
Me_2SiF_2	172·35	CCl_4	176·83
Me_3SiF	173·06	C_6H_{12}	176·88
C_6F_6	173·44	C_6H_6	176·98
CCl_2F_2	173·67	BBr_3	178·0
Et_2O	173·70	$SiBr_4$	178·61

TABLE 63. CONCENTRATION DEPENDENCE OF ^1J FOR SiF$_4$
IN VARIOUS SOLVENTS

Solvent	SiF$_4$ concn. (%)	^1J (c/s)
Et_2O	15	173·70
	50	172·71
C_6H_{12}	15	176·88
	25	176·77
$SiBr_4$	17	178·61
	30	178·37
Me_4Si	15	174·68
	31	174·27
	50	173·78
$CClF_3$ (gas, 28°C)	15	170·78
(liq., −3°C)	15	171·40
(liq., −30°C)	15	172·10

would suggest[262] the possibility of relatively specific F—F interactions. Schaefer[263] is able to account for this strong solvent dependence of J(SiF) in terms of simple dispersion interactions, which act perhaps by lowering ΔE values (mean excitation energies) in the theoretical expression for the Fermi contact term.[264]

12.2. Geminal Coupling Constants (^2J)

By solvent effects upon coupling constants, we mean changes in the magnitude or sign of the coupling of two nuclei brought about by the solvents, without any (or with the minimal) concomitant variation of their geometrical relationship. For this reason, this type of phenomenon is better studied on the so-called "rigid" molecules. At the time of writing, the fairest practice is to present the raw data alone, providing the reader with a rather indigestible fare. Alternatively, despite the paucity of results, we could discuss this topic within the apparently logical framework of the determination of the signs of coupling constants via the direction of the solvent shift, the corresponding working hypothesis about the origin of this effect then being the reaction field.

In effect, since the first experiments or theoretical calculations, presented in Section 4, showing that the chemical shifts were a suitable probe to test the reaction field concept, it has been thought that coupling constants ought to be affected in a similar way. Let us consider, for instance, the one bond $^1J_{CH}$ coupling constant. The sign is known, absolutely, to be positive. The magnitude of this carbon-13 hydrogen coupling parameter increases from methane to chloroform, for example, when electron-withdrawing groups are being substituted. Relative to methane, the electronic distribution of the C—H bond in chloroform will be characterized by some amount of dipole moment going from a slightly more positive hydrogen to a slightly more negative carbon atom, since this is the direction of the molecular dipole moment. The reaction field would be expected to increase this tendency, hence the magnitude of the positive ^1J, if this ad-hoc operational postulate is made in accordance with the results of Evans.[249] This crude model ignores the second order terms by assuming that, in a way similar to that for chemical shifts, the contribution of the reaction field squared in a manner similar to Buckingham's[10,265] electric field effect is very small. Hence, by analogy, a ^2J negative coupling constant is expected to be made more negative by an increase in the dielectric constant of the medium, while a ^2J positive coupling constant would be correspondingly decreased. The data of Smith and Cox,[266] who introduced this fascinating formalism, further illustrated by the examples of Bell and Danyluk,[267] are shown in Table 64.

TABLE 64. SOLVENT VARIATION OF COUPLING CONSTANTS FOR STYRENE OXIDE

Solvent	ϵ (35°C)	$^2J_{AB}$	$^3J_{AC}$	$^3J_{BC}$ (± 0.1 c/s)
C_6H_{12}	1·99	6·00	2·38	3·93
CCl_4	2·20	5·85	2·40	3·94
C_6H_6	2·26	5·81	2·42	4·11
PhMe	2·35	5·79	2·39	3·99
$CDCl_3$	4·55	5·55	2·55	4·10
C_5H_5N	12·3	5·69	2·42	4·11
PhCOMe	16·99	5·67	2·41	4·07
o-Nitrotoluene	25·15	5·68	2·46	4·06
$PhNO_2$	32·22	5·53	2·49	4·06
$MeNO_2$	35·0	5·42	2·55	4·17
CD_3CN	35·1	5·40	2·56	4·16
Me_2SO	46	5·31	2·43	4·13
Neat	—	5·63	2·48	4·08

While the variations of the ^3J vicinal coupling constants are irregular and small, of the order of the probable error limits, the $^2J_{AB}$ geminal coupling constant which, in the case of epoxides, is well known to be positive,[268] indeed decreases markedly, from 6·00 c/s in cyclohexane to 5·31 c/s in dimethyl sulphoxide, when the bulk dielectric constant increases. In a similar way, the $^2J_{HF}$ coupling constant in CHFClBr, whose sign would be expected to resemble that for $CHFCl_2$ and also be positive, decreases with increasing solvent polarity as shown by Evans'[249] measurements (see Table 65).

TABLE 65. SOLVENT VARIATION OF THE H—F
COUPLING PARAMETER IN CHFClBr

Solvent	ϵ	$^2J_{HF}$ (c/s)
C_6H_{12}	1·99	52·1$_0$
Me_2CO	20·7	51·1$_8$
Et_2O	4·34	51·1$_2$
NEt_3	2·42	50·8$_3$
$MeNO_2$	35·0	51·5$_6$
MeCN	35·1	51·4$_3$
Me_2SO	46	50·6$_7$

In this case of CHFClBr, the variations are not as smooth as with styrene oxide, despite the omission of the aromatic solvents data in the former case that could have indicated specific associations in the latter. Furthermore, $^2J_{HF}$ in a related molecule, CHF_3, has a smaller value for a 5% solution in cyclohexane (79·23 ± 0·10 c/s), by 0·49 ± 0·17 c/s, than for the pure liquid. Nevertheless, there is additional supporting evidence for the Smith–Cox[266] Bell–Danyluk[267] empirical rule. The geminal coupling constant, also positive, in formaldoxime (I) decreases from 9·95 c/s in dibutyl ether to 7·63 c/s in heavy water, while the corresponding variation for its methyl ether (II) is from 9·22 c/s in cyclohexane to 6·96 c/s in water.[269] Conversely, the geminal coupling constants for vinyl bromide[270] and vinyl acrylate increase with the

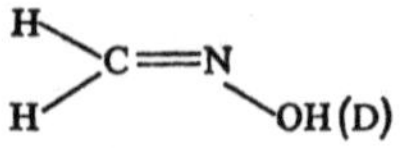

I 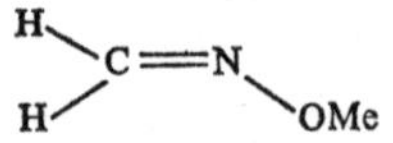   II

dielectric constant of the medium (see Tables 66 and 67). The results obtained by Brügel[271] for vinyl acrylate show a marked variation for the coupling parameters in the acryl group, mainly the geminal coupling, while the couplings for the vinyl group remain practically constant; there is no correlation whatsoever with the solvent dielectric constant.

TABLE 66. SOLVENT VARIATION OF THE GEMINAL COUPLING CONSTANT FOR $H_2C{=}CHBr$

Solvent	$^2J_{ab}$	$^3J_{ac}$	$^3J_{bc}$
Neat	-1.25	$+15.05$	$+6.65$
Hexamethylphosphotriamide (0·1 M)	-1.70	$+14.80$	$+6.75$

TABLE 67. SOLVENT DEPENDENCE OF COUPLING CONSTANTS
FOR VINYL ACRYLATE $H_2C{=}CHCOOCH{=}CH_2$

Solvent	ϵ	Acryl group			Vinyl group		
		$^2J_{ab}$	$^3J_{ac}$	$^3J_{bc}$	$^3J_{ab}$	$^3J_{ac}$	$^3J_{bc}$
		(±0.15 c/s)					
Neat		2·27	10·21	16·72	1·48	6·35	13·97
C_6H_{12}	1·99	1·57	10·77	17·35	1·41	6·36	14·05
CCl_4	2·20	2·45	10·37	16·58	1·43	6·35	13·87
Dioxane	2·22	2·49	10·23	16·56	1·48	6·32	13·95
C_6H_6	2·26	1·28	10·54	17·51	1·53	6·37	14·04
C_5D_5N	12·3	1·34	10·65	17·16	1·63	6·37	13·95
Me_2CO	20·7	1·35	10·60	17·40	1·63	6·27	13·85
MeCN	35·1	1·78	10·73	17·08	1·63	6·37	13·95
Me_2NCHO	37·6	1·57	10·56	17·21	1·63	6·37	14·04
Me_2SO	46	1·31	10·80	17·09	1·60	6·30	13·90

The geminal coupling constant J_{AB} in 1-phenylethyl benzyl ether $PhCH_AH_BOCHMePh$ shows an apparently significant solvent dependence, high magnitudes being associated with low values of the solvent dielectric constant—from 11·5–12·3 in chloroform, carbon tetrachloride, cyclohexane, n-pentane, to 10·2 in dimethyl formamide and 9·3 in nitromethane.[235] Here, not only the usual ^2J solvent dependence for molecules with a fixed geometry would be operative, it may also be that the conformational dependence of the $^2J_{HH}$ coupling phenomenon upon the relative positions of the methylene protons with respect to the adjacent centre of unsaturation[272] plays an important role.[235]

Small variations of the ^2J coupling constant have been measured by Hutton and Schaefer[231] for 1-phenylcyclopropylcarboxylic acid: -3.89 (6 mole % in CS_2), -3.99 (8 mole % in C_6H_6), and -4.03 (7 mole% in $CHCl_3$), when the trans vicinal coupling constant appears to change only by 0·08 c/s. Much more significant changes are reported in another article by the same authors,[273] the geminal coupling in 1-chloro-2-methyl-1,2-dicyclopropane carboxylic acid varies from 6.39 ± 0.05, for the saturated solution in an equimolecular benzene–acetone mixture, to 6.97 ± 0.05, for the formic acid-saturated solution, corresponding to an increase in the dielectric constant of about 40.

Sumners, Piette and Schneider[274] have measured the following coupling constants in formamide, which may reflect the variations of the contributions of the various resonance forms, hence the geometry (see Table 68). Martin and Besnard[275] report a maximum of 8% variation for proton phosphorus coupling constants in organophosphorus compounds, when the solvent is changed between carbon tetrachloride, chloroform, cyclohexane or the neat solution.

TABLE 68. SOLVENT DEPENDENCE OF H—H COUPLING CONSTANTS IN FORMAMIDE $H_XCONH_AH_B$

Solvent	ϵ	$^2J_{AB}$	$^3J_{AX}$	$^3J_{BX}$
		(c/s)		
Me_2CO	20·7	2·9	1·7	13·5
H_2O	80·4	2·3	2·1	13·6
Neat	109	2·4	2·1	12·9

Finegold[56] finds variations by as much as 5% for the geminal couplings in 1,2-dichloro- and 1,2-dibromopropane (see Table 69) (cf. the results of Ng, Tang and Sederholm[189] on the $^2J_{FF}$ coupling in $CF_2BrCFBrCl$). In this work (see Table 70) of Bates, Cawley and Danyluk[276] on vinyl silanes, the

TABLE 69. GEMINAL COUPLINGS FOR 1,2-DICHLORO- AND 1,2-DIBROMOPROPANE[a]

Solvent	ϵ	MeCHClCHHCl	MeCHBrCHHBr
		(± 0.05 c/s)	
Neat		−10·97	−9·88
CCl_4	2·20	−10·82	−9·63
C_6H_6	2·26	−11·12	−9·89
$CHCl_3$	4·55	−11·07	−10·14
Me_2CO	20·7	−11·22	—
MeCN	35·1	−11·27	−10·09
Me_2SO	46	−11·07	−10·09

[a] Mole fraction of the solute kept between 0·077 and 0·231.

values of the geminal couplings follow approximately the increase of the solvent dielectric constant; oxygenated solvents, such as tetrahydrofuran and dioxane, are notable exceptions. The sensitivity to solvent change depends upon the polarity of the solute: for the more polar trichloro and bromo compounds at the top of Table 70, the change from carbon tetrachloride to acetonitrile is 25–30% of the coupling parameter, it is only 6%—*in the opposite direction*—for the tetraphenyl molecule at the bottom of Table 70. These J_{gem} variations could not be induced by temperature, the

TABLE 70. SOLVENT DEPENDENCE OF $^2J_{HH}$ COUPLING PARAMETERS IN SUBSTITUTED VINYL SILANES

Solvent	ϵ	$^2J_{HH}$ (± 0.07 c/s)
	$Cl_3SiC(Cl){=}CH_2$	
C_6H_{12}	1·99	1·86
CCl_4	2·20	1·98
Dioxane	2·22	2·65
Neat		2·04
C_6H_6	2·26	2·44
Tetrahydrofuran		2·60
$C_6H_{12} + Me_2CO$ (2 : 1)	4·16	2·42
Me_2CO	20·7	2·65
MeCN	35·1	2·53
	$Ph_3SiBrC{=}CH_2$	
CCl_4	2·20	1·33
Neat		1·56
Me_2CO	20·7	1·77
MeCN	35·1	2·08
	$Ph_3SiPhC{=}CH_2$	
CCl_4	2·20	2·83
Me_2CO	20·7	2·79
MeCN	35·1	2·65

coupling parameters in a given solvent remaining constant between -60 and $+50°C$. The authors interpret their results as consistent with a decrease in electronegativity of the central silicon atom upon a change in coordination with the ligand from fourfold to sixfold,[277] which is known to markedly affect the magnitude of $^2J_{HH}$ in vinyl metallic compounds.[278] If the dioxane and tetrahydrofuran molecules, which may readily donate electrons into the vacant d orbitals of silicon, are suitable ligands, one does not clearly understand the phenomenon for the other solvents.

For α-chloroacrylonitrile (see Table 71), the $^2J_{12}$ parameter, which is negative,[279] at infinite dilution varies between 1·96 (cyclohexane) and 3·24 (dimethyl sulphoxide), and the values obtained for the different solvents plot reasonably well against the $(\epsilon - 1)/(2\epsilon + 2.5)$ dielectric function (so does the internal chemical shift parameter H_1—H_2).[280–1] McLauchlan, Reeves and Schaefer[279] have conducted a similar study, but they varied the temperature also (see Table 72). These latter authors[279] also find reasonable reaction field correlations for J_{12} and H_1—H_2, but different from those of Watts and Goldstein[281]; appreciable temperature changes are found only for the "more inert" methyl cyclohexane and toluene solutions, which also correspond to lower values of the dielectric constant, for which the reaction field picture is expected to be the most realistic.[279] Similar results are obtained

TABLE 71. SOLVENT DEPENDENCE OF $^2J_{HH}$ IN α-CHLORO-
ACRYLONITRILE (Cl)(CN)C=CH₂

Solvent	ϵ, 35°C	$^2J_{HH}$ ($\pm$0·05 c/s)[a]
TMS	1·89	1·96
C₆H₁₂	1·97	1·96
Mixt. A[b]	3·02	2·16
CHBr₃	4·23	2·46
CHCl₃	4·59	2·42
EtI	7·42	2·52
EtBr	8·78	2·56
Mixt. B[c]	10·63	2·75
(n-Pr)₂CO	11·72	2·94
Me₂CO	19·75	3·07
MeOH	30·61	2·93
Me₂NCHO	35·05	3·19
MeCN	35·10	3·00
MeNO₂	35·10	3·04
Me₂SO	45	3·24

[a] Extrapolated to infinite dilution.
[b] 0·4622 mole CHBr₃ + 0·5378 mole C₆H₁₂.
[c] 0·6381 mole EtBr + 0·3719 mole (n-Pr)₂CO.

TABLE 72. TEMPERATURE AND SOLVENT DEPENDENCE OF THE PROTON COUPLING CONSTANTS
AND CHEMICAL SHIFTS IN α-CHLOROACRYLONITRILE

Solvent	Temp.	^2J ($\pm$0·03 c/s)	H₁ (ppm)	H₂ (ppm)	$Z(\epsilon)$[a]
MeC₆H₁₁	−58°C	2·60	5·868	6·006	0·169
„	40°C	2·40	5·874	6·974	0·153
„	72°C	2·36	—	—	0·148
PhMe	−80°C	2·84	4·657	4·697	0·209
„	40°C	2·61	5·166	5·222	0·187
„	72°C	2·53	5·234	5·292	0·181
MeCN	−40°C	3·04	6·351	6·479	0·477
„	40°C	3·00	6·273	6·389	0·468
„	72°C	2·97	6·241	6·351	0·462
Me₂NCHO	−58°C	3·11	6·798	6·992	0·478
„	40°C	3·15	6·153	6·701	0·468
„	72°C	3·13	6·462	6·624	0·463

[a] $Z(\epsilon) = \dfrac{\epsilon - 1}{2\epsilon + 2 \cdot 5}$.

for the α-bromoacrylonitrile, where $^2J_{HH}$ varies between 1·99 (cyclohexane) and 3·16 (dimethyl formamide). This behaviour is confined to the 1,1-disubstituted ethylene system, for vinyl derivatives (X = F, Cl, Br, I, CN) only show slight variations, whether for the $^3J_{HH}$ or $^2J_{HH}$ parameters, between cyclohexane and dimethyl formamide[281] (see Table 73).

TABLE 73. SOLVENT EFFECTS ON COUPLING PARAMETERS OF VINYL COMPOUNDS

Solute	$^3J_{cis}$		$^3J_{trans}$		^2J	
	C_6H_{12}	Me_2NCHO	C_6H_{12}	Me_2NCHO	C_6H_{12}	Me_2NCHO
$H_2C{=}CHF$	4·70	4·63	12·68	12·47	−3·06	−3·39
$H_2C{=}CHCl$	7·16	6·95	14·78	14·64	−1·28	−1·67
$H_2C{=}CHBr$	7·27	7·00	15·10	14·87	−1·59	−2·05
$H_2C{=}CHI$	7·83	7·78	15·82	15·82	−0·88	−1·52
$H_2C{=}CHCN$	11·65	11·81	17·89	17·88	1·20	0·96

With the exception of the *cis* vicinal coupling in the cyano derivative, all coupling parameters are reduced going from cyclohexane to dimethyl formamide; conversely, the magnitudes of the H—F couplings in vinyl fluoride increase, in the same change of solvent, from 84·6 to 86·6, from 51·6 to 55·6, and from 19·5 to 21·5 c/s, with a very small concentration effect.[281] Ng, Tang and Sederholm[189] have further studied the effect of solvents on ^{19}F spin–spin coupling constants (see Table 74). In this case, the geminal

TABLE 74. EFFECT OF SOLVENTS ON F—F COUPLINGS OF BROMOTRIFLUOROETHYLENE[a]

Solvent	J_{13}	J_{12}	J_{23}
		($\pm0\cdot10$ c/s)	
CS_2	71·7	56·4	123·1
$SOCl_2$	72·1	55·8	122·4
$CFCl_3$	73·3	56·8	123·4
Et_2O	73·8	56·0	123·1
Dioxane	73·8	55·1	122·6
Me_2CS	74·3	53·7	122·5
Me_2CO	74·9	54·9	122·5
MeCHO	74·2	55·1	122·6
MeOH	74·5	55·7	122·9
EtOH	73·9	55·9	122·9
AcOH	74·3	56·0	122·7
Ac_2O	74·4	55·4	122·6
MeCN	74·6	54·9	121·9

[a] At 20°C, concentration of 50% per volume.

coupling constant $^2J_{13}$ increases in absolute magnitude with solvent polarity, while the two vicinal couplings are simultaneously decreased. At −105°C, the same trends are apparent, with the exception of J_{12} for which supplementary reductions of the absolute magnitude of the coupling occur. These authors[189] have also studied the solvent variation of $^3J_{FF}$ in CF_3CFCl_2,

M

with three equally populated rotamers at any temperature, and of the $^2J_{FF}$ and $^3J_{FF}$ couplings in $CF_2BrCFBrCl$, and they observe variations up to 5%.

12.3. Vicinal Coupling Constants (^3J)

Apart from the examples already quoted, and where the authors failed to detect any significant variation,[260, 281] only very small changes were detected for *para*-nitroanisole by Hutton and Schaefer[87] between cyclohexane and acetonitrile for the sum of the ortho and para coupling constants.

We have also found a very small variation for *cis*-1-bromo-2-ethoxyethyl-ene, from 4·25 ± 0·1 c/s in carbon tetrachloride, to 4·04 ± 0·1 c/s in dimethyl sulphoxide,[282] which makes us currently reinvestigate our report[283] of a 30% variation in *cis*-1-chloro-2-ethoxyethylene.

13. AROMATIC SOLVENT INDUCED SHIFTS (ASIS): BENZENE

13.1. Specificity

In 1964 a preliminary communication of Bhacca and Williams[284] kindled a renewed interest for the solvent shifts induced by aromatic solvents. We defer until later discussion of some of the anterior studies dealing with simple polar molecules in benzene, since the immense merit of Bhacca and Williams'[284] work consists of the application to the steroid field. This also is not without precedence, for a note by Slomp and MacKellar, in 1960,[285] directed atten-tion to the considerable shifts of the methyl resonances for these compounds when the solvent is changed from deuterochloroform to pyridine, and emphasized the utility of the pyridine solvent to bring about separation of overlapping bands and other selective effects.

Bhacca and Williams,[284] by examining a series of steroidal ketones, showed that the change from deuterochloroform to benzene produces important solvent shifts (ASIS) whose sign and magnitude are characteristic of the geometrical relationship between the relevant proton and the carbonyl group. The 18-Me and 19-Me resonances of 5α-androstane, the parent hydrocarbon (I), are invariant in the change of solvent, they respectively occur at 0·69 and 0·79 ppm in deuterochloroform, and at 0·71 and 0·79 ppm in benzene solution, whereas typical 5α-androstanones give rise to the shifts listed in Table 75. If the examples of methyl groups adjacent to the carbonyl are selected, the generalization is that an axial methyl is shielded by 0·2–0·3 ppm, while an equatorial methyl is deshielded by 0·05–0·10 ppm. The pertinent data are presented in Table 76.

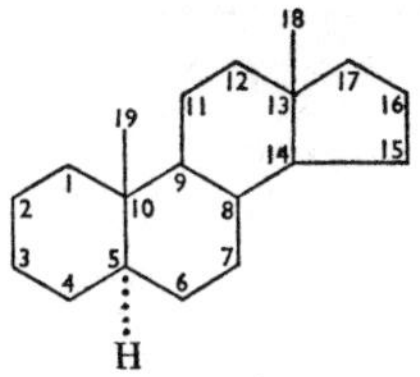

I

TABLE 75. ASIS FOR SOME 5α-ANDROSTANONES

Keto group at position	ASIS (CDCl₃, C₆H₆)	
	Me-19 (ppm)	Me-18 (ppm)
1	0·30	0·00
2	0·16	0·06
3	0·37	0·10
6	0·12	0·06
7	0·32	0·08
11	−0·14	0·11
12	0·24	0·26
15	0·10	0·22
16	0·15	0·32
17	0·12	0·22
5α-14β-androstan-15-one	0·01	0·25
5α-pregnan-20-one (II)	0·08	0·03

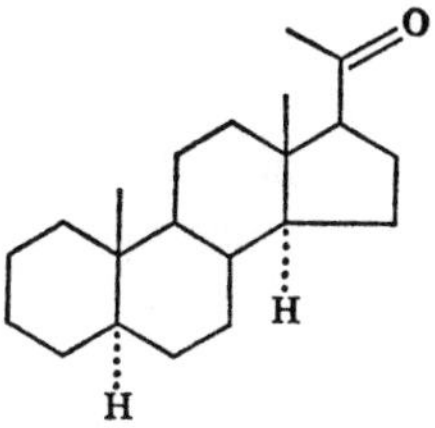

II

13.2. *The Additivity of* ASIS

A very remarkable property of ASIS, also discovered by Williams and Bhacca[286] in the cases of steroids plus benzene, is their additivity. For instance, the benzene shifts incurred by steroidal di- and polyketones can be calculated as the sum of the separate shifts of the 18-Me and 19-Me resonances due to the isolated carbonyl groups of the monoketones (see Table 77). Fétizon and Gramain[287] have used another approach. These authors postulate that the resonances of the 18- and 19-methyl groups in a polysubstituted steroidal ketone, or, more generally, in a polysubstituted

TABLE 76. ASIS FOR METHYL GROUPS ADJOINING THE CARBONYL (HEXATOMIC RINGS)

Compound:	III	IV	V	VI	VII	VIII	
	0·30	—	—	—	0·23	0·12	$\Delta^{CDCl_3}_{C_6H_6}$ axial α-Me
	—	—	−0·06	−0·07	−0·10	−0·07	$\Delta^{CDCl_3}_{C_6H_6}$ equatorial α-Me

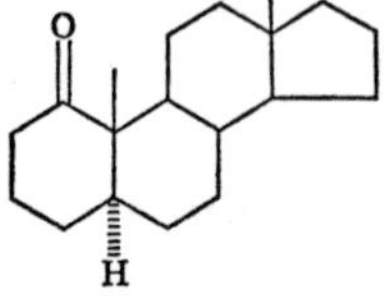

III 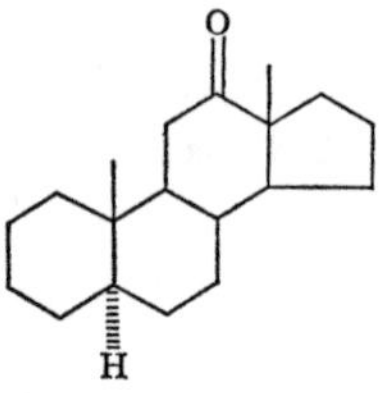   IV 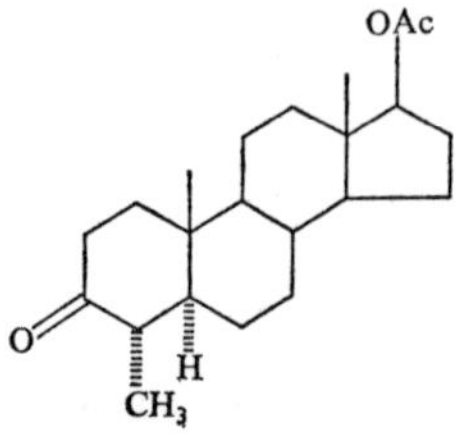   V

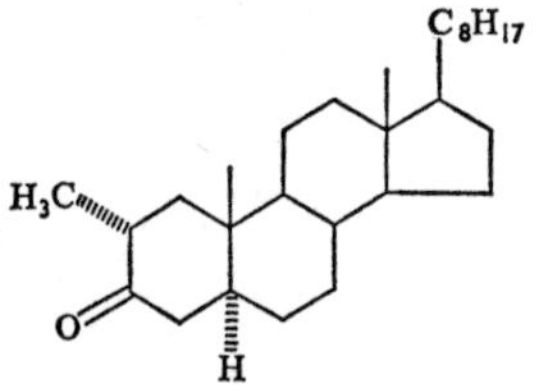

VI 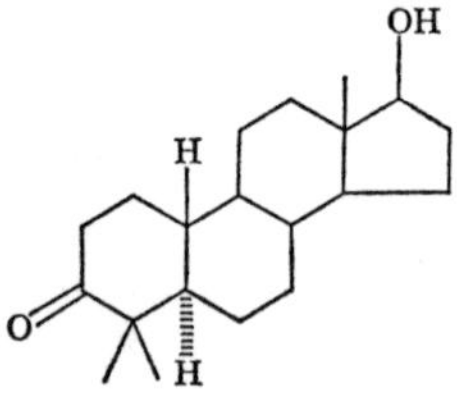   VII 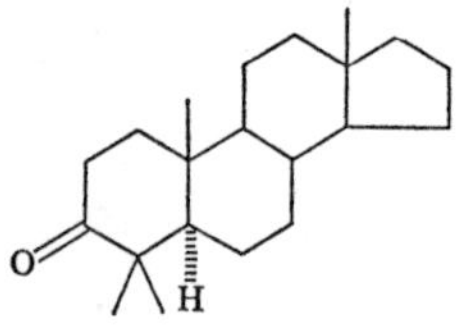   VIII

steroid, can be calculated *in any solvent including benzene* as a sum of increments in the manner proposed by Zürcher.[288-90] They then proceed to
demonstrate that this is indeed so for benzene, for which the following
increments are found, and compared with the corresponding constants for
deuterochloroform (see Table 78). It is therefore obvious that the additivity
of shifts, between deuterochloroform and benzene, reported by Williams
and Bhacca[286] for steroidal polyketones, is obtained *if and only if* the
18-Me and 19-Me resonances of the hydrocarbon skeleton are identical in
the reference ("inactive") and in the aromatic ("active") solvent. This is not
always the case, for there exists in the 13α-Me series a 3·6 c/s discrepancy[287]
between the skeleton resonance of Me-18 in deuterochloroform and in
benzene. In the 5α series, where Williams and Bhacca[286] had observed the
additivity of solvent shifts, the skeletal shifts are indeed the same, as these
authors had the lucidity and caution to check carefully.[291] According
to this argument, the Δ value is not the difference in the *total* chemical shift,
for a given solute's proton(s) between deuterochloroform and benzene, but
the difference, identical or not to the former, between the *incremental* shifts
for the keto group in a given position in deuterochloroform and in benzene.
Evidently, these are complementary viewpoints, with the latter carefully

TABLE 77. BENZENE SHIFTS FOR STEROIDAL DIKETONES

| Compound | | (CDCl$_3$, C$_6$H$_6$), ppm. | |
		ASIS obs	ASIS calc
I	19—Me	0·02	0·02
	18—Me	0·17	0·17
	1—H	−0·29	−0·30
II	19—Me	0·23	0·23
	18—Me	0·20	0·21
III	19—Me	0·49	0·49
	18—Me	0·29	0·32
IV	19—Me	0·45	0·45
	18—Me	0·10	0·13
	19—Me	0·20	0·22
V	18—Me	0·40	0·44

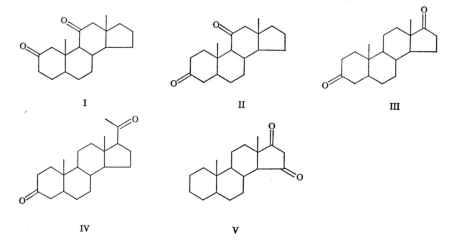

I   II   III

IV   V

avoiding to discuss, or even inquire into, the origin of the benzene shifts. It should be nevertheless noted at this point that Zürcher himself[292] has provided an interpretation of the influence of substituents upon the chemical shifts of the 18-Me and 19-Me protons. He could very successfully relate the chemical shift increments of the tertiary methyl protons, in deuterochloroform, at each position of the steroid frame, with the algebraic sum of an electric and a magnetic term, the anisotropic susceptibility and the dipole moment, respectively, for the preferred staggered conformation of the methyl

TABLE 78. SUBSTITUENT EFFECTS ON 18-Me AND 19-Me
PROTON CHEMICAL SHIFTS[a]

Substituents	CDCl₃		C₆H₆	
	18-Me	19-Me	18-Me	19-Me
17-keto	−6·0	+5·1	+8·4	+9·6
3β-OH	−0·4	−2·0	+0·5	+3·2
3β-OAc	−0·5	−3·0	+1	+6·2
3-keto	−1·3	−14·5	+1·2	+7·7

[a] c/s at 60 Mc/s, to be substracted from 13α-androstane values: 52·0 (18-Me) and 42·7 (19-Me) in CDCl₃; 55·6 (18-Me) and 42·6 (19-Me) in C₆H₆.

groups. The electric field term is the larger of the two. Since, according to these ideas, the magnetic susceptibility term is fairly independent of solvent, and the electric (then reaction field) term is not expected to vary much between deuterochloroform and benzene (which have rather similar dielectric constants) one is forcefully led again to the concept of, for example a dipole-induced dipole association of solute and solvent.

The great merit of the treatment of Fétizon and Gramain[287] is the recognition that the additivity of ASIS exists whatever the nature of the polar substituents introduced into the steroid frame, and need not be confined to the sole keto groups. In this particular and rather restricted case of the steroids, additivity is just a consequence of the "equivalent position" notion, which had been fruitfully introduced by Zürcher.[288-90] Nevertheless, *there are other examples of such an additivity*, when a molecule containing several polar groups, even as heteroatoms, is dissolved in benzene. The diazines are one such example, the observed ASIS for pyridine, pyridazine, pyrimidine, and pyrazine (ppm; carbon tetrachloride, benzene), from the determination of Murrell and Gil[84] are diagrammatically given in Table 79, and compared with the ASIS calculated from pyridine. Here independent "contributions" of both nitrogen heteroatoms to the ASIS of the ring protons are assumed. This comparison is all the more valid since the steric factor is kept constant in these planar, unsubstituted molecules. Murrell and Gil[84] do not comment upon this interesting phenomenon, which may have gone unnoticed by them because of their emphasis upon the values of the internal chemical shifts H₄—H₃, etc., from solvent to solvent. This property of the cyclazines is not without analogy, for the author was able to calculate with good accuracy the ¹³C—H coupling constants for these molecules as the sum of two independent parameters, an angular term dependent upon the endocyclic interatomic angle, and a polar term which had substantial value only for the positions adjacent to (ortho) the nitrogen—or, more generally for all the

TABLE 79. OBSERVED AND CALCULATED VALUES OF THE ASIS FOR THE CYCLAZINES

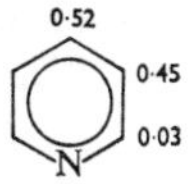

0·52
0·45
0·03

Calculated	Observed	Δ
0·97	1·10	+0.13
0·48	0·43	−0.05
0·90	0·93	+0.03
0·55	0·50	−0.05
0·06	−0·08	−0.14
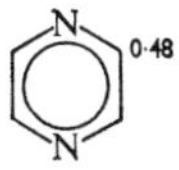 0·48	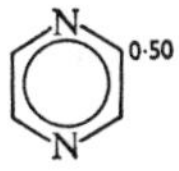 0·50	+0.02

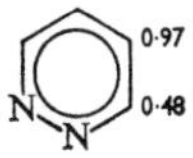

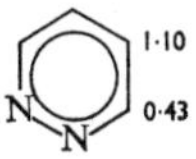

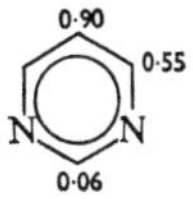

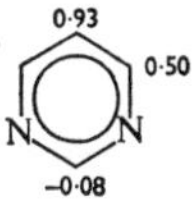

aromatic heterocycles, the heteroatom.[194] This comparison cannot be stretched very far, however, since there exist substantial ASIS in the meta and para, and not only ortho, positions for pyridine itself, in a way reminiscent of the earlier, and probably erroneous[194] increments proposed by Malinowski et al.[293] Nevertheless, this parallel suggests very nearly identical electronic distributions around nitrogen for the cyclazines, in marked contrast with the guesses made in a number of theoretical calculations.[294] A related observation was made by Coppens, Nasielski and Sprecher[148] for the hydrogen bonding of methanol with the monoaza-aromatics (see Section 6). However, there is a marked contrast between the results of Williams and Bhacca[286] or Fétizon and Gramain,[287] for the steroid series, with those of Diehl[90, 295-6] for the aromatic series, where additivity is *also* observed for the effect of substituents, as well as for the effect of solvents. Both types of results may well be correlated along with those of Murrell and Gil,[84] for the diazines, where only the additivity of ASIS can be shown to be approximately true, there being no additivity of substituents, even in the crudest sense: it is well known that the chemical shifts in the cyclazine series represent a complex and unexplained phenomenon.[294]

Diehl[90, 295] has also reported upon the additivity of solvent effects for substituted benzenes. This additivity is obtained either for acetone or benzene, but we shall concentrate here on the latter case. It means that the total ASIS at a certain position of a substituted benzene is simply the sum of the individual ASIS relative to each substituent. For example, in the case of the 1,3-disubstituted benzene VI, the solvent shifts (relative to an internal reference) for the 2 and 5 protons, are in the Diehl formalism:

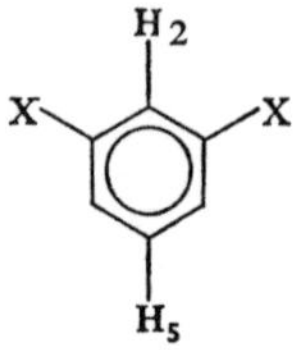

$$L_2 = 2L_{o,\mathrm{X}}$$
$$L_5 = 2L_{m,\mathrm{X}}$$

VI

Similarly, for the symmetrical *para*-disubstituted benzene VII:

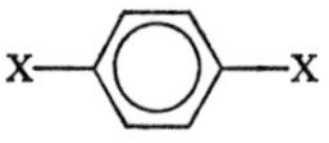

$$L = L_{o,\mathrm{X}} + L_{m,\mathrm{X}} = \tfrac{1}{2}(L_2 + L_5) \text{ (VI)}$$

VII

Therefore, additivity leads to the prediction of the ASIS for the para-disubstituted molecule VII as the average of the ASIS at positions 2 and 5 for the meta-disubstituted molecule VI. This is realized with an excellent approximation, as shown in Table 80 for a few examples. The individual

TABLE 80. PREDICTION OF ASIS FOR MOLECULES VII FROM THE ASIS OF MOLECULES VI

Substituents X,X	Calcd (ppm)	Obsd (ppm)
Br, Br	$+0.25$	$+0.34$
Cl, Cl	$+0.33$	$+0.42$
OMe, OMe	$+0.40_5$	$+0.39$
NO_2, NO_2	-0.40	-0.36
COOMe, COOMe	-0.84_5	-0.81

values of the L_X parameters are given in Table 81. For instance, one calculates the following ASIS for 1-nitro-3-chlorobenzene: 0.35 (H-2); 0.87 (H-5); 0.52 (H-6). The observed values are: 0.31 (H-2); 0.84 (H-5); 0.50 (H-6). There are departures from additivity in ortho-disubstituted benzenes. For 1,2-dichloro-benzene and 1,2,3-trichlorobenzene, the difference between the calculated and observed values can amount to as much as 30% of the total ASIS. This is to be attributed to the interaction between the substituents.

This additivity of the benzene shifts in the substituted benzene series is

TABLE 81. INCREMENTAL ASIS IN THE BENZENE SERIES (ppm)

Substituent X	L_o	L_m	L_p
NH$_2$	+0·07	−0·10	
OH	+0·16	+0·10	
OMe	−0·07	−0·05	−0·07
F	+0·06	+0·23	
Me	−0·02	−0·05	
Cl	+0·09	+0·26	+0·26
Br	+0·09	+0·27	+0·27
I	+0·09	+0·09	
CN	+0·37	+0·53	
NO$_2$	+0·26	+0·61	

to be compared with the additivity of substituent shifts in the same series, also established by Diehl.[296] It should be clear that the former results from spectral comparison of a same molecule between the hexane and benzene solutions, while the latter is apparent when the spectra of molecules differently substituted are compared, in hexane solution. Evidently, and this recalls the remarks of Fétizon and Gramain[287] formulated for steroids, the implications are that substituent effects would also be additive in benzene, so that both viewpoints are in fact complementary, and equivalent in a purely formal way.

13.3. Conformational Analysis

In two papers,[297–8] we have studied the ASIS for cyclohexanones, methyl-substituted in the α, β and γ positions. The variations can be interpreted as generated by a conformational change since the results are self-coherent within each series of compounds. However, it should be pointed out that these Δ also depend upon the ease of formation of the postulated 1 : 1 complex (cf. the following paragraphs) between the solute and a benzene molecule. And this in turn depends on the steric hindrance of the solute and therefore on its degree of substitution. The results are summarized in Table 82. The observed ASIS ($\Delta_{C_6H_6}^{CCl_4}$, $\Delta_{C_6H_6}^{CDCl_3}$) are well consistent with those of Bhacca and Williams[284, 299] obtained in the steroid field. As already stressed (Section 8), use of the hydrogen bonding chloroform solvent leads to a $\Delta_{C_6H_6}^{CDCl_3}$ greater than $\Delta_{C_6H_6}^{CCl_4}$ by 1–5 c/s for protons α to an hydrogen bonding site, such as the carbonyl group for all the cyclohexanones discussed, and as the oxygen heteroatom in the tetrahydropyranone series XI to XIV. Whenever possible, carbon tetrachloride or cyclohexane ought to be preferred as "inert" reference solvents for such studies.

We have indicated[297] a flattened chair conformation for molecule IV on the basis of the deviation of the ASIS (1·8 c/s) from the value calculated

(5·4 c/s) assuming interconversion between true chair forms, as in II; the incremental ASIS for an axial methyl group (15·7 c/s) and an equatorial methyl group (−4·8 c/s) were from the observation on III which exists in a practically non-deformed chair form. In a similar way, it could be shown that I exists essentially as an equatorial conformer. This treatment cannot be extended identically to molecules V to XII; these are more liable to experimental error, since the experimental ASIS are now of the order of 22 c/s (equatorial methyl) and 12 c/s (axial methyl), which differ much less than the above values for the α position. Nevertheless, they are consistent with a slightly deformed chair for VII, and a flattened chair for VIII, whose ASIS

TABLE 82. ASIS FOR METHYL SUBSTITUTED CYCLOHEXANONES AND TETRAHYDROPYRANONES

Molecule	Proton group	$\Delta_{C_6H_6}^{CCl_4}$	$\Delta_{CCl_4}^{CDCl_3}$
		(c/s at 60 Mc/s)	
I	Me(d)	−1·3	4·0
II	Me(s)	5·2	2·9
III	Me(d): *eq.*	−4·0	3·2
	Me(s): *eq.*	−5·7	3·4
	Me(s): *ax.*	15·7	2·3
IV	Me(s)	1·8	3·0
V	Me(t)	21·7	−0·5
VI	Me(s)	16·7	−0·8
VII	Me(t): *eq.*	18·4	0·1
	Me(s): *eq.*	19·3	0·1
	Me(s): *ax.*	11·6	0·2
VIII	Me(s)	13·6	−0·3
IX	Me(d)	21·4	0·0
X	Me(s)	23·0	0·1
XI	Me(d)	21·1	4·0
XII	Me(s)	15·8	3·2
XIII	Me(d)	15·1	2·3
	Me(s)	13·0	4·5
	Me(s)	12·7	3·5
XIV	Me(s)	14·4	0·9

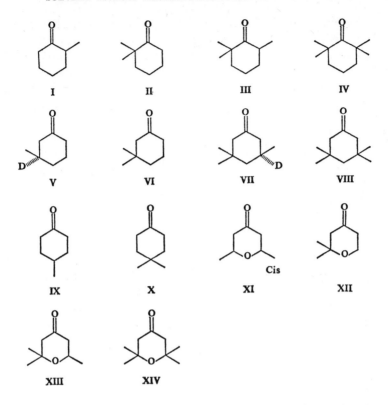

(13·6 c/s) deviates appreciably from the expected value for a chair (17 c/s, as in VI). The molecules VIII and XIV have probably twist chair conformations. If the precision of this method, which cannot possibly be a set of accidental coincidences, is shown by further studies to be sufficient in standardized conditions, it may be of use for conformational investigations.

13.4. *On the Origin of the* ASIS *Phenomenon*

In the preceding paragraphs it has been shown that, by selecting rigid polar solutes like steroidal ketones, the chemical shift modifications are quite characteristic of the position of the affected proton in the molecule. Application of these results to mobile systems provided a new method of conformational analysis.

It is now necessary to investigate the factors responsible for this phenomenon, and to discuss the appropriate experiments. These can be characterized *a priori* as ambiguous, because of the circularity of the argument. As we shall see, a 1 : 1 specific association is the simplest to imagine, and it is a very convenient *model* to consider as a gross first approximation. The features of this model, which ought to make possible a check on its real existence, are:

(a) Its geometry, by which we mean the relative orientation of the solute and solvent molecules in the postulated complex. Obviously, this construction which is meant to accommodate the experimental data presupposes what it is unable to demonstrate, the existence in solution of a 1:1 complex with a definite geometry.

(b) Similarly, the stoichiometry of the complex is important, since dilution curve experiments could be expected to show whether the observed dependence fits better a 1:1 or a 2:1 or a 1:2, or more generally a *m* : *n* association between *m* molecules of solute and *n* molecules of solvent. The mathematical treatment then postulates, in the absence of any knowledge concerning the chemical shift for the pure complex, that it is independent of the concentration, and this hypothesis cannot be justified.

(c) The enthalpy of formation, as well as the entropy of formation of the complex, can be "measured" by variable temperature techniques, plotting log K at a given temperature as a linear function of $1/T$, whose slope and intercept readily yield the desired information. The assumption here is that of the stoichiometry, which will be believed for instance to correspond to the 1:1 interaction.

(d) Finally, the time factor is an important consideration, and we shall discuss upper and lower limits for the duration of such associations between polar solutes and aromatic solvents. At this point, one question has been and may be asked: what is the use? One may also ask whether this question is legitimate. Science proceeds through the elaboration of appropriate models. The function of a model should be distinguished from the function of a theory. A theory is comprehensive, synthetic, by nature while a model is the intellectual isolation of a concept which by the process of analogy is compared with another phenomenon out of the realm of the physical sciences. A model does not serve to the understanding of a phenomenon, which is the function of theory only. The function of a model is threefold, namely to suggest new experiments, i.e. to make testable predictions, to enable the usage of an appropriate mathematical tool for their treatment, and finally to have didactic value. As far as the ASIS phenomenon is concerned, no theory is yet available, so that the approach either is through a model, or altogether empirical. A choice between the two is a matter of intellectual honesty, and use of a model at least allows one to record the steps of the argument. In this particular case of the ASIS, we shall first examine carefully the implications of the 1:1 complex model, before concluding upon the merits of that approach or of a completely empirical approach.

The reader further interested in these methodological problems might care to look at references 300–2.

13.5. *The Steric Effect*

To test the existence of a $1:1$ complex between a polar solute and an aromatic solvent, it has been proposed that hindrance of this association by steric congestion of either member of the pair would be expected to lead to decreased solvent shifts.

Diehl,[90] in his excellent pioneering account of benzene solvent shifts, has established that there is a reduction in their magnitude by a factor of about 0·7 upon the ring protons meta or para to the substituent in mesitylenes and durenes, compared with the monosubstituted benzenes (see Table 83). The

TABLE 83. BENZENE SOLVENT EFFECTS (L') FOR THE META POSITION OF I COMPARED WITH THE META POSITION OF II

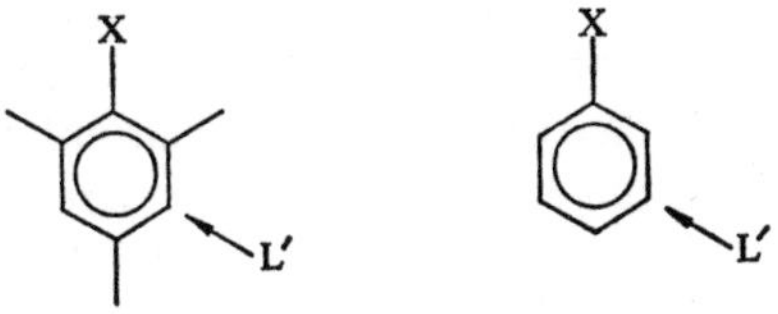

Substituent X	L' (I)	L' (II)
NO$_2$	+37	+61
F	+10	+23
Cl	+13	+26
Br	+13	+27
I	+14	+26
OH	−3	+10
H	−5	0
NH	−8	−10

qualitative explanation of this result, when the steric bulk of the *solute* is increased, is that the solvent benzene molecule may not approach the influenced protons as closely as in the monosubstituted benzene.

One may doubt whether another case of steric hindrance of the solute can be established from the results of Murrell and Gil[84] upon the methyl pyridines (see Table 84).

In our opinion, since the magnitudes of the ASIS are all decreased by the

TABLE 84. ASIS FOR THE METHYL PYRIDINES (CCl₄, C₆H₆). ppm

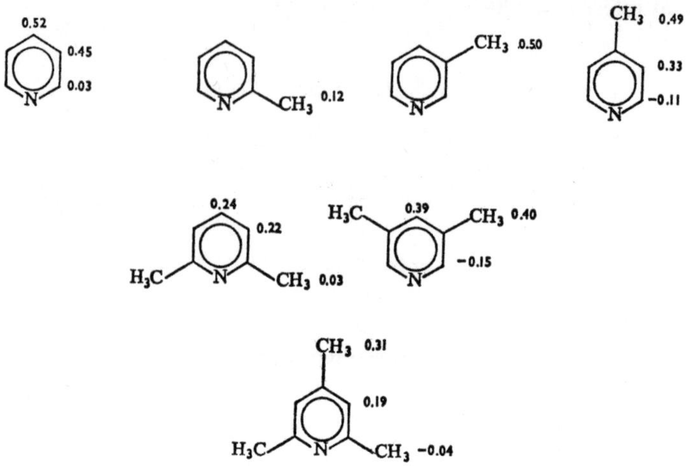

introduction of the methyl substituents (which simultaneously may reflect the reduced electron acceptor demands of nitrogen towards the solvent benzene molecules in a charge-transfer association) it is readily apparent that the simple steric argument is not tenable here. We shall now discuss examples where methyl groups are substituted into benzene, the electron donor molecule, thus increasing this particular feature, and where, by a definite paradox, reductions of the ASIS are again observed, and are not to be attributed to steric compression in the complex. For instance, Hatton and Richards[303] obtained the sequence of solvent shifts for dimethyl formamide (III) shown in Table 85. Their interpretation of these results favoured a collision complex, since a steric effect is operative.

TABLE 85. SOLVENT SHIFTS FOR DIMETHYL FORMAMIDE

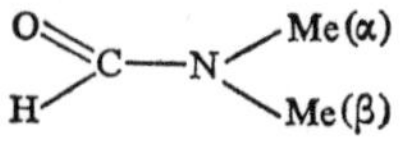

III

Solvent	Molar vol. of solvent M/d	Me(α)-Me(β), c/s
Benzene	90·12	20·1
Toluene	106·3	15·6
p-Xylene	123·3	15·1
t-Butylbenzene	154·9	14·5
Mesitylene	139·0	12·9

Brown and Stark[304] have also noted the decrease of the ASIS for the organometallic examples they studied, in the sequence:

benzene > toluene > mesitylene (see Table 86).

TABLE 86. SOLVENT SHIFTS FOR A NUMBER OF HALIDES

Compound	ASIS (CCl_4, ppm/TMS)		
	Benzene	Toluene	Mesitylene
$MeSnCl_3$	1·432	1·327	1·126
Me_2SnCl_2	0·825	0·763	0·635
Me_3SnCl	0·402	0·368	0·312
$MeSnBr_3$	1·228	1·118	1·042
$MeSnI_3$	1·02	0·90	0·86
Me_4Sn	0·091	0·032	0
$MeSiCl_3$	0·74	0·67	0·59
Me_2SiCl_2	0·396	0·367	0·311
Me_3SiCl	0·244	0·224	0·201
$MeCCl_3$	0·594	0·588	0·502
Me_2CCl_2	0·407	0·350	0·290
Me_3CCl	0·226	0·224	0·208
Me_4C	0·018	0·018	0

Brown and Stark[304] have remarked upon the contradiction between their results (Table 86) and the magnitude of equilibrium constants which, for many solutes, increase in the order:

benzene < toluene < mesitylene;[305]

they ascribe this to predominant dipole induced–dipole interaction, diminished in toluene and in mesitylene by "steric hindrance which impedes simultaneous packing of solvent molecules about a particular solute". Another related observation is that of Ledaal,[306] the shielding of various aromatic solvents on acetonitrile going from 0·72 ppm in benzene, down to 0·47 ppm in mesitylene, through 0·75 and 0·51 ppm, in toluene and m-xylene, respectively. Again, there is a substantial reduction of the solvent shift with increasing solvent bulk.

This steric hindrance interpretation is in fact erroneous. To start with a notorious counter-example, and to quote from the work of Connolly and McCrindle:[307] "In the ketone IV derived from caryophyllene alcohol, the geminal methyls are behind the plane but well above the carbonyl group. From models it appears that the methyl groups strongly hinder the approach of benzene molecules to the rear of the carbonyl in the formation of a preferred collision complex of the type previously postulated. The same is true of the geminal methyls in camphor. In spite of this, the methyls show appreciable

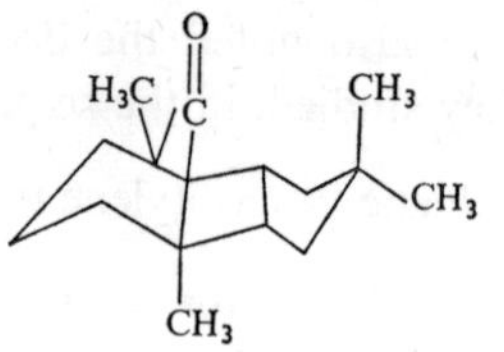

IV

upfield shifts in both cases". It was also somewhat surprising to find benzene shifts for simple methyl cyclohexanones[297-8] almost identical to those established by Williams and Bhacca[284, 286, 299, 308] for the much more hindered steroidal ketones they studied. The solvent shift appears to be characteristic of the precise geometrical position of a given proton, or methyl group, with respect to the carbonyl, irrespective of the steric environment of the former or the latter.

There are also examples where the ASIS are slightly increased with the donor aptitude of the solvent towards polar solutes, as the benzaldehydes studied by Klinck and Stothers,[309] which resemble the benzalmalononitriles examples of Weinberger, Heggie and Holmes[310] (see Table 87). In these, no steric effect is evident, polar factors appear to predominate.

TABLE 87. PROTON SHIFTS OF SOME PARA-SUBSTITUTED BENZALDEHYDES IN AROMATIC SOLVENTS[a]

Solvent	Formyl protons (para-substituent)				Methyl protons NMe$_2$
	NO$_2$	CHO	H	NMe$_2$	
CDCl$_3$	611·2	608·8	600·8	583·8	183·5
Benzene	563·4	573·1	579·3	589·9	142·0
Toluene[b]	562·4	570·4	576·2	584·5	144·2
Mesitylene	561·1	566·6	572·2	579·0	150·3
1,3,5-tri-i-Propylbenzene	569·8	—	578·9	579·2	157·5

[a] c/s from TMS at 60 Mc/s; concn. 5 mole %.
[b] Or toluene-d$_8$.

It was therefore deemed worthwhile to inquire into the possibility of other factors contributing to the experimental observations: Weimer and Prausnitz[311] have measured the equilibrium constants for the complexes between aromatic hydrocarbons and aliphatic ketones, by the classical Benesi–Hildebrand[96, 312] ultraviolet spectrophotometry procedure, the data being reduced with Drago's equation[313] and obtained the results shown in Table 88 for acetone, nitroethane and nitromethane; the NMR method of Hanna and Ashbaugh[108] has been used by Foster and Fyfe[107] to obtain the

TABLE 88. EQUILIBRIUM CONSTANTS FOR CHARGE-TRANSFER COMPLEXES OF POLAR ACCEPTORS WITH POLYMETHYLBENZENES (M^{-1})

Acceptor	Benzene	Toluene	p-Xylene	Mesitylene
Acetone[a]	0.07 ± 0.04	—	0.25 ± 0.06	0.31 ± 0.20
Nitroethane[a]	0.020 ± 0.010	0.022 ± 0.010	0.054 ± 0.012	0.093 ± 0.020
Nitromethane[a]	0.016 ± 0.010	—	0.055 ± 0.010	0.048 ± 0.020
p-Dinitrobenzene[b]	0.16_8	0.17_0	0.21_9	0.25_9
s-Trinitro-benzene[b]	0.31_5	0.42_3	0.57_7	0.79_7

[a] At 25°C, in n-hexane.
[b] At 33.5°C, in carbon tetrachloride.

complementary data for *sym*-dinitro- and trinitro-benzene, with zero permanent dipole moments (see Table 88). Despite the experimental imprecision, it is evident that the strength of the complex increases:

(a) in the inverse order of the NMR shifts of Hatton and Richards,[303] or of Ledaal;[306]

(b) but, as expected, with the increasing *donor* ability of the aromatic, in the sequence: benzene < toluene < *p*-xylene < mesitylene.

The apparent contradiction has been pointed out and studied in detail by Sandoval and Hanna,[314] who have reinvestigated by NMR the molecular complexes of dimethyl formamide with aromatic donors, the subject of the previous study by Hatton and Richards.[303] It was indeed found (Table 89)

TABLE 89. EQUILIBRIUM QUOTIENT Q FOR THE COMPLEXES OF DIMETHYL FORMAMIDE WITH POLYMETHYLBENZENES

Donor	Max Δ_α^i, c/s	Max Δ_β, c/s	Q_α, kg mole^{-1}	Q_β, kg mole^{-1}
Benzene	30.5	13.6	0.126	0.130
Toluene	27.8	13.5	0.136	0.140
p-Xylene	26.0	15.3	0.202	0.20
Mesitylene	25.7	13.0	0.21	0.21
Durene	18.6	10.6	0.28	0.28

that the equilibrium quotient Q (it reflects, but is not identical to, the equilibrium constant K which demands knowledge of the respective activities), obtained according to the procedure of Hanna and Ashbaugh,[108] is very nearly the same if either the solvent shift of the α-Me or the β-Me is used for computation. This calibrates the method, then used to demonstrate a marked *increase* of Q with the donor force of the methyl-substituted aromatic compounds. It is found that the infinite dilution shift in the pure complex is decreased, and that not only does it not reflect a concomitant decrease of the equilibrium quotient, caused by the steric bulkiness of the solvent molecules, but that, on the contrary, this association is normally strengthened by the additional methyl substituents. At this point, speculation upon the origin of this phenomenon is not on firm ground, since the origin of the parent phenomenon, the ASIS, is not really known with precision. It can be, as Diehl[90] pointed out by way of commenting on his results for the substituted mesitylenes and durenes, that their molecular dipole moment is decreased compared with that of the substituted benzenes; but this effect would account for only about one-third of the reduction. It may be, as suggested by Sandoval and Hanna,[314] that the ring current of benzene is substantially reduced upon methyl substitution because *any* substituent would cause this electron flow to be decreased;[315] however, the hypothesis of a ring current in benzene itself has been strongly criticized by Musher,[316] and Dailey[317] has indicated the discrepancy between the values of the anisotropy in the diamagnetic susceptibility and the experimental chemical shifts due to ring currents, which are theoretically predicted to be directly proportional. The controversy between Elvidge and Jackman,[318] and Abraham *et al.*[319] has emphasized the difficulty in evaluating the ring current of an aromatic molecule from the ^{1}H resonance shifts.

Another suggestion of Sandoval and Hanna[314] to account for the decreased solvent shift is that of an increased paramagnetic shift as previously established by Hanna and Ashbaugh[108] for the tetracyanoquinone complexes.

One may note finally the analogy with the results of Voigt,[320] who measured the gas-to-solution ultraviolet absorption band shifts of the complexes between tetracyanoethylene and aromatic hydrocarbons: 2·12 (benzene); 1·65 (*p*-xylene); 1·55 (mesitylene); 1·41 (durene); and 0·99 × 10^{-3} cm^{-1} (hexamethylbenzene). This is in reverse order to the strength of these complexes and opposite to the shift predicted by theory, but very closely resembling in magnitude the ratio of the measured heats of formation of these complexes, in carbon tetrachloride at 20°C:[321] respective values of 2·14 and 2·3 are obtained for the ratios of $\Delta\nu$'s and ΔH_f^0's for the benzene–hexamethylbenzene pair, 1·41 and 1·42 for the benzene–durene pair. Nevertheless, this

V

paradoxical reduction of ASIS with increasing electron donor ability of the aromatics is observed only for dipolar molecules. Symmetrical polar solutes, with a zero permanent dipole moment, have a normal behaviour. Such is the case of *para*-benzoquinone (V), studied for the first time by Schneider,[322] who measured an 0·50 ppm ASIS from cyclohexane to benzene (5 mole % solutions), and 0·965 ppm to 1-methyl naphthalene. The nature of the association is believed to be of the charge-transfer type. More recently, Perkampus and Krüger[323] measured the following charge-transfer shifts for this molecule (see Table 90).

TABLE 90. CHARGE-TRANSFER SHIFTS OF QUINONE V WITH AROMATICS[a]

Electron donor	Δ (ppm)	Electron donor	Δ (ppm)
Benzene	+0·06	1,2,4-Trimethylbenzene	+0·09
o-Xylene	+0·10	Durene	+0·13
m-Xylene	+0·09	Hexamethylbenzene	+0·25
Toluene	+0·07	Naphthalene	+0·17
p-Xylene	+0·09	1-Methyl naphthalene	+0·20
2-Methyl naphthalene	+0·20		

[a] These figures are obtained for carbon tetrachloride solutions, with 25 mole % of quinone and 50 mole % of the aromatic compound.

13.6. *The Geometry of the Complex*

At the time these results were obtained, there was general agreement to interpret the empirical observations of this type, when polar solutes are dissolved in benzene, by the existence in solution of a discrete complex between *one* molecule of the polar solute and *one* molecule of benzene. This association would result from the polarization of the π-electrons in the aromatic nucleus by the permanent dipole moment of the dissolved molecule. The geometry generally postulated for these 1:1 complexes recognizes that the solute's dipole moment lies in a plane parallel to that of the aromatic ring, so that in general the planes of both solvent and solute molecules are parallel. The aromatic molecules are twice as polarizable in the molecular plane ($\alpha_{\parallel} = 12\cdot3$ A^3) than normal to it ($\alpha_{\perp} = 6\cdot35$ A^3).[324] The supplementary argument of Brown and Stark[304] that, for the solutes of type $(CH_3)_nMX_{4-n}$ "a closer approach to the solute molecules can be made along the six-fold axis than in the molecular plane, because of the hindrance presented by hydrogens or methyl groups", omits consideration of the actual width of the pi-electronic distribution, which is quite large. In any case, the solute protons tend to be located above the plane of the benzene ring and adjacent to the axis of symmetry. Interelectronic repulsion is efficacious to separate as much as possible the negative end of the solute from the electron rich benzene ring, which means that, since both molecules are constrained to lie in parallel planes, the electron attracting groups of the solute will be anchored away from the six-fold axis of symmetry. An alternative model considers the possibility of charge-transfer complex formation between the polar solute, e.g. a ketone, and the aromatic ring, with its somewhat stringent geometric requirements. The result is similar to the consequences of the dipole induced–dipole model, the preferred family of conformers for the association will avoid the nodal plane of the aromatic nucleus, devoid of electrons for donation, but will instead stick the solute next to the axis of symmetry. The overlap, which necessarily goes with some charge transfer, between the donor and the acceptor molecular orbitals, pulls the aromatic π-electrons in slight excess towards the face of the ring in contact with the acceptor solute, thus making less likely an association through the other face with a second acceptor molecule. The preferred stoichiometry is again therefore that of a 1 : 1 complex.

We shall first discuss in some detail the geometry attributed by Schneider[322] to the acetonitrile–benzene system. The argument is very well conducted, and introduces clearly the different concepts. The initial observation is the 56·8 c/s (at 60 Mc/s) shift to high field that acetonitrile (H_3C—$C{\equiv}N$) undergoes when it is changed from a dilute solution in neopentane (CMe_4) to a dilute solution in benzene, neopentane being the internal reference in both cases (the separate chemical shifts are −49·8 c/s in neopentane, and +7·0 c/s in benzene). It is postulated that neopentane is an inert solvent, and that the non-specific van der Waals interactions of acetonitrile and neopentane will be very weak and comparable to the non-specific van der Waals interactions of acetonitrile with benzene. Now, it must be recognized that benzene is a most remarkable solvent molecule, since it is associated with the property of magnetic anisotropy. This, which can be conceived independently of the ring current model,[316] implies large upfield shifts (shielding effect) for protons situated above the aromatic plane, and near the axis, and large downfield shifts (the so-called deshielding effect) for protons in the plane of the aromatic ring. The ring current picture accounts for these results by the hypothesis of a circular conducting loop, constituted by the distribution of π-electrons, induced into a current by the application of the constant field H_0, whose effect will be to create a secondary magnetic field, collinear with, opposed to, and evidently much smaller than, H_0. The corresponding magnetic lines of force, which reproduce the empirical anisotropy in the chemical shifts, have been mapped, using a semi-empirical method, by Johnson and Bovey:[325] these authors have provided the contours of equi-shielding, in the space surrounding the benzene ring, measured with the cylindrical coordinates ρ and z from its centre. Their calculation is based upon the "ring current" hypothesis, which may have no sound physical basis,[316] but had nevertheless considerable success for elementary intramolecular discussions.[326] The next step of Schneider's argument is to rationalize the large upfield shifts he observes for the methyl resonance by the above model of a dipole induced–dipole force leading to non-randomization of the relative molecular orientation of solute and solvent, with regard to the neopentane reference. He believes that the actual shifts due to the mechanism of complex formation itself are very small, the anisotropy of the solvent serves to amplify them. Because of the very small difference in dielectric functions between neopentane and benzene the reaction field term like the van der Waals term can be neglected. To confirm this reasoning, Schneider[322] also carefully checked that the observed shift does not arise (instead of molecular complex formation) from molecular shape factors, due to the mixing of the disk-shaped benzene with the rod-like acetonitrile. The similar experiments he conducted with non-polar hydrocarbon solutes, either linear like ethane, dimethylacetylene and allene, or planar like ethylene and naphthalene, gave essentially zero shifts between the neopentane and the benzene solutions, demonstrating that the dipolar nature of the solute is a necessary condition. The geometry of the complex between dimethyl formamide (I) and benzene has been studied by Hatton and Richards;[303] they observed unequal shifts for the α-Me and the β-Me, the methyl group labelled α being that trans to the carboxylic

Ia   Ib

oxygen. The upfield ASIS is greater for the α-Me than for the β-Me, the respective values according to Schneider[322] are 46·0 and 17·1 c/s (at 60 Mc/s). The two methyl groups are non-equivalent, diastereotopic in the sense of Mislow and Raban,[232] because of the relatively high barrier to internal rotation in amides arising from contribution of the dipolar form Ib, with a formal negative charge on the oxygen, a positive charge on nitrogen, and partial double-bond character for the carbon–nitrogen bond. This dipolar resonance form Ib is responsible for the preferred orientation of the N,N-dimethylformamide over benzene, the positively charged nitrogen opposing the aromatic electrons. The conformation of the system is not known with more precision, and it cannot be known with more precision, given the information which has been used. As Baldwin[327] has remarked, five parameters are needed to define the coordinates of the associating molecules: two parameters, ρ and z in cylindrical coordinates, locate the barycentre of the solute relative to that of benzene, and three others are necessary to fix the orientation of the molecule about that point. In this case of dimethyl formamide, where there are only two chemical shifts involved (namely the two ASIS) an infinite number of geometries would simultaneously fit the experimental data with the Johnson and Bovey[325] tables for the magnetic anisotropy of the benzene ring; three degrees of freedom remain for the molecule. Similar geometries would be expected for the complexes of other amides with benzene. The ASIS for the methyl group of the *trans* isomer of N-methyl formamide IIa is larger than that for the *cis* isomer IIb, which

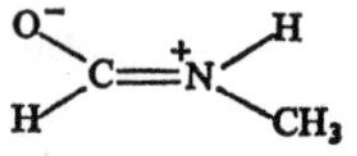

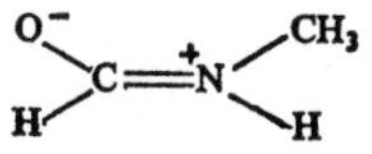

predominates (92%).[328] Hatton and Richards,[329–30] Moriarty and Kliegman[331–2] have performed similar studies for dimethyl acetamide[329] and other N-methyl amides[330] or N-methyl lactams.[331] In the case of the N-methyl lactams, the results are shown diagrammatically in Table 91. It is seen that the geometry for the complex must be at variance with the previous cases of the amides,[303, 328–31] for the ASIS are almost equal for the N–CH_3 and the H_2C—CO protons, and much smaller than for the N—CH_2 protons, in contradistinction with the case of the acetamides.[331] Furthermore, it

TABLE 91. ASIS (CCl_4, C_6H_6: ppm) for the Five- to Nine-membered
N-Methyl Lactams

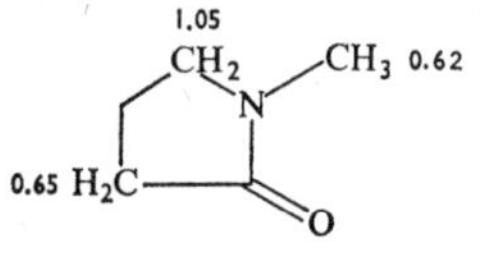

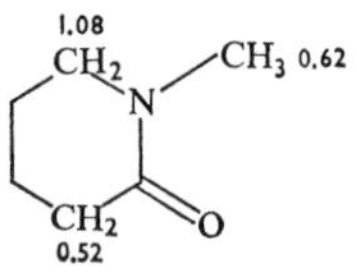

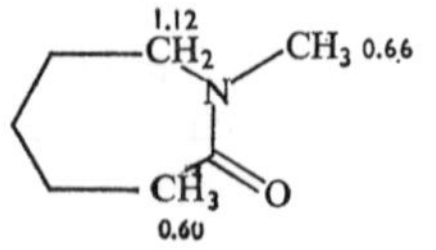

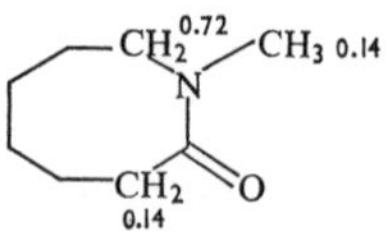

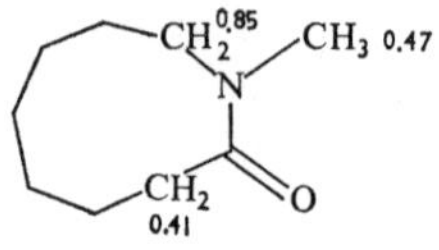

should be noted that there is a loose empirical correlation between the magnitude of the ASIS and the "absolute" chemical shifts of the protons in carbon tetrachloride, which vary between 3·32 and 3·50 ppm (N—CH₂), 2·75 and 2·97 ppm (N—CH₃), and 2·23 to 2·47 ppm (H₂C—CO). It would not be surprising if the electronic properties of the N-methyl lactam solute were to correlate with the stability of its complex with benzene. Moriarty and Kliegman[332] also use the intensities of the two separate N—CH₃ peaks they observe in the eleven-membered and thirteen-membered cases, and which they identify by the observation that the ASIS for the s-*trans* conformation is twice that for the s-*cis* conformation, to evaluate the respective ratios s-*cis*/s-*trans* = 55/45 and 40/60. For all these lactams, the ¹³C—H coupling constant relative to the N-methyl group has the value 137–8 c/s, the same as for N,N-dimethyl acetamide (138 c/s), which is taken to indicate identical positive charges on nitrogen,[332] and also implies identical rotational barriers, according to the correlation established by Neuman and Young.[333] Baldwin[327] has re-assigned the NMR spectrum of mesityl oxide for which a variety of methods (infra-red, ultraviolet, Raman and dipole moment determinations) indicate the s-*cis* conformation IIIa, in conflict with the picture by Hatton and Richards[330] of the s-*trans* conformation IIIb forming a stereospecific complex with benzene.

β—trans   β—cis

IIIa   IIIb

Baldwin[327] used the comparison with the NMR spectra of 1,1,1-trideutero-4-methyl-3-penten-2-one, *cis* and *trans*-3-penten-2-one. According to Timmons' data,[334] the ASIS offer an independent confirmation of the s-*cis* conformation IIIa for mesityl oxide: there is a characteristic difference between the positive ASIS for the resonance of a β-*cis* substituent in the s-*trans* diastereomer, and the negative ASIS for the analogous resonance in the s-*cis* diastereomer, and this second category includes mesityl oxide. There is some discrepancy between the ASIS reported by Hatton and Richards,[330] which are quoted by Baldwin,[327] and the new measurements of Timmons:[334] these are respectively, 0·24 and 0·17 ppm (H₃C—C = O), 0·40 and 0·31 ppm (β-*trans*-CH₃), and 0·00 against 0·03 ppm (β-*cis*-CH₃). This difference is not altogether surprising since, if benzene was the common aromatic solvent, Hatton and Richards[330] used the pure liquid instead of a reference solvent, while the Timmons study is based upon carbon tetra-chloride;[334] as this more recent work indicates, the resonances of mesityl oxide shift by an average of 0·05 and up to 0·13 ppm when the concentration changes from 1% to 25%.

Arriving at the end of this paragraph, the reader may have better understood why we did not follow the customary practice of providing pictures of the solute–benzene association. Nevertheless, such simple pictures have met with astonishing success, given their crude character (cf. important reserva-tions to be made subsequently), as is illustrated in Fig. 25 taken from our work on norbornenes:[335] the solvent shifts are calculated using the Johnson–Bovey[325] tables for cis-endo-dichloronorbornene (IV).

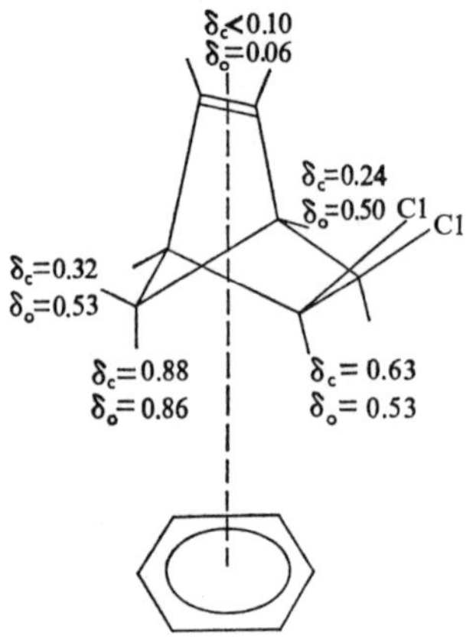

FIG. 25. Suggested geometry for the association between compound IV and benzene, as deduced from the solvent shifts. The observed (δ_o) and calculated (δ_c) chemical shifts refer to CCl_4 solution. (Laszlo and Schleyer.[335])

A similar geometry is proposed by Subramanian, Emerson and LeBel[336] in their more recent study of 5,6-dihalo-2-norbornenes. Williams and Wilson[308] have defined an average position of the aromatic ring, *perpendicular and not parallel*, to the plane of the carbonyl double bond for 11-androstanone, resulting from the steric constraints on the face of such a steroid skeleton.

13.7. Correlation of the ASIS with the Solute Dipole Moment

In 1962 Schneider[322] tested his hypothesis of a dipole induced–dipole mechanism for the ASIS phenomenon, by means of the expected linear dependence between the ASIS for polar solutes of the type CH_3X and their dipole moment μ, and corrected for the molecular volume V to allow for the varying distance between the centres of gravity of the solute and of the benzene molecule. The simplest function of this type is the ratio μ/V. The plot obtained by Schneider[322] indeed shows some degree of correlation, with the notable exception of chloroform, due to hydrogen bonding with the aromatic π electrons. This procedure hinges upon the estimation of the quantity V.

Brown and Stark[304] have encountered much success using this suggestion of Schneider,[322] and developed a fairly accurate technique for measuring molecular dipole moments: they considered solutes for which the molecular volume is fairly large and constant so that they could omit this very term V, which cannot be known with precision. The ASIS which they obtained, together with values of the permanent dipole moment μ, are summarized in Table 92. It is indeed proved that the solutes with zero dipole moments

TABLE 92. CORRELATION OF THE ASIS WITH THE
SOLUTE DIPOLE MOMENT

Compound	ASIS (CCl$_4$, C$_6$H$_6$: ppm)	$\mu(\pm 0.2 \text{ D})$
MeSnCl$_3$	1·432	3·6
MeSnBr$_3$	1·228	3·2
MeHgBr	1·158	3·1
MeHgI	1·104	2·9
MeSnI$_3$	1·02	2·6
MeSiCl$_3$	0·74	1·9
MeCCl$_3$	0·594	1·5
MeI	0·667	1·48
Me$_2$SnCl$_2$	0·825	4·2
Me$_2$SiCl$_2$	0·396	2·3
Me$_2$CCl$_2$	0·407	2·2
Me$_2$Hg	0·153	0
Me$_3$SnCl	0·402	3·5
Me$_3$CCl	0·226	2·2
Me$_3$SiCl	0·244	2·1
Me$_4$Sn	0·091	0
Me$_4$C	0·018	0

suffer negligible ASIS, the quality of this approximation seems to depend upon the atomic polarizabilities, as expected. Three separate correlation lines are obtained for the monomethyl, dimethyl and trimethyl derivatives, with slopes respectively 0·384, 0·188 and 0·115 ppm/D; to explain these results the authors[304] suggest convincingly that the corresponding number of *contacts* (with the solvent, and in a sense similar to that of the theory by

Satake *et al.*;[112] see Section 6.1) is 3·0, 1·5 and 1·0. Accordingly, the slopes of the lines, which pass through the origin, are in the ratio 3·3 : 1·6 : 1·0, a quite remarkable result, as the whole of this empirical method which meets with amazing success, given all the neglected factors such as consideration of the actual molecular sizes, geometries and polarizabilities. Its limits can be appreciated when one considers the example of a smaller, rod-like, solute, acetonitrile, whose ASIS (1·016 ppm) is somewhat smaller than the prediction based upon the value of its dipole moment (3·5 D). As emphasized by Brown and Stark:[304] "this method would possess the singular advantage that it could be applied to the components of mixtures, for estimation of dipole moment, for distinguishing isomers on the basis of polarity, and for other like purposes".

Diehl[90] has successfully correlated the ASIS at the meta position of substituted benzenes with the substituent dipole moment, the slope being −0·14 ppm/D. An equivalent treatment, also proposed by Diehl,[295] relates the same L'_m ASIS with the composite Hammett parameter $\sigma_I + \frac{1}{2}\sigma_R$, with a slope 0·74 ppm/Hammett unit; there is much more scatter when the L'_o ASIS for the ortho position is plotted instead.

This type of relationship appears to be general, and the ASIS reported for MeX (X = CN, NO₂, Br, I) and H₂C=CHX (X = CN, NO₂, Br) by Schneider[322] (Table 93) vary in the same way. In the latter case of the vinyl

TABLE 93. ASIS FOR ALKYL-X, VINYL-X, AND PHENYL-X SOLUTES[a]

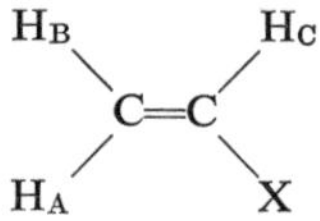

Substituent X	CH₃X	Hₐ	H_B	H_C	C₆H₅X (ortho protons)[b]
CN	0·95	0·70	0·84	0·80	0·37
NO₂	0·94₅	0·62	0·97	0·68	0·26
Br	0·49	0·22	0·42₅	0·37	0·09
I	0·51	—	—	—	0·09

[a] ppm; neopentane[322] or hexane[295] to benzene.
[b] The constant solvent shift due to benzene (0·38 ppm) has been omitted, the values given are the L'_o values.[295]

derivatives, the ASIS for the β-proton *cis* to the electronegative substituent (Hₐ) closely follows the L'_0 values, which are the ASIS reported by Diehl[295] for the ortho protons, in the analogous position of substituted benzene rings. Further research is to be done in this area, where only few results exist.

It should be noted finally that Nakagawa and Fujiwara[337] have also correlated the ASIS (CCl_4/C_6H_6) for the methyl group of meta- and para-substituted toluenes with the Hammett σ constant, and that Seyden-Penne, acting upon a suggestion of the author that the cyano group would give larger ASIS than the carbonyl group, could verify it in the cyclopropane series.[338]

13.8. *Dilution Curve Experiments*

Foster and Fyfe[107] have developed a technique for the determination of the association constants of the organic charge-transfer complexes by nuclear magnetic resonance. Given the equilibrium:

$$\text{acceptor} + \text{donor} \rightleftharpoons \text{complex} \tag{67}$$

the association constant is:

$$K = \frac{(AD)}{(A)(D)} \tag{68}$$

where (A), (D) and (AD) are respectively the concentrations of the acceptor, the donor and the complex. Making the usual assumption that the observed chemical shift δ^A_{obs} is a weighted average of the chemical shifts for molecule A, free, δ^A_0, and in the pure complex, δ^A_{AD}, one gets:

$$\varDelta = \delta^A_{obs} - \delta^A_0 = \frac{(D)}{1 + (D) \cdot K}(\delta^A_{AD} - \delta^A_0) = \frac{(D)}{1 + (D)K} \cdot \varDelta_0 \tag{69}$$

where $\varDelta$ is the observed shift resulting from the association, and $\varDelta_0$ is the chemical shift in the pure complex. It follows that:

$$\frac{1}{\varDelta} = \frac{1}{K\varDelta_0} \cdot \frac{1}{(D)} + \frac{1}{\varDelta_0}, \tag{70}$$

an equation identical to that used by Hanna and Ashbaugh,[108] which it is more convenient to consider in the form:

$$\frac{\varDelta}{(D)} + \varDelta \cdot K = \varDelta_0 \cdot K. \tag{71}$$

It is seen that, if the complex exists with the postulated 1:1 stoichiometry, the plot of $\varDelta/(D)$ against $\varDelta$ should be a straight line which gives the equilibrium constant K directly as the negative gradient. As noted by Fort,[339] the reverse of this statement is not necessarily true, the linear relationship is not a proof of complex formation. This technique has been applied for instance to 5α-androstan-6-one, out of a series of steroidal ketones for which the heat of complex formation with toluene was determined by variable temperature studies (see Section 13.9). A straight line is obtained, with a slope measuring the apparent equilibrium constant $K = 0.20$ M^{-1}, at 33°C, in fair agreement with the results of the alternative temperature method.[340] Tyrrell, in his investigation of complex formation between propargyl chloride and benzene,[341] used an equivalent approach with satisfactory results since the equilibrium constant derived from the methylene protons resonances is $K = 1 \cdot 01 \pm 0 \cdot 02$ M^{-1}, against $K = 0 \cdot 96 \pm 0 \cdot 06$ M^{-1} from independent consideration of the ethynyl proton. Anderson[342] applied the procedure of Tyrrell[341] to his study of dioxanes in benzene solution. Linear plots are often obtained when the ASIS are plotted for a given solute–solvent system vs. their concentration ratio,[339] the mole fraction of the aromatic solvent[298] or solute.[337] These latter two are not always equivalent, Nakagawa and Fujiwara[337] have practised the curious procedure of correcting the mole fraction of the solute by counting one methyl group or one phenyl group of the solute as an entire molecule, so that their mixing curves for acetone in benzene, or *ortho*-toluidine in pyridine vary between 0 and 0·5 mole fraction, instead of the normal 0 to 1 range. Very similar curves are obtained when the chemical shift of the methyl groups for acetone, 3,3-dimethyl cyclohexanone,

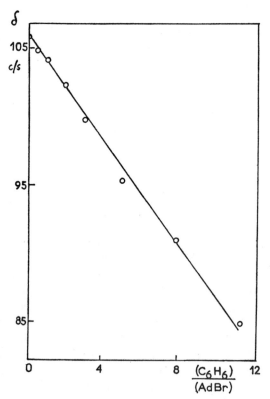

Fig. 26. Titration curve of bromoadamantane by benzene in carbon tetrachloride. (Fort.[339])

and *cis*-2,6-dimethyl tetrahydropyranone is plotted against the mole fraction of benzene : positive deviations (*ca.* 1–5 c/s) from the straight line expected for unassociated ideal binary mixtures[103] occur in each case. The precision of the measurements was not sufficient to apply reliably the curves calculated by Mavel,[103] and to determine the values of *K* by this method. Also it does not necessarily follow from these curves, as already stressed above, that the associations or complexes present in these solutions are similar. The straight line obtained by Fort[339] for bromoadamantane and benzene in carbon tetrachloride solution is shown in Fig. 26. It should be emphasized at this point

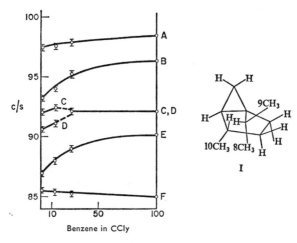

FIG. 27. ¹H Chemical shifts for the individual methyl group bands of compound I in 0·15 M concentration in carbon tetrachloride–benzene mixtures. C, F = 10-methyl; A, D and B, E = 8, 9-methyl groups.

that the most reliable procedure, i.e. the titration curve of the polar solute–aromatic solvent complex in *dilute* solution with an inert solvent, is seldom applicable. The shifts induced become too small to be measured accurately, so that it is more practical to examine the direct mixture of the pure solute with the aromatic donor molecule.

The interest of this technique is considerable for the determination of the structure of natural products. Hashimoto and Tsuda[343] have identified the methyl groups of triterpenes by such a continuous change of solvent from deuterochloroform to benzene, both the maximum shift—*which may occur at some concentration other than infinite dilution*—and the concavity of the mixing curve being characteristic of substituents. In fact the maximum shift is closely related to the ASIS, and the latter to the equilibrium constant. A very illustrative example (Fig. 27) is that of thujane I, in which the doublet components for secondary methyl groups at positions 8, 9 and 10 are paired according to identity of their mixing curves, by Dieffenbacher and Philipsborn.[344] These authors have further studied the ASIS in this hydrocarbon series, and found that it differs very much in magnitude from isomer to isomer.[345]

The conclusion from this Section, as indicated in the introduction to the ASIS phenomenon (section 13.3), is that the dilution experiments cannot demonstrate the existence of a 1:1 complex, but that the apparent association constants thus measured for an assumed complex are in good agreement with independent evidence that these are relatively weak complexes.

13.9. *Variable Temperature Determination of Enthalpy of Formation*

To restate it briefly, the working hypothesis is that of a 1:1 complex, according to the equation:

$$\text{solvent} + \text{ketone}\dagger \rightleftharpoons \text{complex} \qquad (72)$$

† or another polar solute.

The prediction can therefore be made, to test the quality of this model, that by lowering the temperature, the equilibrium (72) would be shifted and the ASIS resulting from complex formation enhanced. In addition, the temperature variation of the equilibrium constant K for the equilibrium (72) would permit an evaluation of the enthalpy of formation of the complex. Obviously, benzene is not a suitable solvent for such variable temperature studies because its melting point is 5°C. For this reason, the experiments which will be described here have been generally carried out in toluene, which produces at ambient temperatures ASIS very similar in magnitude to the ASIS caused by benzene itself. Before deducing from the temperature curves the constant K for the above equilibrium (72) at various temperatures, it is necessary to answer two questions. The first one, that it might be that the temperature dependence of the chemical shifts is an *intra*molecular effect, independent of any association. For this reason, the temperature curves for the 8-Me, 9-Me and 10-Me groups of camphor, in toluene solution, are presented in Fig. 28. This Fig. 28 is to be compared with Fig. 18 where similar curves were presented for camphor in carbon disulphide solution[200]: in the present case, the variations have an opposite sign, the 8-Me and 9-Me resonances being displaced towards a higher field when the temperature is lowered, and they are of a much larger magnitude, suggesting that an association indeed occurs in the toluene case. The second question is relative to the proof for the 1:1 stoichiometry for this association, dealt with in Section 13.8 where we discuss dilution curve experiments. For the calculation of thermodynamic parameters on the basis of an assumed 1:1 complex, Abraham[346] has suggested the following method. If, in a dilute solution, a fraction p of the solute is in the complex form, then equation (73) holds, where K is the equilibrium constant

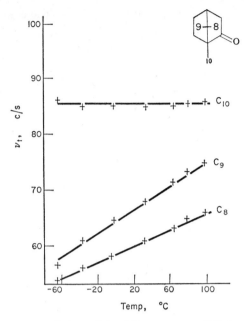

FIG. 28. Temperature variation of the ^1H resonances of the methyl groups of camphor dissolved in toluene. (Laszlo and Williams.[340])

for equation (72), and ΔS and ΔH are the entropy of formation and heat of formation of the complex, respectively:

$$K = \frac{p}{1-p} = \exp\left(\frac{\Delta S}{R}\right) \exp\left(-\frac{\Delta H}{RT}\right). \tag{73}$$

The fraction of solute molecules complexed, p, is given at any temperature, (t) by:

$$p = \frac{\delta_t - \delta_0}{\delta_c - \delta_0} \tag{74}$$

where δ_t = the observed chemical shift at temperature t; δ_0 = the chemical shift of the uncomplexed solute; and δ_c = the proton resonance in the pure complex.

If the position of the resonance in an "inert" solvent such as carbon tetrachloride is taken to give δ_0, and this is arbitrarily taken as zero, then equation (74) simplifies to:

$$p = \frac{\delta_t}{\delta_c}. \tag{75}$$

Estimates of the δ_c quantities can be made by measuring δ_t as a function of temperature, and extrapolating to 0°K. At this point, all of the solute molecules are assumed to be complexed and $\delta_{t=0} = \delta_c$. The equilibrium constant at any temperature is then (equations (73) and (75)):

$$K = \frac{p}{1 - p} \tag{76}$$

and ΔH may be obtained as usual from the slope of a plot of log K against $1/T$. In this procedure, the two assumptions which may be criticized concern δ_0 and δ_c. The concentration dependence of the chemical shift in the inert solvent makes it necessary to use 1–2 mole % solutions, and to check that in this concentration range the dilution shift is less than the error in extrapolation. As for the extrapolated shift δ_c in the pure complex, the good straight line which was obtained by Fort[339] for the t-BuBr/C$_6$H$_5$Cl system is shown in Fig. 29.

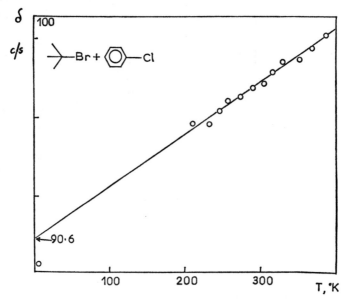

FIG. 29. Extrapolation of the association shift to 0°K. (Fort.[339])

In another example, that of 5α-androstan-11-one (I), studied by Laszlo and Williams,[340] the enthalpy of formation has been determined as $\Delta H = -0.65 \pm 0.15$ kcal mole^{-1}. This result is practically independent of the value of $\delta_c(12\,\alpha\text{-}H) = +120 \pm 20$ c/s, for the values between 110 and 130 c/s, the calculated entropy of formation, on the contrary, being obviously sensitive to this parameter. This example of 11-androstanone (I) is favourable since the solvent shifts of no less than five proton types: 18-Me, 19-Me, 12α-H, 12β-H and 1β-H, variously oriented with respect to the 11-carbonyl

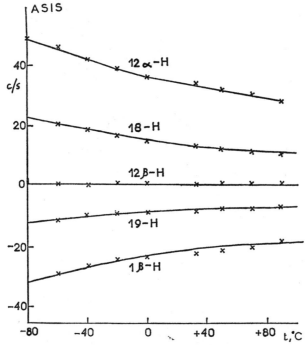

group, can be followed (Fig. 30).The largest toluene ASIS are seen to be relative to the 12α-H, for which δ_c is given above, and for 1β-H ($\delta_c = -85 \pm 15$ c/s). The plot of log K, based on the 1β-proton, is seen to give a line of the same slope as for the 12α-proton (Fig. 31); the best fit of log K versus $1/T$ to a common line for both corresponds to the respective values of $\delta_c = 110$; -70; $120 : -77$; and $130 : -83$, corresponding to the above value of ΔH.

FIG. 30. Temperature effects on ${}^1\mathrm{H}$ chemical shifts for 5α-androstan-11-one in toluene. (Laszlo and Williams.[340])

Despite this good agreement, the results may be partly fortuitous, for the following reasons. The circular dichroism curve of I is temperature dependent.[347] This observation has been explained in terms of a conformational

N

equilibrium in ring A, whereby the interaction between the 1β-proton and the 11-keto group may be relieved. This hypothesis is supported by variable temperature studies on the NMR spectrum of I in carbon tetrachloride solution; the chemical shift of the 1β-H resonance changes from 2·46 ppm at + 65°C to 2·41 ppm at − 20°C, whereas those of the 19-H resonance (0·99 ppm at + 65°C and 0·98 ppm at − 20°C) and the 12-H resonances are only slightly altered. The observation that the chemical shifts of the 19-H, 12α-H and 12β-H resonances do show a *small*, but significant temperature variation in carbon tetrachloride solution is compatible with a very weak interaction between carbon tetrachloride and the solute molecule[342] (see Section 2.4).

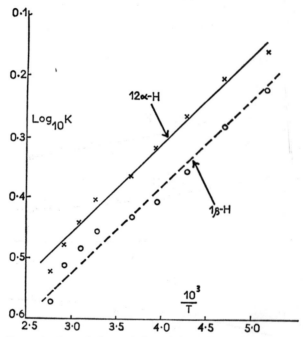

FIG. 31. Determination of the enthalpy of formation of the complex formed between 5α-androstan-11-one and toluene. (Laszlo and Williams.[340])

However, an *intrinsic* effect of temperature is not excluded (see Section 7). The appreciable change in the chemical shift of the 1β-H resonance with temperature *in an "inert" solvent* suggests that the average orientation of the 1β-proton with respect to the 11-keto group changes with temperature, i.e. ring A is not conformationally homogeneous in I. Nevertheless, the notion that this technique of Abraham[346] is very reliable is reinforced by the fact that values of ΔH derived from different ASIS for protons at various positions of the solute do converge. Another example is provided by the adamantyl halides studied by Fort[339] where the γ and δ resonances separately lead to the same value of the enthalpy of formation. The values of the enthalpy of formation of the postulated 1:1 complex for a variety of solutes and solvents are summarized in Table 94.

TABLE 94. ENTHALPY OF FORMATION FOR THE COMPLEX BETWEEN POLAR SOLUTES AND AROMATIC SOLVENTS

Solute[a]	Solvent	$-\Delta H°$, kcal mole^{-1}	Ref.
CH_3I	Toluene	$1·3 \pm 0·5$	346
CHI_3	Toluene	$1·6 \pm 0·2$	346
Pyrimidine	Toluene	$1·2 \pm 0·4$	84
Acetonitrile	Toluene	1–2	348
p-Benzoquinone	Toluene	1–2	348
Me_2NCHO	Toluene	1–2	348
Molecule I	Toluene	$0·65 \pm 0·15$	340
t-BuCl	Benzene	$1·19 \pm 0·25$	339
t-BuBr	Benzene	$1·23 \pm 0·25$	339
t-BuI	Benzene	$1·00 \pm 0·25$	339
AdCl	Benzene	$1·09 \pm 0·25$	339
AdBr	Benzene	$0·82 \pm 0·25$	339
AdI	Benzene	$0·92 \pm 0·25$	339
t-BuCl	PhF	$0·86 \pm 0·25$	339
t-BuBr	PhF	$0·90 \pm 0·25$	339
t-BuI	PhF	$0·88 \pm 0·25$	339
AdCl	PhF	$1·07 \pm 0·25$	339
AdBr	PhF	$0·96 \pm 0·25$	339
AdI	PhF	$1·01 \pm 0·25$	339
t-BuCl	PhCl	$0·93 \pm 0·25$	339
t-BuBr	PhCl	$0·83 \pm 0·25$	339
t-BuI	PhCl	$0·84 \pm 0·25$	339
AdCl	PhCl	$1·08 \pm 0·25$	339
AdBr	PhCl	$1·00 \pm 0·25$	339
AdI	PhCl	$1·03 \pm 0·25$	339

[a] Ad = adamantyl.

The conclusions to be drawn from Table 94 are:
(a) The values indicated for the enthalpies of formation are to be regarded only as an order of magnitude 0·5–2 kcal mole^{-1}, since there are several possible perturbing phenomena, e.g. simultaneous occurrence of 1 : 1, 1 : 2, etc., complexes. There may also be temperature shifts associated with dispersion forces, which may be unequal for the tetramethylsilane reference and the temperature dependent resonances followed.
(b) The interactions are in any event *very weak*, and as stressed by Fort[339]

N§

"variations with substituents, etc., may be very significant on a percentage basis, like 10 or 15%, but it is impossible to measure enthalpies of the order of 1 kcal mole^{-1} with such an accuracy". No relation is seen between the observed enthalpies and the dipole moments, in a related series, which may be due to the small range of solute dipole moments considered: for instance, the tert-butyl and adamantyl halides studied by Fort[339] all have molecular dipoles between 2·10 and 2·30 D.

(c) From the measured free energies of formation ($K = 0\cdot3$–$1\cdot5$ M^{-1} at 25°C), it is readily apparent that only a minority of molecules is complexed at ambient temperatures (still in the hypothesis of a 1:1 complex). Then, the calculations which implicitly assume total complexation to account for the ASIS observed by the Johnson–Bovey tables, as performed by Laszlo and Schleyer[335] or Williams and Wilson[308] (see Section 13.6) are simply not justified and their success is all the more surprising. This is really anomalous since the shifts for a number of protons differently located in the molecule are obtained to a good approximation. One unsatisfactory way out is to dismiss the whole problem by implying that the solute and the benzene complexed molecule just move a little closer.

(d) The relative invariance of the measured heats of formation (cf. (b) above, however) may argue against the formation of a 1:1 complex, and be taken to indicate a much looser, more long range type of interaction.[339] Clearly more data is necessary to clarify this important point.

13.10. *The Time Scale*

There is yet another topic related to the microscopic reality of such complexes. First, one remark, based precisely upon the use of nuclear magnetic resonance in such studies, has to be recalled. The "classical" spectroscopic methods, infra-red or ultra-violet, utilize optical transitions, which are very fast and obey the Franck–Condon principle. Which means that when these physical methods demonstrate the existence of any complex, by charge-transfer, by hydrogen bonding, etc., through the simultaneous observation of a "free" and a "bonded" absorption, only a lower limit of its life-time is defined, of the order of 10^{-14}–10^{-12} sec. It is known that, in the liquid phase, the time a molecule spends in a given orientation, between two collisions, is of the order of 10^{-10}–10^{-11} sec, and this is the true minimum duration for any complex. The NMR measurement is much slower and, for separate "free" and "complexed" resonances to be present in a spectrum, the life-time of the complex has to be longer than the reciprocal of their frequency difference. In the examples we have discussed here, the ASIS were of the order of 0–60 c/s the reciprocal of this shift fixes a limit of 10^{-2} sec for the observation of individual bands. These have not been observed, in the case of 11-androsta-none;[340] for instance, the exchange of benzene molecules between the solvent and the complex was fast enough at − 70°C to escape detection, and only sharp, average peaks were observed. The life-time of such an association, between one molecule of the solute and one molecule of the solvent, is there-fore very small, certainly much smaller than 10^{-2} sec, and there is no such thing as the *permanence* of these 1:1 complexes in solution. This serves to explain the results obtained by Diehl.[90] This author was able to split the solvent shifts brought about by benzene on the protons of substituted ben-zenes into two parts: first, there are the specific shifts for protons ortho, meta or para to the substituent X in a monosubstituted benzene; second, a constant contribution for all ring positions, originating in the large benzene magnetic anisotropy. For such a planar molecule, if it is assumed "that each surface element has the same probability of seeing solute protons",[90] these experience a large upfield shift, evaluated at 0·38 ppm by Diehl,[90] for benzene compared with 5 mole % benzene in hexane, and as 0·70 ppm by Barbier, Delmau and Béné[349] for the similar absolute high field ASIS of TMS from carbon tetra-chloride to benzene. Obviously, "if we had an appreciable concentration of molecules in a *permanent* (our addition, and italics) complex form, the protons of the substituted benzene would, contrary to our experimental results, no longer have a chance to experience an average local magnetic field of the solvent benzene surface".[90] These considerations clearly show the operational definition of a complex for Moriarty and Kliegman:[332] "Com-plex in the present context implies an association between solvent and solute resulting in a discrete polymolecular assemblage which exists at least for the duration of the experiment used for its detection" (footnote 7 of their paper[332] is incorrect; it has been corrected in the full paper).

13.11. *Conclusion About the Origin of* ASIS

The following summarizes the empirical observations about ASIS:
1. Polar solutes only; incremental shifts depending upon the situation of the protons with respect to the polar groups.
2. Remarkable additivity, which implies a very specific and *local* property.
3. A paradox that the observed ASIS do not vary as the apparent enthalpy of the complex, but in the opposite direction: substitution of methyl groups into the solvent molecule increases the strength of the association but diminishes the ASIS; actual shifts correlate with dipole moments, while the apparent enthalpies of formation are invariant.
4. Benzene molecules "bind" with positive centres, avoiding the negative parts of the solute.
5. "Chemical" nature (see Table 96).
6. Small life-time.

Speculation about the origin of ASIS will require independent, complementary and confirmatory evidence from other physical methods, which at the time of writing is scarce. No 1:1 complex with a congruent melting point could be detected by thermal analysis between benzene and any of the following ketones: acetone, acetophenone, benzophenone and camphor. In each case, a single two-component eutectic was detected.[350] The same is true for the tert-butylbromide–benzene system, for which Fort[339] constructed the phase diagram. The infra-red carbonyl band of the ketone solutes is little affected by the change from an inert solvent to benzene, less than when it is partially hydrogen bonded [351] (see Table 95).

TABLE 95. SOLVENT INDUCED FREQUENCY SHIFTS OF THE INFRA-RED CARBONYL BAND

Solvent	$\nu(C=O)$ cm^{-1}			
	Acetone	Mesityl oxide	Acetophenone	Benzophenone
n-Hexane	1727	1702	1701	1673
CCl$_4$	1723	1697$_5$	1693	1669
C$_6$H$_6$	1721$_5$	1695	1693	1667
CHCl$_3$	1715$_5$	1690	1687	1662

More solid indications are provided by the observation,[352] reported by Murrell and Gil,[84] that the dipole moment of pyridine is larger by 0·12 D in carbon tetrachloride than in benzene, and by the preliminary results of Golfier:[353] owing to the sensitivity of the optical rotation to distant perturbations, and considering the precision possible in its determination, it could be expected that the molecular rotation would vary with the solvent. Indeed, it is found that for 17-androstanone (I) away from the ultraviolet absorption bands of either the solute or the solvent (533 mμ), the molecular rotations

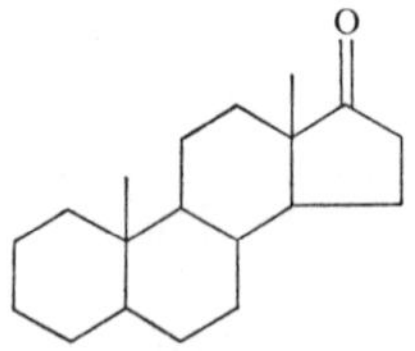

I

((α)$_{363}$) were 625 $\pm$ 2° in carbon tetrachloride solution, 558 $\pm$ 2° in benzene, and 569 $\pm$ 2° in pyridine. A number of other aromatic solvents bearing polar substituents were also used, but they differed little from benzene itself, which produces the biggest part of the effect. Anderson, Cambio and Prausnitz[354] have used ultraviolet spectrophotometry in the classical manner to measure the constants for the associations between acetone and benzene, or cyclo-hexanone and benzene, as $0 \cdot 10 - 0 \cdot 20$ M^{-1} at 25°. These are in very good agreement with the results of variable temperature or dilution curve experiments (discussed in the previous paragraphs) which define by NMR a similar order of magnitude for the strength of the association.

A confusion has progressively crept into the preceding paragraphs devoted to this phenomenon of ASIS: we have been writing either about a dipole induced–dipole interaction, or about a charge transfer, which are not synonymous; with respect to the geometry of the resulting complex, the former image requires the dipolar axis of the solute to lie in a plane parallel to that of the aromatic ring, while the latter corresponds to the dipolar axis being coincident with the six-fold symmetry axis. This distinction, which could not be resolved —it may be that certain solute–solvent systems pertain to one type, and others to the second, or that even both be simultaneously active—lies between complexes of a *chemical* nature. A more primitive distinction is the classical one between a "physical" and a "chemical" association, as it has been formulated by Mulliken,[355–6] Hildebrand,[359] and a number of other authors.[357–8] An attempt is made to summarize the distinctions in Table 96.

TABLE 96. ARE THE ASIS A "PHYSICAL" OR A "CHEMICAL" PHENOMENON?

Nature of the assoc.:	PHYSICAL	CHEMICAL
Criteria	all mols affected	coexistence of a "bonded" and a "free" form.
Heat of mixing ΔH_m	>0 (endothermic)	<0 (exothermic)
Volume of mixing Δv_m	>0 (expansion)	<0 (contraction)
,	proportional to:	
	the product of the volume fractions	the product of the mole fractions.
		a 1:1 complex if: * between molecules, not between charged particles. * ΔH_{assoc} independent of temperature.

By this token, there is no doubt that the observations are best accounted for by a "chemical" association, specially as the recourse to an internal reference in order to measure the chemical shifts removes all of the possible "physical" perturbations. Here arises a small problem in nomenclature: how does one name these species? The terms "contact complex" and "collision complex" have been employed, by a generalization of the findings of the optical spectro-scopists. The collision complexes, as they are defined by Ketelaar,[360] are not stable entities in the thermodynamic sense, their life-time being short as compared with the average time between two collisions, and they occur in compressed gases for not too high densities, where they are detected by the observation of simultaneous vibrational transitions in the mixture. The con-tact complex, according to Orgel and Mulliken,[356] Bayliss and McRae,[357] is a description of the electronic charge transfer between molecules, which can accompany a random encounter, without implying a true complex for-mation. Bayliss and McRae[357] speak of the resulting "Franck–Condon strain", since the solvation cage around the excited solute molecule are still appropriate to the ground state, the solvent molecules not having time to rearrange themselves during the very short time of an optical transition. This last term obviously cannot be used to describe the NMR situation since it has just been pointed out that the time scale is much longer for the relevant transition to occur. Huong, Lascombe and Josien[361] prefer the term "contact complex" rather than "collision complex" after Horak, or than "van der Waals complex", after Bratoz, to describe their infra-red observations for phenols diluted in non-polar solvents. To close this discussion about the origin of the ASIS phenomenon, the author wishes to strongly affirm his contention that the approach through the working hypothesis of a $1:1$ dipole-induced dipole complex, with the geometry of parallel planes where the aromatic pi-electrons would tend to avoid the negative end of the solute's dipole, has been right and fruitful. This model, if it is too simple to account for all the observations (it is probable that admixing of a quantum mechanical term to the simple electrostatic, as for the hydrogen bond, will provide a correct description), or to reflect the complexity of the ordering process taking place in the liquid around the polar solute, has led to predictions, which have generally been shown to be true. It has led also to interesting new experiments, which have provided additional questions to answer and it has suggested a number of useful correlations or techniques, like an unusual method for measuring dipole moments, or the additivity of the ASIS, etc. The merits, if any, of a purely empirical attitude cannot be clearly seen, *a posteriori*. This $1:1$ complex model has done no harm to science and the caution with which it has been put forward and discussed by the authors is indeed most remarkable. Now, it is clearly obsolete, and a more sophisticated tool will have to be constructed.

To quote Hoyle:[362] "It is difficult to know how to deal with an opponent who when he loses an argument insists that if only things were made more complicated he would win it. I suppose that in the next few years we shall just have to plough through the agony of dealing with the more complicated situation. . . ."

13.12. *Structural Applications*

Williams and Bhacca[286] deduced the preferred orientation of the 17β-acetyl group in 5α-pregnan-20-one from the solvent shifts quoted in Table 75, it is consistent with dipole moment data[363] and ORD and CD evidence.[364] Bowie *et al.*[365] have measured the ASIS ($CDCl_3$, C_6H_6) for a number of quinones; in the anthraquinone I series, they are 0·06–0·17 ppm for 1-methyl

I

groups, 0·52–0·60 ppm for the 2- and 3-methyl groups. Solvent shifts support the previously assigned differences in stereo-chemistry and conformation between isoleutherin, eleutherin and a related system derived from the aphins. By measuring the ASIS for a large variety of monocyclic and polycyclic ketones, Connolly and McCrindle[307, 366] have established a correlation which is obeyed by all the examples yet reported, in the literature: a reference plane drawn through the carbon of the carbonyl group at right angles to the carbon oxygen bond separates the anterior oxygen-containing space, for which negative ASIS are the rule, from the rear zone where positive ASIS occur. Protons situated in this plane suffer no change from the inert to the benzene solvent; the maximum shift is observed at some distance of this reference plane, and the ASIS then decrease with the distance to the carbonyl group. Di Maio, Tardella and Iavarone[367] also report ASIS for lactones, also studied by Connolly and McCrindle,[366] the results of the former study are presented in Table 97. The results depend mainly upon the spatial relationship of the methyl and the polar lactonic group in the solute, *irrespective of the rest of the molecule.*[366-7] This is shown diagrammatically for the examples of the *cis* (II) and *trans* (III) ring junctions:

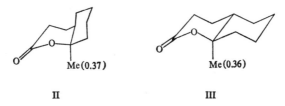

II   III

For 2-alkyl-1,3-dioxanes, the following ASIS (from carbon tetrachloride to benzene) have been established by Anderson[342] (see Table 98).

TABLE 97. ASIS FOR METHYL RESONANCES OF δ-LACTONES[367]

Compound	Asis (CDCl$_3$/C$_6$H$_6$: ppm)	Compound	Asis (CDCl$_3$/C$_6$H$_6$: ppm)
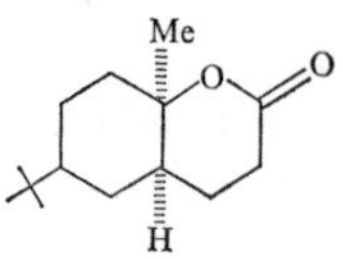 Me / H structure	+0.38	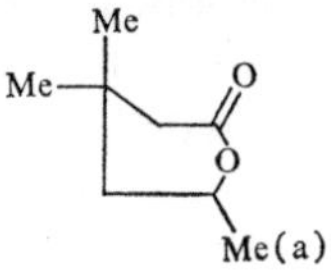 Me—, Me(a) structure	(a) 0.27[a]
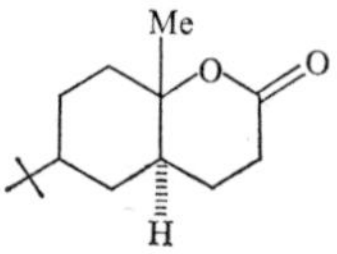 Me / H structure	+0.39	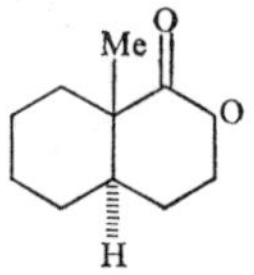 Me structure	0.27
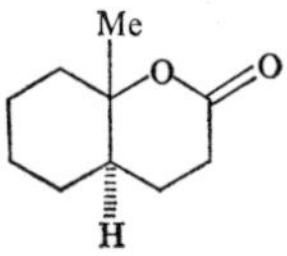 Me / H structure	+0.36[b]	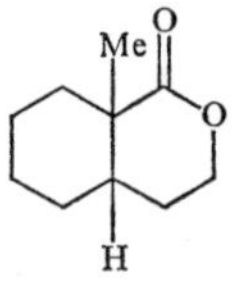 Me structure	0.22
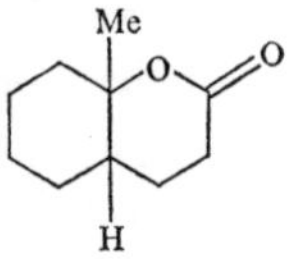 Me / H structure	+0.37[c]	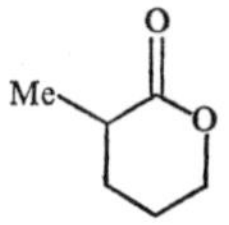 Me structure	0.14
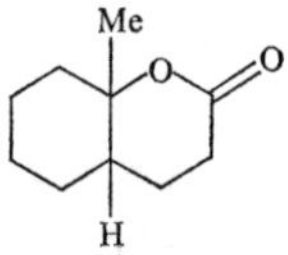 Me / H structure	+0.28		

[a] Equilibrium constant $K = 0.65 \pm 0.03$.   [b] $K = 0.67 \pm 0.03$.   [c] $K = 0.64 \pm 0.03$. Determined by the method of Tyrrell.[341]

Anderson[342] explains the difference between the *gem*-dimethyl (C–5) series and the parent monosubstituted series, by the benzene molecule forming a 1:1 complex with the dioxane solute (this point was checked by means of a Tyrrell plot[341]). This is formed on either side of the molecular surface, or at a greater distance from the dipole of the solute in the former than in the latter case, since all the ASIS are reduced by 0·15–0·30 ppm. It may well be that not only the geometry of the complex depends upon the presence of axial methyl groups on either side of the solute, but that the solute itself responds to their introduction by making slight conformational changes. This study postulates

TABLE 98. ASIS FOR 1,3-DIOXANES (ppm)

Compound	R	Ha	Hb	Hc	Hd	He
(1,3-dioxane: He, Hd, Hc, Ha, Hb)	t-Bu	−0·01	+0·32	+0·24	+0·59	+0·24
	i-Pr	−0·02	+0·28	+0·19	+0·55	—
	Et	0	+0·27	+0·19	+0·55	—
	Me	+0·05	+0·28	+0·18	+0·54	+0·29
(dioxane)		+0·02	+0·28	+0·19	—	—
(Me(e), Me(d), Hc; Ha, Hb)	t-Bu	−0·06	+0·17	+0·07	+0·36	0
	i-Pr	−0·04	+0·16	+0·02	+0·35	−0·01
	Et	−0·03	+0·13	+0·05	+0·34	−0·01
	Me	−0·04	+0·09	+0·01	+0·33	−0·03
(Me, Me)		−0·07	+0·13	+0·20	—	—
(Me, Me, Me)		—	+0·13	+0·20	—	—
(Me(d))		−0·03	+0·21ᵃ	+0·13		

ᵃ Mean value.

identical conformations and conformer populations in the "inactive" and "active" solvents, which seems reasonable. It should be remarked that these ASIS have been measured for relatively high solute concentrations (0·4 M) which may be quite different from the dilution values.

Seyden-Penne, Strzalko and Plat[368] could, empirically but unambiguously, relate the ASIS for five pairs of methyl cyclopropanes disubstituted by a methylenearyloxy and by an amido grouping with their configurations IV and V.

Detailed consideration of the effect of substituents in the aromatic ring shows that both the $ArOCH_2$ and the $CONH_2$ substituents are active towards the aromatic solvent.

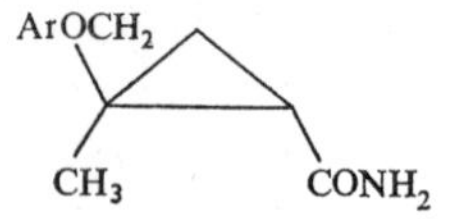

$$\Delta_{C_5H_5N}^{CDCl_3} = 0.27 \text{ ppm}$$

$$\Delta_{C_5H_5N}^{CDCl_3} = 0.20 - 0.27 \text{ ppm}$$

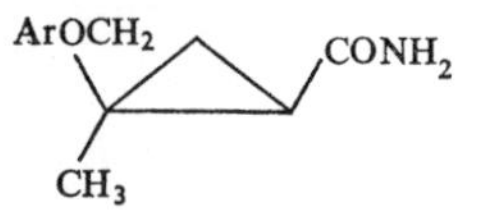

IV

V

14. POLAR SOLUTES IN HETEROCYCLIC AROMATIC SOLVENTS

The data available are not yet systematic enough to enable clear-cut relationships—which may be absent because of the duality of factors, the aromaticity and the heteroatom—to be established, as in the benzene case. Williams,[369] who studied the chemical shifts induced by pyridine in ketones relative to carbon tetrachloride, proposes another reference plane, parallel to that of Connolly and McCrindle[307] and passing through the α carbon atoms C-2 and C-6. There have also been some studies[370-1] of steroids in pyridine solutions. Connolly and McCrindle[372] have done an extensive survey of the magnitudes of shifts produced by aromatic solvents upon 2-methylcyclopentanone, 2,2,6,6-tetramethylcyclohexanone, camphor, pulegone and fenchone, from which it appears that pyrrole produces the largest shifts.

15. CHIRAL SOLVENTS*

Pirkle[373] has used a chiral solvent to distinguish the spectra of enantiomers. The ^{19}F resonance of the racemic fluoroalcohol I shows a doublet in carbon tetrachloride ($J_{HF} = 6.7$ c/s) or in racemic α-phenetylamine (II). In the

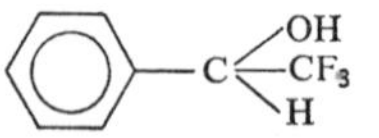

I

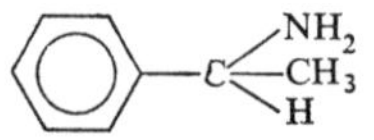

II

chiral l-amine, two sets of doublets of equal intensity, separated by 2.0 c/s, are observed. Measurement of the relative intensities allowed the determination of the optical purity of a sample of I($[\alpha]_D + 2.19°$) produced by asymmetric induction to be $53 \pm 0.6\%$. The specific rotation of the dextro-rotatory enantiomer is calculated to be $36.5 \pm 7.5\%$, in agreement with the maximum possible rotation of this enantiomer of $41.2°$. Table 99 summarizes the different possibilities.

* Optically active solvents.

TABLE 99. NON-EQUIVALENCE OF NMR SPECTRA OF ENANTIOMERS IN CHIRAL SOLVENTS

	rac. mixt. $(\pm)$ or $(+, -)$	rac. mixt. $(\pm)$ or $(+, -)$		rac. mixt. $(\pm)$ or $(+, -)$
Solute:				
Solvent:	0 achiral	LD racemic		L chiral
		slow exchange	fast exchange	
Spectra: (of enantiomers)	identical	different[a] (+)L ⋮ (−)D ⎵⎵⎵⎵ (+)D ⋮ (−)L	average to the same	different (+)L } (−)L

[a] ·········· enantiomeric relationship; ⁓⁓⁓ diastereomeric relationship.

16. DETERMINATION OF SOLVATION NUMBERS OF CATIONS

We shall not provide here a detailed discussion of how the structure of water or water mixtures is perturbed by dissolved electrolytes, but summarize in this section only the experiments where exchange between solvent and solvation shell was sufficiently slow to allow two distinct resonances to be observed for the solvent protons.

In 1960 Jackson, Lemons and Taube[374] reported that when $Co(ClO_4)_2$ is added to aqueous solutions of certain diamagnetic salts, it becomes possible to distinguish the bonded $H_2^{17}O$ resonances from the water molecule in the first coordination sphere, and the free $H_2^{17}O$ resonances in the bulk solvent. The paramagnetic cobalt salt shifts the free H_2O signal to lower field, whereas the signal of the bound H_2O is not shifted appreciably. For instance, with Al^{3+},[375] the difference in the chemical shift of the bound and the free water is 430 ppm, equivalent to $8 \cdot 2 \times 10^4$ ppm for a 1:1 mole ratio of Co^{2+} to moles of free water, while the paramagnetic shift of the bound water is only 60 c/s (at 5·415 Mc/s), relative to pure water. In this study, Connick and Fiat,[375] using 11·5% ^{17}O enriched water, showed that the coordination number of Al^{3+} is 6, that of Be^{2+} being 4. Another technique was successful with 1·5% ^{17}O enriched water.[376] The same paramagnetic salt method was used by Swinehart and Taube[377] to determine the solvation number of Mg^{2+} in methanolic solution as 6, by the joint observation of a free and a bound peak at −75°C using a solution of $Mg(ClO_4)_2$ in methanol. For the

Li^+ ion, in the system $LiClO_4/MeOH$, the exchange appears to be too fast as no separate bands could be detected for the complex. Thomas and Reynolds[378] have used the same procedure of the intensity ratio comparison between the bulk solvent peak, and the coordinated solvent peak, the latter being proportional to the concentration of added salt, to determine within experimental error the coordination number of Al^{3+} in dimethyl sulphoxide as 6 at 20 and at 40°C. In all these experiments, the actual solvation number is somewhat smaller, by a few per cent, due to residual water molecules and/or to the existence of discrete ion pairs. Thomas and Reynolds[378] further defined the thermodynamic parameters for the exchange of solvent molecules with the solvent cage, using the temperature dependence of the exchange line width variation:

$$\Delta H^{\ddagger} = 20 \pm 1 \text{ kcal mole}^{-1},$$
$$\Delta S^{\ddagger} = 3 \cdot 7 \pm 2 \cdot 5 \text{ cal mole}^{-1} \text{ deg}^{-1}.$$

Luz and Meiboom[380–1] have studied the solvation of Co^{2+} and Ni^{2+} in methanol–water mixtures. For $Co(ClO_4)_2$ in acidified methanol, separate peaks are observed below $-50°C$ for the free and bonded MeOH molecules, again a case of slow exchange. The coordination number for Co^{2+} is 6; from the peak frequencies the values of the hyperfine interaction between the bonded MeOH protons and the Co^{2+} ion are $+8 \cdot 8 \times 10^5$ c/s (hydroxyl protons), and $+3 \cdot 9 \times 10^5$ c/s (methyl protons) (in this latter case, a value of $7 \cdot 9 \times 10^5$ c/s is obtained with the Ni^{2+} cation). The water molecules tend to disrupt this ordering. With paramagnetic salts, as used by preceding authors, separate resonance signals are more easily observed because displacements are very large. Fratiello, Schuster and Miller[382] could also differentiate between bulk and complexed solvent molecules using *non-paramagnetic salts*, like Be(II), Al(III), Ga(III), Si(IV), Ti(IV) and Sb(V), in solvents such as N,N-dimethyl formamide, N-methyl formamide, N-ethyl formamide and ethyl formate. For instance, with $SbCl_5/Me_2NCHO$ complex peaks appear for the aldehydic and methyl protons, respectively displaced downfield by 56 c/s (for either methyl group). These figures are consistent with complex formation at the oxygen atom of the solvent molecule, for which the resonance form $Me_2N^+CHO^-$ is therefore confirmed. The coordination number of Sb(V) was not determined in this initial work, which contained a number of other interesting observations. For instance, whereas the coalescence temperature for the two methyl groups of the bulk solvent is lower than 100°C, the corresponding signals for the complex remain unchanged as a pair at this temperature. This indicates that in the complex, the increased double bond character of the carbon–nitrogen bond prevents internal rotation. It has already been pointed out that the methyl resonances are displaced by different amounts upon complex formation, also the ^4J coupling constant between the methyl and aldehyde protons almost doubles in magnitude. A good correlation is obtained when, for example, the aldehydic proton separations between the bulk and the complexed amide are plotted as a function of $Z/(r + a)^2$, for an ion of charge Z (for the cations listed above), with an ionic radius r increased by an arbitrary constant $a = 0.2$ Å, chosen to produce a good fit of the data. The success of such an elementary electric field approach has encouraged the authors to perform electron density calculations on these molecules in order to relate them to the observed signal separations.

REFERENCES

1. W. T. RAYNES, A. D. BUCKINGHAM and H. J. BERNSTEIN, *J. Chem. Phys.* **36**, 3481 (1962).
2. L. PETRAKIS and H. J. BERNSTEIN, *J. Chem. Phys.* **37**, 2731 (1962); **38**, 1562 (1963).
3. G. WIDENLOCHER, E. DAYAN and B. VODAR, *Compt. Rend.* **256**, 2584 (1963); **258** 4242 (1964).
4. E. DAYAN and G. WIDENLOCHER, *Compt. Rend.* **257**, 883 (1963).
5. G. WIDENLOCHER and E. DAYAN, *Compt. Rend.* **260**, 6856 (1965).
6. F. H. A. RUMMENS and H. J. BERNSTEIN, *J. Chem. Phys,* **43**, 2971 (1965).
7. S. GORDON and B. P. DAILEY, *J. Chem. Phys.* **34**, 1084 (1961).
8. A. D. BUCKINGHAM, *J. Chem. Phys.* **36**, 3096 (1962).
9. L. PETRAKIS and C. H. SEDERHOLM, *J. Chem. Phys.* **35**, 1174 (1961).
10. A. D. BUCKINGHAM, *Can. J. Chem.* **38**, 300 (1960).
11. J. C. SCHUG, *J. Phys. Chem.* **70**, 1816 (1966).
12. A. A. BOTHNER-BY, *J. Mol. Spectroscopy* **5**, 52 (1960).
13. B. B. HOWARD, B. LINDER and M. J. EMERSON, *J. Chem. Phys.* **36**, 485 (1962).
14. M. J. STEPHEN, *Mol. Phys.* **1**, 223 (1958).
15. T. W. MARSHALL and J. A. POPLE, *Mol. Phys.* **1**, 199 (1958).
16. N. LUMBROSO, T. K. WU and B. P. DAILEY, *J. Chem. Phys.* **67**, 2469 (1963).
17. B. FONTAINE, M. T. CHENON and N. LUMBROSO-BADER, *J. Chim. Phys.* **10**, 1075 (1965).
18. F. LONDON, *Z. Physik.*, **63**, 245 (1930); *Z. Phys. Chem.* (Leipzig) **B11**, 222 (1930).
19. J. A. POPLE, W. G. SCHNEIDER and H. J. BERNSTEIN, *High-resolution Nuclear Magnetic Resonance*, McGraw-Hill Book Co., Inc., New York, 1959, sections 4–8 and 4–9.
20. A. FRATIELLO and D. C. DOUGLASS, *J. Mol. Spectroscopy* **11**, 465 (1963).
21. C. A. REILLY, H. M. MCCONNELL and R. G. MEISENHEIMER, *Phys. Rev.* **98**, 264A (1955).
22. D. J. FROST and G. E. HALL, *Nature* **205**, 1309 (1965).
23. K. FREI and H. J. BERNSTEIN, *J. Chem. Phys.* **37**, 1891 (1962).
24. C. LUSSAN, *J. Chim. Phys.* **61**, 462 (1964).
25. A. A. BOTHNER-BY and R. E. GLICK, *J. Chem. Phys.* **26**, 1647 (1957).
26. A. A. BOTHNER-BY and R. E. GLICK, *J. Mol. Spectroscopy* **5**, 52 (1960).
27. A. A. BOTHNER-BY and C. NAAR COLIN, *J. Amer. Chem. Soc.* **80**, 1728 (1958).
28. R. E. GLICK, D. F. KATES and S. J. EHRENSON, *J. Chem. Phys.* **31**, 567 (1959).
29. R. E. GLICK and S. J. EHRENSON, *J. Phys. Chem.* **62**, 1599 (1958).
30. K. VENKATESWARLU and S. SRIRAMAN, *Bull. Chem. Soc. Japan* **31**, 211 (1958).
31. D. J. FROST and G. E. HALL, *Mol. Phys.*, **10**, 191 (1966).
32. H. A. LAUWERS and G. P. VAN DER KELEN, *Bull. Soc. Chim. Belges* **75**, 238 (1966).
33. C. P. NASH and G. E. MACIEL, *J. Phys. Chem.* **68**, 832 (1964).
34. A. FRATIELLO and E. G. CHRISTIE, *Trans. Faraday Soc.* **61**, 306 (1965).
35. D. C. DOUGLASS and A. FRATIELLO, *J. Chem. Phys.* **39**, 3161 (1965).
36. A. PACAULT, *Principes généraux de la magnétochimie, Revue Scientifique*, 1944; quoted by Lussan.[24]
37. J. REISSE and R. OTTINGER, private communication.

38. R. A. Y. Jones, A. R. Katritzky, J. N. Murrell and N. Sheppard, *J. Chem. Soc.*, 2576 (1962).
39. G. E. Maciel and J. J. Natterstad, *J. Chem. Phys.* **42**, 2752 (1965).
40. A. C. Chapman, J. Homer, D. J. Mowthorpe and R. T. Jones, *Chem. Comm.*, No. 7, 121 (1965).
41. G. Filipovich and G. V. D. Tiers, *J. Phys. Chem.* **63**, 761 (1959).
42. R. W. Taft, E. Price, I. R. Fox, I. C. Lewis, K. K. Andersen and G. T. Davis, *J. Amer. Chem. Soc.* **85**, 709 (1963).
43. G. V. D. Tiers, *J. Phys. Chem.* **62**, 1151 (1958).
44. A. L. McClellan and S. W. Nicksic, *J. Phys. Chem.* **69**, 446 (1965).
45. H. M. Hutton and T. Schaefer, *Can. J. Chem.* **43**, 3116 (1965).
46. F. Hruska, E. Bock and T. Schaefer, *Can. J. Chem.* **41**, 3034 (1963).
47. G. E. Maciel and G. C. Ruben, *J. Amer. Chem. Soc.* **85**, 3903 (1965).
48. H. A. Christ and P. Diehl, *Helv. Phys. Acta*, **36**, 170 (1963).
49. L. J. Bellamy and R. L. Williams, *Trans. Faraday Soc.* **55**, 14 (1959).
50. O. Hassel and C. Romming, *Quart. Rev.* **16**, 1 (1962); P. Groth and O. Hassel, *Mol. Phys.* **6**, 543 (1963); M. Veyret and M. Gomel, *Compt. Rend.* **258**, 4506 (1964).
51. P. Laszlo, L. C. Allen, P. von R. Schleyer and R. M. Erdahl, Abstracts, Fourth Omnibus Conference on the Experimental Aspects of Nuclear magnetic resonance Spectroscopy (4th OCEANS), Mellon Institute, Pittsburgh, Pa., March 2, 1963.
52. G. Suryan, *Proc. Indian Acad Sci.* A**33**, 107 (1951); P. M. Denis, G. J. Bènè and R. C. Extermann, *Arch. Sci. (Geneva)* **5**, 32 (1952); A. L. Bloom and J. N. Shoolery, *Phys. Rev.* **90**, 358 (1953); C. Sherman, *Phys. Rev.* **93**, 1429 (1954); A. M. J. Mitchell and G. Phillips, *Brit. J. Appl. Phys.* **7**, 67 (1956); K. Antonowicz and T. Waluga, *Polon. Sci., Classe III* **5**, 815, 1069 (1957); A. Z. Hrynkiewicz and T. Waluga, *Acta Phys. Polon.* **16**, 381 (1957); S. Forsén and A. Rupprecht, *J. Chem. Phys.* **33**, 1888 (1960); B. L. Donally and G. E. Bernal, *Amer. J. Phys.* **31**, 779 (1963).
53. *Nuclear Magnetic Resonance in a Flowing Liquid*, Aleksandr I. Zhernovoi and Georgh D. Latyshev, Consultants Bureau Enterprises, Inc., New York, 1965.
54. L. W. Reeves, *N.M.R. Measurements of Reaction Velocities and Equilibrium Constants as a Function of Temperature*, Advances in Physical Organic Chemistry, V. Gold, Ed., Vol. 3, Academic Press, 1965, p. 187.
55. H. S. Gutowsky, G. G. Belford and P. E. McMahon, *J. Chem. Phys.* **36**, 3353 (1962).
56. H. Finegold, *J. Chem. Phys.* **41**, 1808 (1964).
57. M. Karplus, *J. Chem. Phys.* **30**, 11 (1959).
58. F. A. L. Anet, *J. Amer. Chem. Soc.* **84**, 747 (1962).
59. R. J. Abraham, L. Cavalli and K. G. R. Pachler, *Nuclear Magnetic Resonance in Chemistry*, Academic Press, 1965, p. 111.
60. D. H. Williams and N. S. Bhacca, *J. Amer. Chem. Soc.* **86**, 2742 (1964).
61. E. I. Snyder, *J. Amer. Chem. Soc.* **88**, 1155, 1161, 1165 (1966).
62. W. F. Trager, B. J. Nist and A. C. Huitric, *Tetrahedron Letters* **33**, 2931 (1965).
63. E. W. Garbisch, Jr., *J. Amer. Chem. Soc.* **86**, 1780 (1964).
64. H. Feltkamp and N. C. Franklin, *Ann. Chem.* **683**, 55 (1965); *J. Amer. Chem. Soc.*, **87**, 1616 (1965).
65. R. U. Lemieux and J. W. Lown, *Can. J. Chem.*, **42**, 893 (1964).
66. A. Allerhand and P. von R. Schleyer, *J. Amer. Chem. Soc.* **85**, 1715 (1963).
67. K. Takahashi, *Bull. Chem. Soc. Japan* **37**, 291 (1964).
68. J. Reisse, J. C. Celotti and R. Ottinger, *Tetrahedron Letters* **19**, 2167 (1966).
69. G. O. Dudek and E. P. Dudek, *J. Amer. Chem. Soc.* **88**, 2407 (1966).
70. J. C. Woodbrey and M. T. Rogers, *J. Amer. Chem. Soc.* **84**, 13 (1962).
71. C. W. Fryer, F. Conti and C. Franconi, *Ricerca scientif.* (2a) **35**, 3 (1965).
72. L. Onsager, *J. Amer. Chem. Soc.* **58**, 1486 (1936).
73. A. J. Dekker, *Physica* **12**, 209 (1946); T. H. Scholte, *Physica* **15**, 437, 450 (1949).
74. W. S. Lovell, Ph.D. dissertation, Princeton, 1963; National Bureau of Standards Report 8457, 20-8-1964.

75. A. D. BUCKINGHAM, T. SCHAEFER and W. G. SCHNEIDER, *J. Chem. Phys.* **32,** 1227 (1960).
76. J. I. MUSHER, *J. Chem. Phys.* **37,** 34 (1962).
77. P. DIEHL and R. FREEMAN, *Mol. Phys.* **4,** 39 (1961).
78. I. G. ROSS and R. A. SACK, *Proc. Phys. Soc. London* **1363,** 893 (1960).
79. P. LASZLO and J. I. MUSHER, *J. Chem. Phys.* **41,** 3906 (1964).
80. P. LASZLO, *Bull. Soc. Chim. France* 1131 (1966).
81. D. DECROOCQ, *Bull. Soc. Chim. France* 127 (1964).
82. P. LASZLO, *Bull. Soc. Chim. France* 85, 1964).
83. J. K. BECCONSALL and P. HAMPSON, *Mol. Phys.* **10,** 21 (1965).
84. J. N. MURRELL and V. M. S. GIL, *Trans. Faraday Soc.* **61,** 402 (1965).
85. D. H. WHIFFEN, *Quart. Rev.* **4,** 131 (1950).
86. C. A. COULSON, *Proc. Roy. Soc.* A **255,** 69 (1960).
87. H. M. HUTTON and T. SCHAEFER, *Can. J. Chem.* **43,** 3116 (1965).
88. I. YAMAGUCHI, Abstracts, *Tokyo NMR Symposium,* 1965, p. M-2-12.
89. P. DIEHL, *Helv. Chim. Acta* **45,** 504 (1962).
90. P. DIEHL, *J. Chim. Phys.* **61,** 199 (1964).
91. J. C. SCHUG and J. C. DECK, *J. Chem. Phys.* **37,** 2618 (1962).
92. W. G. PATERSON and N. R. TIPMAN, *Can. J. Chem.* **40,** 2122 (1962).
93. J. I. MUSHER, *J. Chem. Phys.* **40,** 2399 (1964).
94. R. W. TAFT, E. PRICE, I. R. FOX, I. C. LEWIS, K. K. ANDERSEN and G. T. DAVIS, *J. Amer. Chem. Soc.* **85,** 3146 (1963).
95. J. G. KIRKWOOD, *J. Chem. Phys.* **2,** 351 (1934).
96. G. BRIEGLEB, *Elektronen-Donator-Acceptor Komplexe,* Springer-Verlag, Berlin, 1961.
97. D. J. CRAM and L. A. SINGER, *J. Amer. Chem. Soc.* **85,** 1084 (1963).
98. W. KOCH and HCH. ZOLLINGER, *Helv. Chim. Acta* **48,** 554 (1965).
99. J. F. BIELLMANN, private communication.
100. A. R. KATRITZKY, F. J. SWINBOURNE and B. TERNAI, *J. Chem. Soc.* (B), 235 (1966).
101. M. SAUNDERS and J. B. HYNE, *J. Chem. Phys.* **29,** 1319 (1958).
102. H. S. GUTOWSKY and A. SAIKA, *J. Chem. Phys.* **21,** 1688 (1953).
103. G. MAVEL, *J. Physique Rad.* **21,** 37 (1960).
104. O. REDLICH and A. T. KISTER, *J. Chem. Phys.* **15,** 849 (1947).
105. P. J. BERKELEY, Jr. and M. W. HANNA, *J. Phys. Chem.* **67,** 846 (1963).
106. C. M. HUGGINS, G. C. PIMENTAL and J. N. SHOOLERY, *J. Chem. Phys.* **23,** 1244 (1955).
107. R. FOSTER and C. A. FYFE, *Trans. Faraday Soc.* **61,** 1626 (1965).
108. M. W. HANNA and A. L. ASHBAUGH, *J. Phys. Chem.* **68,** 811 (1964).
109. C. LUSSAN, *J. Chim. Phys.* **60,** 1100 (1963).
110. M. T. CHENON and N. LUMBROSO-BADER, *J. Chim. Phys.* **62,** 1208 (1965).
111. L. A. LaPLANCHE, H. B. THOMPSON and M. T. ROGERS, *J. Phys. Chem.* **69,** 1482 (1965).
112. I. SATAKE, M. ARITA, H. KIMIZUKA and R. MATUURA, *Bull. Chem. Soc. Japan* **39,** 597 (1966).
113. E. A. GUGGENHEIM, *Mixtures,* Oxford University Press, 1952.
114. J. A. BARKER, *J. Chem. Phys.* **20,** 1526 (1952).
115. N. SOUTY and B. LEMANCEAU, *Compt. Rend.* **259,** 4255 (1964).
116. B. B. HAYWARD, C. F. JUMPER and M. T. EMERSON, *J. Mol. Spectroscopy* **10,** 117 (1963).
117. R. KAISER, *Can. J. Chem.* **41,** 430 (1963).
118. C. S. SPRINGER and D. W. MEEK, *J. Phys. Chem.* **70,** 481 (1966).
119. J. FEENEY and L. H. SUTCLIFFE, *J. Chem. Soc.* 1123 (1962).
120. B. LEMANCEAU, C. LUSSAN, N. SOUTY and J. BIAIS, *J. Chim. Phys.* **61,** 195 (1964).
121. J. C. DAVIS, Jr., K. S. PITZER and C. N. R. RAO, *J. Phys. Chem.* **64,** 1744 (1960).
122. B. B. HOWARD, C. F. JUMPER and M. T. EMERSON, *J. Mol. Spectroscopy* **10,** 117 (1963).
123. V. P. BYSTROV and V. P. LEZINA, *Opt. i Spektroskopiya* **16,** 790 (1964); *Chem. Abstr.,* **61,** 4186a (1964).
124. C. LUSSAN, B. LEMANCEAU and N. SOUTY, *Compt. Rend.* **254,** 1980 (1962).
125. W. B. SMITH, *J. Org. Chem.* **27,** 4641 (1962).
126. A. PERRIER-DATIN, P. SAUMAGNE and M. L. JOSIEN, *Compt. Rend.* **259,** 1825 (1964).

127. G. S. KASTLA and K. C. MEDHI, *Indian J. Phys.* **37**, 568 (1963).
128. G. PANNETIER and L. ABELLO, *Bull. Soc. Chim. France* 1645 (1966).
129. J. L. MATEOS, R. CETINA and O. CHAO, *Chem. Comm.* 519 (1965).
130. L. SAVOLEA-MATHOT, *Trans. Faraday Soc.* **49**, 8 (1953).
131. F. HRUSKA, G. KOTOWYCZ and T. SCHAEFER, *Can. J. Chem.* **43**, 3188 (1965).
132. A. L. McCLELLAN and S. W. NICKSIC, *J. Phys. Chem.* **69**, 446 (1965).
133. A. L. McCLELLAN, S. W. NICKSIC and J. C. GUFFY, *J. Mol. Spectroscopy* **11**, 340 (1963).
134. M. A. WHITEHEAD, N. C. BAIRD and M. KAPLANSKY, *Theor. Chim. Acta* **3**, 135 (1965).
135. T. SCHAEFER, private communication.
136. P. J. BERKELEY and M. W. HANNA, *J. Chem. Phys.* **41**, 2530 (1964).
137. P. J. BERKELEY and M. W. HANNA, *J. Amer. Chem. Soc.* **86**, 2990 (1964).
138. A. LOEWENSTEIN and Y. MARGALIT, *J. Phys. Chem.* **69**, 4152 (1965).
139. J. A. POPLE, *Proc. Roy. Soc. (London)*, A **239**, 541, 550 (1957).
140. H. M. McCONNELL, *J. Chem. Phys.* **27**, 226 (1957).
141. J. A. CRESWELL and A. L. ALLRED, *J. Amer. Chem. Soc.* **85**, 1723 (1963).
142. J. A. CRESWELL and A. L. ALLRED, *J. Amer. Chem. Soc.* **84**, 3966 (1962).
143. E. L. ELIEL, N. L. ALLINGER, S. J. ANGYAL and G. A. MORRISON, *Conformational Analysis*, Interscience–Wiley, New York, 1965, chap. 2.
144. W. G. PATERSON and D. M. CAMERON, *Can. J. Chem.* **41**, 198 (1963).
145. A. L. McCLELLAN, S. W. NICKSIC and J. C. GUFFY, *J. Mol. Spectroscopy* **11**, 340 (1963); and references quoted in this article.
146. H. SAITO, K. NUKADA, H. KATO, T. YONEZAWA and K. FUKUI, *Tetrahedron Letters* **2**, 111 (1965).
147. A. MATHIAS and V. M. S. GIL, *Tetrahedron Letters* **35**, 3163 (1965).
148. G. COPPENS, J. NASIELSKI and N. SPRECHER, *Bull. Soc. Chim. Belges* **72**, 626 (1963).
149. I. GRANÄCHER, *Helv. Phys. Acta* **34**, 272 (1961).
150. G. MAVEL, *J. Physique Rad.* **20**, 834 (1959).
151. E. R. LIPPINCOTT and R. SCHRÖDER, *J. Chem. Phys.* **23**, 1099, 1131 (1955).
152. H. F. HAMEKA, *Nuovo Cimento* **11**, 382 (1959); *Mol. Phys.* **1**, 203 (1958); **2**, 64 (1959).
153. A. D. BUCKINGHAM, *Mol. Phys.* **3**, 219 (1960).
154. M. J. STEPHEN, *Mol. Phys.* **1**, 223 (1958).
155. T. W. MARSHALL and J. A. POPLE, *Mol. Phys.* **1**, 199 (1958).
156. L. PAOLONI, *J. Chem. Phys.* **30**, 1045 (1959).
157. N. N. SHAPET'KO, D. N. SHIGORIN, A. P. SKOLDINOV, T. S. RYABCHIKOVA and L. N. RESHETOVA, *J. Struct. Chem. USSR* **6**, 138 (1965).
158. I. GRANÄCHER, *Helv. Phys. Acta* **31**, 734 (1958); P. DIEHL and I. GRANÄCHER, *Helv. Phys. Acta* **32**, 288 (1959); I. GRANÄCHER and P. DIEHL, *Arch. Sci.* **12**, 238 (1959).
159. S. FORSÉN and B. ÅKERMARK, *Acta Chem. Scand.* **17**, 1907 (1963).
160. R. J. OUELLETTE, *J. Amer. Chem. Soc.* **86**, 3089, 4378 (1964).
161. E. D. BECKER, V. LIDDEL and J. N. SHOOLERY, *J. Mol. Spectroscopy* **2**, 1 (1958).
162. R. J. OUELLETTE, G. E. BOOTH and K. LIPTAK, *J. Amer. Chem. Soc.* **87**, 3436 (1965).
163. I. YAMAGUCHI, *Bull. Chem. Soc. Japan* **34**, 353 (1961).
164. I. YAMAGUCHI, *Bull. Chem. Soc. Japan* **34**, 451 (1961).
165. I. YAMAGUCHI, *Bull. Chem. Soc. Japan* **34**, 744 (1961).
166. I. YAMAGUCHI, *Bull. Chem. Soc. Japan* **34**, 1602 (1961).
167. I. YAMAGUCHI, *Bull. Chem. Soc. Japan* **34**, 1606 (1961).
168. B. G. SOMERS and H. S. GUTOWSKY, *J. Amer. Chem. Soc.* **85**, 3065 (1963).
169. H. J. FRIEDRICH, *Angew. Chem.* **16**, 721 (1965).
170. R. W. TAFT, *J. Amer. Chem. Soc.* **74**, 3120 (1952).
171. J. N. MURRELL, V. M. S. GIL and F. B. VAN DUIJNEVELDT, *Rec. Trav. Chim. Pays-Bas* **84**, 1399 (1965).
172. D. COOK, *J. Amer. Chem. Soc.* **80**, 49 (1958).
173. E. D. BECKER, *Spectrochim. Acta* **17**, 436 (1961).
174. M. HOEKE and A. L. KOEVOET, *Rec. Trav. Chim. Pays-Bas* **82**, 17 (1963).
175. T. GRAMSTAD, *Spectrochim. Acta* **19**, 497 (1963).

176. G. E. MACIEL and R. V. JAMES, *Inorg. Chem.* **3**, 1650 (1964).
177. J. B. STOTHERS, *Quart. Rev.* **19**, 144 (1965).
178. G. B. SAVITSKY, K. NAMIKAWA and G. ZWEIFEL, *J. Phys. Chem.* **69**, 3105 (1965).
179. H. C. BROWN and K. ICHIKAWA, *Tetrahedron* **1**, 221 (1957).
180. A. V. KAMERNITZKY and A. A. AKHREM, *Tetrahedron* **18**, 705 (1962); J. C. RICHER, *J. Org. Chem.* **30**, 324 (1965).
181. W. G. DAUBEN, G. J. FONKEN and D. S. NOYCE, *J. Amer. Chem. Soc.* **78**, 2579 (1956).
182. D. HADZI, *J. Chem. Soc.* 5128 (1962).
183. G. E. MACIEL, Abstracts, *Tokyo NMR Symposium*, September 1965, p. M-2-10.
184. C. REICHARDT, *Angew. Chem. Int. Ed.* **4**, 29 (1965).
185. J. J. DELPUECH, *Tetrahedron Letters* **25**, 2111 (1965); *Bull. Soc. Chim. France*, 1624 (1966).
186. R. OTTINGER, G. BOULVIN, J. REISSE and G. CHIURDOGLU, *Tetrahedron* **21**, 3435 (1965).
187. T. G. ALEXANDER and M. MAIENTHAL, *J. Pharm. Sci.* **53**, 962 (1964).
188. P. E. PETERSON, *J. Org. Chem.* **31**, 439 (1966).
189. S. NG, J. TANG and C. H. SEDERHOLM, *J. Chem. Phys.* **42**, 79 (1965).
190. S. NG and C. H. SEDERHOLM, *J. Chem. Phys.* **40**, 2090 (1964).
191. P. C. MYHRE, J. W. EDMONDS and J. D. KRUGER, *J. Amer. Chem. Soc.* **88**, 2459 (1966).
192. K. C. RAMEY and W. S. BREY, Jr. *J. Chem. Phys.* **40**, 2349 (1964).
193. R. THEIMER and J. R. NIELSEN, *J. Chem. Phys.* **30**, 98 (1959).
194. P. LASZLO, *Bull. Soc. Chim. France*, 558 (1966).
195. K. FREI and H. J. BERNSTEIN, *J. Chem. Phys.* **38**, 1216 (1963).
196. J. LASCOMBE, J. DEVAURE and M. L. JOSIEN, *J. Chim. Phys.* **61**, 1271 (1964).
197. S. S. DANYLUK, *J. Amer. Chem. Soc.* **86**, 4504 (1964).
198. D. P. EVANS, S. L. MANATT and D. D. ELLEMAN, *J. Amer. Chem. Soc.* **85**, 238 (1963).
199. N. MULLER and R. C. REITER, *J. Chem. Phys.* **42**, 3265 (1965).
200. M. FÉTIZON, M. GOLFIER and P. LASZLO, *Bull. Soc. Chim. France*, 3486 (1965).
201. P. LASZLO, *Bull. Soc. Chim. France*, 2658 (1964).
202. K. TORI, K. KITAHONOKI, Y. TAKANO, H. TANIDA and T. TSUJI, *Tetrahedron Letters* **11**, 559 (1964).
203. K. TORI, K. AONO, K. KITAHONOKI, R. MUNEYUKI, Y. TAKANO, H. TANIDA and T. TSUJI, *Tetrahedron Letters* **25**, 2921 (1966).
204. J. I. MUSHER, *Mol. Phys.* **6**, 93 (1963).
205. B. B. DEWHURST, J. S. E. HOLKER, A. LABLACHE-COMBIER, M. R. G. LEEMING, J. LEVISALLES and J. P. PETE, *Bull. Soc. Chim. France*, 3259 (1964).
206. M. GORODETSKY and Y. MAZUR, *Tetrahedron Letters*, **4**, 227 (1964).
207. J. C. ROWELL, W. D. PHILLIPS, L. R. MELBY and M. PANAR, *J. Chem. Phys.* **43**, 3442 (1965).
208. A. SAUPE, *Z. Naturforsch.* **20A**, 572 (1965).
209. A. ABRAGAM, *The Principles of Nuclear Magnetism*, Oxford University Press, Oxford, 1961, p. 453.
210. H. SPIESECKE, personal communication, June 1966.
211. L. C. SNYDER and S. MEIBOOM; cf. *Chem. Engng. News*, April 25, 1966, p. 51; and June 13, 1966, p. 9; *J. Amer. Chem. Soc.*, **89**, 1038 (1967).
212. L. C. SNYDER and R. L. KORNEGAY, The determination of molecular structure by computer simulation of NMR spectra in liquid crystal solvents, *ACS Symposium on Ordered Fluids and Liquid Crystals*, Atlantic City, N.J., September 1965.
213. L. C. SNYDER and E. W. ANDERSON, *J. Amer. Chem. Soc.* **86**, 5023 (1964).
214. L. C. SNYDER and E. W. ANDERSON, *J. Chem. Phys.* **42**, 3336 (1965).
215. L. C. SNYDER, *J. Chem. Phys.* **43**, 4041 (1965).
216. R. REAVILL and H. J. BERNSTEIN, personal communication, May 1966.
217. G. ENGLERT and A. SAUPE, *Z. Naturforsch.* **19A**, 172 (1964).
218. A. SAUPE and G. ENGLERT, *Phys. Rev. Letters* **11**, 462 (1963).
219. A. SAUPE, *Z. Naturforsch.* **19A**, 161 (1964).

220. *Chemical and Engineering News*, March 16, 1964, p. 42.
221. G. ENGLERT, *Symposium on High Resolution NMR of Oriented Molecules*, Pittsburgh Conference on Analytical Chemistry and Applied Spectroscopy, 1964.
222. J. NEHRING and A. SAUPE, personal communication, June 1966; *Z. Naturforsch.* (in press).
223. R. G. JONES, R. C. HIRST and H. J. BERNSTEIN, *Can. J. Chem.* **43**, 683 (1965).
224. D. E. MCGREER and M. M. MOCEK, *J. Chem. Ed.* **40**, 358 (1963).
225. D. M. GRANT, R. C. HIRST and H. S. GUTOWSKY, *J. Chem. Phys.* **38**, 470 (1963).
226. G. FARGES and A. S. DREIDING, *Helv. Chim. Acta* **49**, 552 (1966).
227. W. F. TRAGER, B. J. NIST and A. C. HUITRIC, *Tetrahedron Letters* **33**, 2931 (1965).
228. W. F. TRAGER, B. J. NIST and A. C. HUITRIC, *Tetrahedron Letters* **4**, 267 (1965).
229. P. COURTOT, S. KINASTOWSKI and H. LUMBROSO, *Bull. Soc. Chim. France* 489 (1964).
230. H. M. HUTTON and T. SCHAEFER, *Can. J. Chem.* **41**, 2774 (1963).
231. H. M. HUTTON and T. SCHAEFER, *Can. J. Chem.* **41**, 2429 (1963).
232. K. MISLOW and MORTON RABAN, *Stereoisomeric Relationships of Groups in Molecules* (in press); cf. K. MISLOW, *Introduction to Stereochemistry*, W. A. Benjamin, 1965.
233. B. SINGER and K. MISLOW, private communication, August 1964.
234. K. MISLOW, M. A. W. GLASS, H. B. HOPPS, E. SIMON and G. H. WAHL, Jr., *J. Amer. Chem. Soc.* **86**, 1710 (1964).
235. G. M. WHITESIDES, J. G. GROCKI, D. HOLTZ, H. STEINBERG and J. D. ROBERTS, *J. Amer. Chem. Soc.* **87**, 1058 (1965).
236. E. I. SNYDER, *J. Amer. Chem. Soc.* **85**, 2624 (1963).
237. M. RABAN, *Tetrahedron Letters* **27**, 3105 (1966).
238. S. GRONOWITZ, *Arkiv Kemi* **18**, 133 (1962); M. CHABRE, D. GAGNAIRE and C. NOFRE, *Bull. Soc. Chim. France* 108 (1966).
239. J. R. HOLMES, D. KIVELSON and W. C. DRINKARD, *J. Chem. Phys.* **37**, 150 (1962).
240. O. L. CHAPMAN and R. W. KING, *J. Amer. Chem. Soc.* **86**, 1256 (1964).
241. C. P. RADER, *J. Amer. Chem. Soc.* **88**, 1713 (1966).
242. J. J. UEBEL and H. W. GOODWIN, *J. Org. Chem.* **31**, 2040 (1966).
243. A. S. PERLIN, *Can. J. Chem.* **44**, 539 (1966).
244. T. F. PAGE, Jr. and W. E. BRESLER, *Anal. Chem.* **36**, 1981 (1964).
245. B. CASU, M. REGGIANI, G. G. GALLO and A. VIGEVANI, *Tetrahedron Letters* **39**, 2839 (1964).
246. W. B. ANDERSON, Jr. and R. M. SILVERSTEIN, *Anal. Chem.* **37**, 1417 (1965).
247. V. W. GOODLETT, *Anal. Chem.* **37**, 431 (1965).
248. J. FEENEY and A. HEINRICH, *Chem. Comm.* No. 10, 295 (1966).
249. D. F. EVANS, *J. Chem. Soc.*, 5575 (1963).
250. T. YONEZAWA, I. MORISHIMA, M. FUJII and K. FUKUI, *Bull. Chem. Soc., Japan*, **38**, 1224 (1965).
251. I. OMURA, H. BABA, K. HIGASI and Y. KANAOKA, *Bull. Chem. Soc., Japan* **30**, 633 (1957).
252. K. WATANABE, T. NAKAYAMA and J. MOTTL, *J. Quant. Spectrosc. Radiat. Transfer* **2**, 369 (1962).
253. R. HOFFMANN, *J. Chem. Phys.* **40**, 2745 (1964).
254. A. ALBERT, R. GOLDACRE and J. PHILIPPS, *J. Chem. Soc.*, 2240 (1948); J. CLARK and D. D. PERRIN, *Quart. Rev.* **18**, 295 (1964).
255. E. D. BECKER, private communication.
256. J. REUBEN and D. SAMUEL, private communication.
257. G. MARTIN, B. CASTRO and M. MARTIN, *Compt. Rend.* **261**, 395 (1965).
258. W. H. DE JEU, H. ANGAD GAUR and J. SMIDT, *Rec. Trav. Chim. Pays-Bas* **84**, 1621 (1965).
259. J. G. POWLES and J. H. STRANGE, *Mol. Phys.* **5**, 329 (1962).
260. V. S. WATTS, J. LOEMKER and J. H. GOLDSTEIN, *J. Mol. Spectroscopy* **17**, 348 (1965).
261. J. REISSE and G. CHIURDOGLU, *Spectrochim. Acta* **20**, 441 (1964).
262. T. D. COYLE, R. B. JOHANNESEN, F. E. BRINCKMAN and T. C. FARRAR, *J. Phys. Chem.* **70**, 1682 (1966).

263. T. SCHAEFER, private communication, June 1966.
264. L. SALEM *The Molecular Orbital Theory of Conjugated Systems*, W. A. BENJAMIN, New York, 1966, Section 4–7.
265. A. D. BUCKINGHAM, *Trans. Faraday Soc.* **58**, 2077 (1962).
266. S. L. SMITH and R. H. COX, *J. Mol. Spectroscopy* **16**, 216 (1965).
267. C. L. BELL and S. S. DANYLUK, *J. Amer. Chem. Soc.* **88**, 2344 (1966).
268. D. D. ELLEMAN and S. L. MANATT, *J. Mol. Spectroscopy* **9**, 477 (1962).
269. B. L. SHAPIRO, S. J. EBERSOLE and R. M. KOPCHIK, *J. Mol. Spectroscopy* **11**, 326 (1963).
270. G. J. MARTIN and M. C. MARTIN, *J. Chim. Phys.* **61**, 1222 (1964).
271. W. BRÜGEL, private communication.
272. M. BARFIELD and D. M. GRANT, *J. Amer. Chem. Soc.* **85**, 1899 (1963); T. TAKAHASHI, *Tetrahedron Letters*, 565 (1964).
273. H. M. HUTTON and T. SCHAEFER, *Can. J. Chem.* **41**, 684 (1963).
274. B. SUMNERS, L. H. PIETTE and W. G. SCHNEIDER, *Can. J. Chem.* **38**, 681 (1960).
275. G. MARTIN and A. BESNARD, *Compt. Rend.* **257**, 898 (1963).
276. P. BATES, S. CAWLEY and S. S. DANYLUK, *J. Chem. Phys.* **40**, 2415 (1964).
277. C. EABORN *Organosilicon Compounds*, Academic Press, Inc., New York, 1960, p. 92; S. BROWNSTEIN, B. C. SMITH, G. EHRLICH and A. W. LAUBENGAYER, *J. Amer. Chem. Soc.* **81**, 3826 (1959).
278. D. W. MOORE and J. A. HAPPE, *J. Phys. Chem.* **65**, 224 (1961); R. T. HOBGOOD, J. H. GOLDSTEIN and G. S. REDDY, *J. Chem. Phys.* **35**, 2038 (1961).
279. K. A. McLAUCHLAN, L. W. REEVES and T. SCHAEFER, *Can. J. Chem.* **44**, 1473 (1966).
280. V. S. WATTS, G. S. REDDY and J. H. GOLDSTEIN, *J. Mol. Spectroscopy* **11**, 325 (1963).
281. V. S. WATTS and J. H. GOLDSTEIN, *J. Chem. Phys.* **42**, 228 (1965).
282. P. LASZLO and M. C. TROMEUR, unpublished results.
283. P. LASZLO and H. J. T. BOS, *Tetrahedron Letters* **18**, 1325 (1965).
284. N. S. BHACCA and D. H. WILLIAMS, *Tetrahedron Letters* **42**, 3127 (1964).
285. G. SLOMP, Jr. and S. L. MacKELLAR, *J. Amer. Chem. Soc.* **82**, 999 (1960).
286. D. H. WILLIAMS and N. S. BHACCA, *Tetrahedron* **21**, 2021 (1965).
287. M. FÉTIZON and J. C. GRAMAIN, *Bull. Soc. Chim. France*, 2289, 3444 (1966).
288. R. F. ZÜRCHER, *Helv. Chim. Acta* **44**, 1380 (1961).
289. R. F. ZÜRCHER, *Helv. Chim. Acta* **46**, 2054 (1963).
290. N. S. BHACCA and D. H. WILLIAMS, *Applications of N.M.R. Spectroscopy in Organic Chemistry*, Holden-Day Inc., San Francisco, 1964, Chap. 2.
291. ref. (290), p. 164, 1st paragraph.
292. R. F. ZÜRCHER, *Nuclear Magnetic Resonance in Chemistry*, Ed. B. PESCE, Academic Press, New York, 1965.
293. E. R. MALINOWSKI, L. Z. POLLARA and J. P. LARMAN, *J. Amer. Chem. Soc.* **84**, 2649 (1962).
294. ref. (194), footnote (*******), p. 559.
295. P. DIEHL, *Helv. Chim. Acta* **45**, 568 (1962).
296. P. DIEHL, *Helv. Chim. Acta* **44**, 829 (1961).
297. S. BORY, M. FÉTIZON, P. LASZLO and D. H. WILLIAMS, *Bull. Soc. Chim. France*, 2541 (1965).
298. M. FÉTIZON, J. GORÉ, P. LASZLO and B. WAEGELL, *J. Org. Chem.*, **31**, 4047 (1966).
299. D. H. WILLIAMS and N. S. BHACCA, *Tetrahedron* **21**, 1641 (1965).
300. KARL R. POPPER, *Conjectures and Refutations*, Routledge & Kegan Paul, London, 1936.
301. C. G. HEMPEL, *Aspects of Scientific Explanation and Other Essays in the Philosophy of Science*, The Free Press, 1965.
302. E. B. WILSON, *An Introduction to Scientific Research*, Dover Press.
303. J. V. HATTON and R. E. RICHARDS, *Mol. Phys.* **5**, 139 (1962).
304. T. L. BROWN and K. STARK, *J. Phys. Chem.* **69**, 2679 (1965).
305. Z. YOSHIDA and E. OSAWA, *J. Amer. Chem. Soc.* **87**, 1467 (1965); M. R. BASILEA, E. L. SAIER and L. R. COUSINS, *J. Amer. Chem. Soc.* **87**, 1665 (1965).

306. T. LEDAAL, *Tetrahedron Letters* **15**, 1653 (1966).
307. J. D. CONNOLLY and R. MCCRINDLE, *Chem. Ind.* 379 (1965).
308. D. H. WILLIAMS and D. A. WILSON, *J. Chem. Soc.* (B), 144 (1966).
309. R. E. KLINCK and J. B. STOTHERS, *Can. J. Chem.* **40**, 2329 (1962).
310. M. A. WEINBERGER, R. M. HEGGIE and H. L. HOLMES, *Can. J. Chem.* **43**, 2585 (1965).
311. R. F. WEIMER and J. M. PRAUSNITZ, *Spectrochim. Acta* **22**, 77 (1966).
312. H. BENESI and J. H. HILDEBRAND, *J. Amer. Chem. Soc.* **71**, 2703 (1949).
313. R. S. DRAGO and N. J. ROSE, *J. Amer. Chem. Soc.* **81**, 6138 (1959).
314. A. A. SANDOVAL and M. W. HANNA, *J. Phys. Chem.* **70**, 1203 (1966).
315. J. S. MARTIN, *J. Chem. Phys.* **39**, 1728 (1963); R. MCWEENY, *Mol. Phys.* **1**, 811 (1958).
316. J. I. MUSHER, *J. Chem. Phys.* **43**, 4081 (1965); **46**, 1219 (1967).
317. B. P. DAILEY, *J. Chem. Phys.* **41**, 2304 (1964).
318. J. A. ELVIDGE and L. M. JACKMAN, *J. Chem. Soc.* 859 (1961); J. A. ELVIDGE, *Chem. Comm.*, No. 8, 160 (1965).
319. R. J. ABRAHAM, R. C. SHEPPARD, W. A. THOMAS and S. TURNER, *Chem. Comm.*, No. 3, 43 (1965).
320. E. M. VOIGT, *J. Phys. Chem.* **70**, 598 (1966).
321. G. BRIEGLEB, J. CZEKALLA and G. REUSS, *Z. Physik. Chem.* (*Frankfurt*) **30**, 316 (1961).
322. W. G. SCHNEIDER, *J. Phys. Chem.* **66**, 2653 (1962).
323. H. H. PERKAMPUS and U. KRÜGER, *Z. Phys. Chem.*, **48**, 379 (1966).
324. LANDOLT-BORNSTEIN, *Zahlenwerte und Functionen*, Vol. 1, Pt. 3, Springer-Verlag, Berlin, 1951, p. 510.
325. C. E. JOHNSON and F. A. BOVEY, *J. Chem. Phys.* **29**, 1012 (1958).
326. J. S. WAUGH and R. W. FESSENDEN, *J. Amer. Chem. Soc.* **79**, 846 (1957).
327. J. E. BALDWIN, *J. Org. Chem.* **30**, 2423 (1965).
328. L. A. LAPLANCHE and M. T. ROGERS, *J. Amer. Chem. Soc.* **86**, 337 (1964).
329. J. V. HATTON and R. E. RICHARDS, *Mol. Phys.* **3**, 253 (1960).
330. J. V. HATTON and R. E. RICHARDS, *Mol. Phys.* **5**, 153 (1962).
331. R. M. MORIARTY, *J. Org. Chem.* **28**, 1296 (1963).
332. R. M. MORIARTY and J. M. KLIEGMAN, *Tetrahedron Letters* **9**, 891 (1966).
333. R. C. NEUMAN, Jr. and L. B. YOUNG, *J. Phys. Chem.* **69**, 2570 (1965).
334. C. J. TIMMONS, *Chem. Comm.* No. 22, 576 (1965).
335. P. LASZLO and P. VON R. SCHLEYER, *J. Amer. Chem. Soc.* **86**, 1171 (1964).
336. P. M. SUBRAMANIAN, M. T. EMERSON and N. A. LE BEL, *J. Org. Chem.* **30**, 2624 (1965).
337. N. NAKAGAWA and S. FUJIWARA, *Bull. Soc. Chem. Japan* **34**, 143 (1961).
338. J. SEYDEN-PENNE, F. STRZALKO and M. PLAT, *Tetrahedron Letters* **30**, 3611 (1966).
339. R. C. FORT, Abstracts, 152nd ACS National Meeting, New York, September 1966.
340. P. LASZLO and D. H. WILLIAMS, *J. Amer. Chem. Soc.* **88**, 2799 (1966).
341. J. TYRRELL, *Can. J. Chem.* **43**, 783 (1965).
342. J. E. ANDERSON, *Tetrahedron Letters* **51**, 4713 (1965).
343. M. HASHIMOTO and Y. TSUDA, Abstracts, *Tokyo NMR Symposium*, p. M-2-13.
344. A. DIEFFENBACHER and W. VON PHILIPSBORN, *Helv. Chim. Acta* **49**, 897 (1966).
345. W. VON PHILIPSBORN, private communication.
346. R. J. ABRAHAM, *Mol. Phys.* **4**, 369 (1961).
347. K. M. WELLMAN, E. BUNNENBERG and C. DJERASSI, *J. Amer. Chem. Soc.* **85**, 1870 (1963).
348. J. V. HATTON and W. G. SCHNEIDER, *Can. J. Chem.* **40**, 1285 (1962).
349. C. BARBIER, J. DELMAU and G. BÉNÉ, *J. Chim. Phys.* **58**, 764 (1961).
350. J. TIMMERMANS, *Les Solutions Concentrées*, Masson, Paris, pp. 218–19; and private communication from C. BRASSY and F. PINCON.
351. M. ITO, K. INUZUKA and S. IMANISHI, *J. Chem. Phys.* **31**, 1694 (1959).
352. A. N. SHARPE and S. WALKER, *J. Chem. Soc.*, 2974 (1961); 157 (1962).
353. M. GOLFIER, private communication, June 1966.
354. R. ANDERSON, R. CAMBIO and J. M. PRAUSNITZ, *A.I.Ch.E. J.* **8**, 66 (1962).
355. R. S. MULLIKEN, *Rec. Trav. Chim. Pays-Bas* **75**, 845 (1956).
356. B. E. ORGEL and R. S. MULLIKEN, *J. Amer. Chem. Soc.* **79**, 4839 (1958).

357. N. S. BAYLISS and E. G. MCRAE, *J. Phys. Chem.* **58**, 1002 (1954).
358. L. J. ANDREWS and R. M. KEEFER, *Molecular Complexes in Organic Chemistry*, Holden-Day, San Francisco, 1964.
359. J. H. HILDEBRAND and R. L. SCOTT, *The Solubility of Non-electrolytes*, Dover, 1964.
360. J. A. KETELAAR, *Rec. Trav. Chim. Pays-Bas* **75**, 857 (1956).
361. P. V. HUONG, J. LASCOMBE and M. L. JOSIEN, *Compt. Rend.* **259**, 4244 (1964).
362. F. HOYLE, *Galaxies, Nuclei, and Quasars*, Heinemann 1966.
363. N. L. ALLINGER and M. A. DA ROOGE, *J. Amer. Chem. Soc.* **83**, 4256 (1961).
364. K. M. WELLMAN and C. DJERASSI, *J. Amer. Chem. Soc.* **87**, 60 (1965).
365. J. H. BOWIE, D. W. CAMERON, P. E. SCHÜTZ, D. H. WILLIAMS and N. S. BHACCA, *Tetrahedron* **22**, 1771 (1966).
366. J. D. CONNOLLY and R. MCCRINDLE, *Chem. Ind.*, 2066 (1965).
367. G. D. DI MAIO, P. A. TARDELLA and C. IAVARONE, *Tetrahedron Letters* **25**, 2825 (1966).
368. J. SEYDEN-PENNE, T. STRZALKO and M. PLAT, *Tetrahedron Letters* **50**, 4597 (1965).
369. D. H. WILLIAMS, *Tetrahedron Letters* **27**, 2305 (1965).
370. B. HAMPEL, *Tetrahedron* **22**, 1601 (1966).
371. C. F. H. GREEN, J. E. PAGE and S. E. STAMFORTH, *J. Chem. Soc.*, 7328 (1965).
372. J. D. CONNOLLY and R. MCCRINDLE, *J. Chem. Soc.* (C), 1613 (1966).
373. W. H. PIRKLE, *J. Amer. Chem. Soc.* **88**, 1837 (1966).
374. A. JACKSON, J. LEMONS and H. TAUBE, *J. Chem. Phys.* **33**, 553 (1960).
375. R. E. CONNICK and D. N. FIAT, *J. Chem. Phys.* **39**, 1349 (1963).
376. M. ALEI and J. A. JACKSON, *J. Chem. Phys.* **41**, 3402 (1964).
377. J. H. SWINEHART and H. TAUBE, *J. Chem. Phys.* **37**, 1579 (1962).
378. S. THOMAS and W. L. REYNOLDS, *J. Chem. Phys.* **44**, 3148 (1966).
379. N. A. MATWIYOFF, *Inorg. Chem.* **5**, 788 (1966).
380. Z. LUZ and S. MEIBOOM, *J. Chem. Phys.* **40**, 1058 (1964).
381. Z. LUZ and S. MEIBOOM, *J. Chem. Phys.* **40**, 1066 (1964).
382. A. FRATIELLO, R. SCHUSTER and D. P. MILLER, *Mol. Phys.*, **11**, 597 (1966).

The following references have appeared after the completion of the manuscript, and are listed with the number of the relevant section between parentheses:

(4) T. MATSUO and Y. KODERA, *J. Phys. Chem.* **70**, 4087 (1966).
    F. HRUSKA, D. W. MCBRIDE, and T. SCHAEFER, *Can. J. Chem.* **45**, 1081 (1967).
    G. KOTOWYCZ and T. SCHAEFER, *Can. J. Chem.* **45**, 1093 (1967).
    H. M. HUTTON and T. SCHAEFER, *Can. J. Chem.* **45**, 1111 (1967).
(5) R. FOSTER and C. A. FYFE, *J. Chem. Soc.* **B 926** (1966).
    M. R. CRAMPTON and V. GOLD, *J. Chem. Soc.* **B 23** (1967).
    R. FOSTER and C. A. FYFE, *Nature* **213**, 591 (1967).
(6) N. MULLER and O. R. HUGHES, *J. Phys. Chem.* **70**, 3975 (1966).
    N. MULLER and P. SIMON, *J. Phys. Chem.* **71**, 568 (1967).
    R. E. GLICK, W. E. STEWART and K. C. TEWARI, *J. Chem. Phys.* **45**, 4049 (1966).
    S. F. TING, S. M. WANG, and N. C. LI, *Can. J. Chem.* **45**, 425 (1967).
(7) A. W. DOUGLAS and D. DIETZ, *J. Chem. Phys.* **46**, 1214 (1967).
(9) R. A. BERNHEIM and B. J. LAVERY, *Bull. Amer. Phys. Soc.* **11**, 172 (1966).
    B. P. DAILEY, C. S. YANNONI, and G. P. CAESAR, Abstracts, 153rd ACS Meeting, Miami, April 9–14, 1967, R-78.
(10) R. FREEMAN and N. S. BHACCA, *J. Chem. Phys.* **45**, 3795 (1966).
(12) R. H. COX and S. L. SMITH, *J. Mol. Spectroscopy*, **21**, 232 (1966).
    V. S. WATTS and J. H. GOLDSTEIN, *J. Phys. Chem.* **70**, 3887 (1966).
    S. L. SMITH and A. M. IHRIG, *J. Mol. Spectroscopy*, **22**, 241 (1967).
    S. L. SMITH and A. M. IHRIG, *J. Chem. Phys.* **46**, 1181 (1967).
(13) C. GANTER, L. G. NEWMAN, and J. D. ROBERTS, *Tetrahedron* Suppl. 8 (II), 507 (1966).
    C. R. NARAYANAN and N. K. VENKATASUBRAMANIAN, *Tetrahedron Letters*, **47**, 5865 (1966).

E. Rahkamaa, *Suomen Kemi.* **B39**, 272 (1966).

J. Ronayne and D. H. Williams, *Chem. Comm.*, 712 (1966).

C. R. Kanekar, G. Govil, C. L. Khetrapal and M. M. Dhingra, *Proc. Ind. Acad. Sci.* **477**, 315 (1966).

F. R. McDonald, A. W. Decora, and G. L. Cook, 18th Conference on Analytical Chemistry and Applied Spectroscopy, Pittsburgh, March 6, 1967.

H. M. Fales and K. S. Warren, *J. Org. Chem.* **32**, 501 (1967).

D. W. Boykin, Jr., A. B. Turner and R. E. Lutz, *Tetrahedron Letters* **9**, 817 (1967).

H. C. Brown and A. Suzuki, *J. Amer. Chem. Soc.* **89**, 1933 (1967).

(14) Gurudata, R. E. Klinck, and J. B. Stothers, *Can. J. Chem.* **45**, 213 (1967).

(15) J. C. Jochims, G. Taigel and A. Seeliger, *Tetrahedron Letters* **20**, 1901 (1967).

(16) S. Nakamura and S. Meiboom, *J. Amer. Chem. Soc.* **89**, 1765 (1967).

T. J. Swift, W. G. Sayre and S. Meiboom, *J. Chem. Phys.* **46**, 410 (1967).

J. F. Hinton and E. S. Amis, *Chem. Comm.* **2**, 100 (1957).

NAME INDEX

SUBJECT INDEX

413

CONTENTS OF PREVIOUS VOLUMES

VOLUME 1

VOLUME 2